Lecture Notes in Computer Science 6020

Commenced Publication in 1973
Founding and Former Series Editors:
Gerhard Goos, Juris Hartmanis, and Jan van Leeuwen

Editorial Board

David Hutchison
Lancaster University, UK

Takeo Kanade
Carnegie Mellon University, Pittsburgh, PA, USA

Josef Kittler
University of Surrey, Guildford, UK

Jon M. Kleinberg
Cornell University, Ithaca, NY, USA

Alfred Kobsa
University of California, Irvine, CA, USA

Friedemann Mattern
ETH Zurich, Switzerland

John C. Mitchell
Stanford University, CA, USA

Moni Naor
Weizmann Institute of Science, Rehovot, Israel

Oscar Nierstrasz
University of Bern, Switzerland

C. Pandu Rangan
Indian Institute of Technology, Madras, India

Bernhard Steffen
TU Dortmund University, Germany

Madhu Sudan
Microsoft Research, Cambridge, MA, USA

Demetri Terzopoulos
University of California, Los Angeles, CA, USA

Doug Tygar
University of California, Berkeley, CA, USA

Gerhard Weikum
Max-Planck Institute of Computer Science, Saarbruecken, Germany

Jean-Marc Ogier   Wenyin Liu
Josep Lladós (Eds.)

Graphics
Recognition

Achievements, Challenges, and Evolution

8th International Workshop, GREC 2009
La Rochelle, France, July 22-23, 2009
Selected Papers

 Springer

Volume Editors

Jean-Marc Ogier
Université de La Rochelle, France
E-mail: jean-marc.ogier@univ-lr.fr

Wenyin Liu
City University of Hong Kong, China
E-mail: csliuwy@cityu.edu.hk

Josep Lladós
Universitat Autònoma de Barcelona, Bellaterra, Spain
E-mail: josep@cvc.uab.es

Library of Congress Control Number: 2010928223

CR Subject Classification (1998): I.4, I.2.10, H.3, I.5, H.4, I.4.6

LNCS Sublibrary: SL 6 – Image Processing, Computer Vision, Pattern Recognition, and Graphics

ISSN   0302-9743
ISBN-10   3-642-13727-X Springer Berlin Heidelberg New York
ISBN-13   978-3-642-13727-3 Springer Berlin Heidelberg New York

springer.com

© Springer-Verlag Berlin Heidelberg 2010
Printed in Germany

Typesetting: Camera-ready by author, data conversion by Scientific Publishing Services, Chennai, India
Printed on acid-free paper   06/3180

Preface

This book contains refereed and improved papers presented at the 8th IAPR Workshop on Graphics Recognition (GREC 2009), held in La Rochelle, France, July 22–23, 2009. The GREC workshops provide an excellent opportunity for researchers and practitioners at all levels of experience to meet colleagues and to share new ideas and knowledge about graphics recognition methods. Graphics recognition is a subfield of document image analysis that deals with graphical entities in engineering drawings, sketches, maps, architectural plans, musical scores, mathematical notation, tables, diagrams, etc. GREC 2009 continued the tradition of past workshops held in the Penn State University, USA (GREC 1995, LNCS Volume 1072, Springer Verlag, 1996); Nancy, France (GREC 1997, LNCS Volume 1389, Springer Verlag, 1998); Jaipur, India (GREC 1999, LNCS Volume 1941, Springer Verlag, 2000); Kingston, Canada (GREC 2001, LNCS Volume 2390, Springer Verlag, 2002); Barcelona, Spain (GREC 2003, LNCS Volume 3088, Springer Verlag, 2004); Hong Kong, China (GREC 2005, LNCS Volume 3926, Springer Verlag, 2006); and (GREC 2007, LNCS Volume 5046, Springer Verlag, 2008).

The program of GREC 2009 was organized in a single-track 2-day workshop. It comprised several sessions dedicated to specific topics. For each session, there was an invited presentation describing the state of the art and stating the open questions for the session's topic, followed by a number of short presentations that contributed by proposing solutions to some of the questions or by presenting results of the speaker's work. Each session was then concluded by a panel discussion. Session topics included structural approaches for recognition and indexing, techniques towards vectorization, sketching interfaces, on-line processing, symbol and shape segmentation, description and recognition, historical documents analysis, indexing, spotting, and performance evaluation and ground truthing. In addition, a panel discussion on the state of the art and new challenges was organized as the concluding session of GREC 2009.

Continuing with the tradition of past GREC workshops, the program of GREC 2009 included graphics recognition contests. In particular, two contests were held: an arc segmentation contest, organized by Hasan S.M. Al-Khaffaf and Abdullah Zawawi Talib, and a symbol recognition contest, organized by Philippe Dosch, Ernest Valveny and Mathieu Delalandre. In these contests, for each contestant, test images and ground truths were prepared in order to have objective performance evaluation conclusions on their methods.

After the workshop, all the authors were invited to submit enhanced versions of their papers for this edited volume. The authors were encouraged to include ideas and suggestions that arose in the panel discussions of the workshop. Every paper was evaluated by two or three reviewers. At least one reviewer was assigned from the workshop attendees. Papers appearing in this volume were selected, and most of them were thoroughly revised and improved, based on the reviewers' comments. The structure of this volume is organized in seven sections, reflecting the workshop session topics.

We want to thank all paper authors and reviewers, contest organizers and participants, and workshop attendees for their contributions to the workshop and this volume. In particular, we gratefully acknowledge Karl Tombre for leading the panel discussion and the group of the University of La Rochelle for their great help in the local arrangements of the workshop.

The 9th IAPR Workshop on Graphics Recognition (GREC 2011) is planned to be held at Seoul, Korea.

April 2010
Jean-Marc Ogier
Liu Wenyin
Josep Lladós

Organization

General Chair

Jean-Marc Ogier

Program Co-chairs

Liu Wenyin
Josep Lladós

Local Arrangements Chairs

Jean-Marc Ogier
Mickael Coustaty
Nathalie Girard

Program Committee

Sergei Ablameyko, Belarus
Sébastien Adam, France
Gady Agam, USA
Dorothea Blostein, Canada
Thomas Breuel, Germany
Luigi Cordella, Italy
Bertrand Coasnon, France
David Doermann, USA
Philippe Dosch, France
Georgy Gimelfarb, New Zealand
Alexander Gribov, USA
Pierre Heroux, France
Joaquim Jorge, Portugal
Young-Bin Kwon, Korea
Sergei Levachkine, Mexico
Howard Leung, China

Rafael Lins, Brazil
Gerd Maderlechner, Germany
Umapada Pal, India
Tony Pridmore, UK
Jean-Yves Ramel, France
Gemma Sánchez, Spain
Zhengxing Sun, China
Eric Saund, USA
Antoine Tabbone, France
Chew-Lim Tan, Singapore
Karl Tombre, France
Lu Tong, China
Ernest Valveny, Spain
Toyohide Watanabe, Japan
Su Yang, China

Additional Referees

Karell Bertet, France
Mathieu Delalandre, Spain
Muriel Visani, France

Patrick Franco, France
Jean-Christophe Burie, France

Sponsoring Institutions

University of La Rochelle
Town of La Rochelle
French Department of Charente-Maritime
Region Poitou-Charentes
European Union
IAPR TC10 - International Association for Pattern Recognition
CNRS - French National Research Center
GDR I3 - French Research Group of the CNRS
Jouve Company
Sood Company
Aproged - Association des Professionnels pour l'économie numérique

Table of Contents

Use of Perceptive Vision for Ruling Recognition in Ancient Documents

Aurélie Lemaitre[1], Bertrand Coüasnon[2], and Jean Camillerapp[2]

[1] Université de Rennes 1, Campus de Beaulieu, F-35042 Rennes
[2] INSA, Avenue des Buttes de Coësmes, F-35043 Rennes
UMR IRISA, Campus de Beaulieu, F-35042 Rennes
Université Européenne de Bretagne, France

Abstract. Rulings are graphical primitives that are essential for document structure recognition. However in the case of ancient documents, bad printing techniques or bad conditions of conservation induce problems for their efficient recognition. Consequently, usual line segment extractors are not powerful enough to properly extract all the rulings of a heterogeneous document. In this paper, we propose a new method for ruling recognition, based on perceptive vision: we show that combining several levels of vision improves ruling recognition. Thus, it is possible to put forward hypothesis on the nature of the rulings at a given resolution, and to confirm or infirm their presence and find their exact position at higher resolutions.

We propose an original strategy of cooperation between resolutions and present tools to set up a correspondence between the elements extracted at each resolution. We validate this approach on images of ancient newspaper pages (dated between 1848 and 1944). We also propose to use the extracted rulings for the structure analysis of newspaper pages. We show that using more reliable extracted rulings simplifies and improves document structure recognition.

1 Introduction

Rulings are a base for the analysis of strongly structured documents like forms, tables or newspapers [3]. However, in the case of ancient or damaged documents, the recognition of rulings is complex: old printing techniques may produce irregular, speckled or dashed lines; bad method of conservation causes smearing ink and folding of the paper; the digitalization step can cause skew and curvature. Some examples of rulings that are difficult to detect in ancient documents are presented on figure 1.

The classical methods of the literature, based on projection or Hough transform are not always convenient: for example they do not easily deal with curvature and skew, and they use global parameters for a whole page. Consequently, other methods have been proposed in the literature. Gatos *et al.* present in [2] a specific process for the extraction of lines, based on gray scale transformation of the binary image. However, they require an a priori known length and width in order to determine the pixels that belong to the final ruling. Hadjar *et al.*

J.-M. Ogier, W. Liu, and J. Lladós (Eds.): GREC 2009, LNCS 6020, pp. 1–11, 2010.

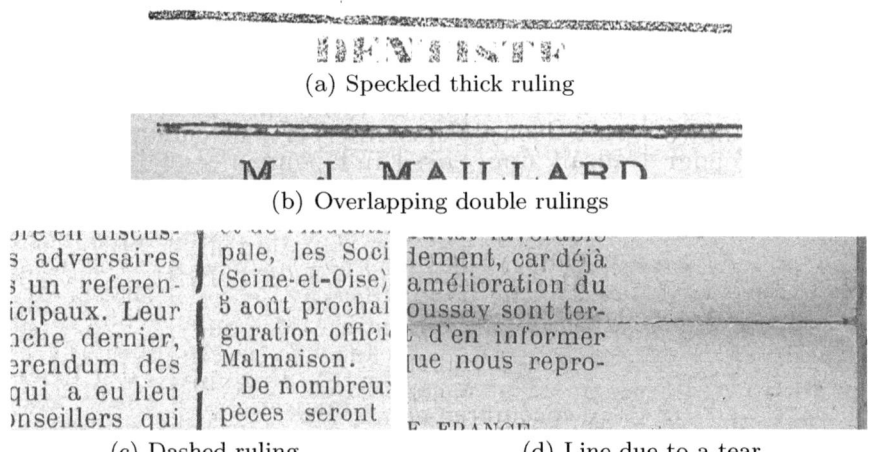

(a) Speckled thick ruling

(b) Overlapping double rulings

(c) Dashed ruling   (d) Line due to a tear

Fig. 1. Examples of rulings that are difficult to detect in ancient documents

in [4] try to detect discontinuous rulings, with an approach based on connected components. Nevertheless, they need to know a maximum distance to group components into a single ruling. A threshold is also required in the work of Liu *et al.* presented in [8]. In all these methods dedicated to damaged documents, a strong a priori knowledge is used to answer the problem of the variation of the thickness of the rulings in the same document, and to deal with over segmented rulings. Xi *et al.* [9] propose a method based on curvelets, using different resolutions of an image. However, their method is not able to deal with curvature or with slope bigger than 2 degrees, that frequently occurs in digitized documents.

Hori and Doerman present in [5] a multi-level analysis for form recognition. We follow this idea in order to solve more generally the problem of ruling recognition. We propose a new approach based on a mechanism used by the human eye: the perceptive vision. Indeed, in order to understand a document, our brain combines the visions at various levels of perception. In the case of rulings, we show that combining the analysis at various levels of resolution makes more accurate their extraction. Moreover, this method is able to extract rulings whatever their thickness, and even if they are damaged, without dedicated knowledge on their nature.

In this paper, we first present intuitively how using several levels of perception of an image can improve ruling recognition. It enables to deduce the perceptive mechanism that we implemented for ruling recognition. In section 4, we show the interest of the perceptive vision for ruling recognition with experiments on old newspapers. At last, thanks to the good obtained results, we propose an application for newspaper structure recognition.

2 Intuitive Approach

For the perceptive vision of rulings, we choose to combine three levels of perception. Indeed, we have experimentally noticed that using three levels of perception enables to obtain differences between visions that are significant but not too important. Thus, we build a multiresolution pyramid, by recursive low-pass filtering and sub-sampling. It recursively divides the dimensions of the images by 4.

On the three obtained resolutions, we apply an efficient line segment extractor, based on Kalman filtering [7]. Since our visual perception of the document varies, the detected line segments can differ depending on the resolution.

High resolution (initial image, ex: 300 dpi). When watching a document with a detailed local vision, that is to say at high resolution, we can see: pieces of double rulings (figure 1(b)), pieces of thin rulings (figure 1(c)) and elements due to noise. Thick rulings may not be perceptible because the white speckle takes too much importance.

Medium resolution (image divided by 4, ex: 75 dpi). When watching this document at medium resolution, we can see: pieces of double rulings (figure 1(b)), pieces of thin rulings (figure 1(c)), pieces of thick rulings (figure 1(a)), elements due to noise (figure 1(d)), and straight parts of capital letters.

Low resolution (image divided by 16, ex: 20 dpi). When watching this document with a global vision, that is to say at low resolution, we can see: thick rulings (figure 1(a)) that are dark enough, double rulings that appear as a single ruling (figure 1(b)), and some bold text lines that also appear as line segments. Thin rulings are too bright to appear at this level.

These different perceptions are gathered in table 1. We distinguish two kinds of rulings: the "true" rulings that are real structural elements (thick, thin, and double ruling) and the "false" ones that are text lines, pieces of characters and more globally noise.

Table 1. Perception of line segments at each resolution

Resolution	Low	Medium	High
Thin ruling	No	Yes	Yes
Thick ruling	Yes	Yes	No
Double ruling	Yes	Yes	Yes
Text lines	Yes	No	No
Characters	No	Yes	No
Noise	No	No	Yes

The table 1 shows that the perception of each element at the three resolutions can determine the kind of studied ruling. Thus, we deduce a strategy of analysis to recognize each kind of ruling. We now present this mechanism in details.

3   Principles of Implementation of the Perceptive Vision of Rulings

In this part, we present the elements that are required to implement a perceptive method of rulling. We present our strategy for combining the line segments that are detected at different resolutions. Then, we explain how we have treated two essential points that are the correspondance between resolutions, and the precise adjustment of the position of the rulings.

3.1   Strategy of Perceptive Cooperation

The perceptive strategy consists in putting forward hypothesis on the presence of a ruling, based on the perception at lower resolutions. This hypothesis may be confirmed or infirmed by the presence of elements on upper resolutions. This strategy is presented on figure 2.

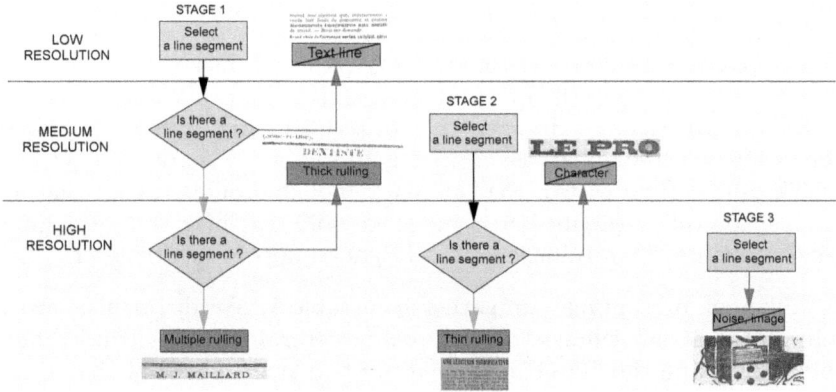

Fig. 2. Strategy for the combination of visions

The full analysis is realized in three stages: first the study of low resolution lines and their presence at highest resolutions (medium and high), then the study of the remaining medium resolution lines and their presence at high resolution, and at last the remaining line segments at high resolution.

First, we select a line segment at low resolution. If it has not any corresponding element at medium resolution, it may be a text line. Else, the presence of corresponding elements at high resolution makes it possible to confirm the hypothesis of a thick or a multiple line. We apply the same kind of mechanism to recognize the thin lines that must have corresponding elements at both medium and high resolutions. This strategy enables to eliminate line segments that are detected only at medium or high resolution.

It is important to see that the position of the line segment at the lowest resolution determines the localization of the search zone at the other resolutions. Moreover, the vision at the lowest resolution gives knowledge on the studied line: curvature, skew, length, thickness, that are used to determine how to gather the pixels at the highest resolution and to constitute the final ruling.

3.2 Correspondence between Resolutions

The strategy of perceptive cooperation requires to compare some line segments extracted at various resolutions. In order to easily realize this comparison, we use a single coordinate system for all the elements of each resolutions of the image. Thus, the figure 3 represents the three sets of line segments that are extracted at three resolutions, but that are stored in the same coordinate system.

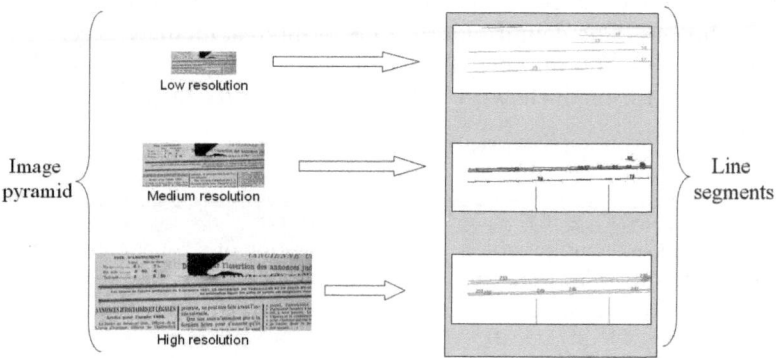

Fig. 3. Pyramid of images and associated extracted line segments stored in the same coordinate system

3.3 Adjustment of Position

Using a single coordinate system is necessary but not sufficient to find the exact position of a ruling. Indeed, the difference of scale between resolutions may cause errors of quantification. An example is presented on figure 4: a line segment has been detected at low resolution (figure 4(a)); if we directly transpose it in the upper resolution (figure 4(b)), its position does not exactly match with the present black pixels of the image.

In order to solve this problem, we propose to introduce a new concept, the *abstract line*, and a new tool of *positioning*. The abstract line is a polygonal

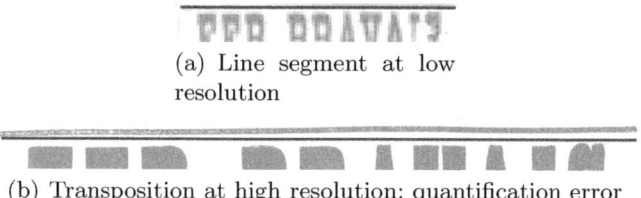

(a) Line segment at low resolution

(b) Transposition at high resolution: quantification error

Fig. 4. Example of quantification error

approximation of a line segment. Thus, it can be manipulated independently of any resolution. This abstract line is mainly used by the positioning tool that aims at adjusting precisely the position of a line, according to the presence of black pixels.

An example of use is presented on figure 5, and enables to treat the quantification error presented on figure 4. First, the abstract line is used to define a zone of interest (figure 5(b)). Indeed, at this step, even if the position of the abstract line need to be precised, it contains interesting information about the thickness, the slope and the curvature of the searched ruling, provided by the perception at lower resolutions. The definition of a zone of interest enables to select a set of relevant black pixels (figure 5(c)). Then, the role of the positioning tool consists in adjusting the position of the abstract line according to the present black pixels (figure 5(d)).

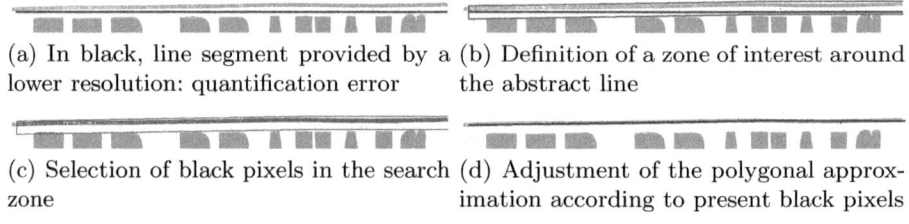

(a) In black, line segment provided by a lower resolution: quantification error

(b) Definition of a zone of interest around the abstract line

(c) Selection of black pixels in the search zone

(d) Adjustment of the polygonal approximation according to present black pixels

Fig. 5. Principle of positioning tool

To sum up, this positioning tool enables to correct small quantification errors and to provide a precise localization of the abstract lines. It is one of the key elements that enables to progressively adjust the position of the ruling, taking into account the different levels of resolution.

4   Evaluation of the Interest of Perceptive Vision

In order to show the interest of the perceptive vision, we apply our method for ruling recognition in ancient newspapers. More precisely, we propose to compare our perceptive approach with a monoresolution method.

4.1   Base of Evaluation

We applied our method for the recognition of heterogeneous rulings in pages of old newspapers, dated between 1848 and 1944, provided by the *Archives Départementales des Yvelines*. Some examples of studied images are presented on figure 6. We manually labeled a ground truth of 4967 rulings in 157 pages of newspapers

The studied images have an initial size of 5400*8000 pixels. The medium resolution images have a size of 1350*2000 pixels. The low resolution images have a size of 330*500 pixels.

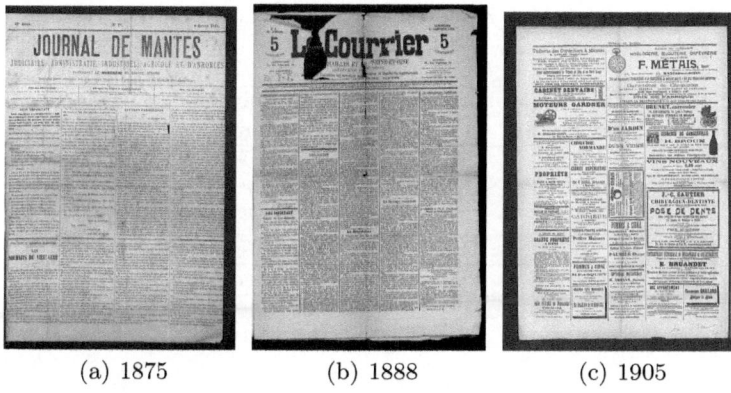

(a) 1875   (b) 1888   (c) 1905

Fig. 6. Examples of studied old newspaper pages

4.2 Metric

In order to evaluate the recognition rate, we have to make a correspondence, for each image, between the expected rulings of the ground truth, and the rulings found by the method. We classify the results according to four categories:

- *total recognition* is the number of expected rulings that have been properly recognized,
- *partial recognition* is the number of expected rulings that have been partially recognized, that is to say the recognized ruling is too short or too long,
- *omission* is the number of rulings of the ground truth that have not been detected,
- *noise* is the number or recognized ruling that does not correspond to any ruling of the ground truth.

4.3 Using Only One Resolution

Our first experiment consists in extracting the rulings using only one resolution of the image. Thus, we apply our line segment extractor at this given resolution and try to gather them into rulings, without using dedicated thresholds. We propose to compare the obtained results for three different resolutions. This experiment is illustrated on figure 7. The obtained results are presented on table 2.

Thanks to high resolution (figure 7(a)), we obtain 69.1% of complete recognition of rulings. Thus, this resolution enables the recognition of many rulings but is not able to deal with discontinuities: many rulings are only partially recognized, some thick speckled lines are forgotten. On the opposite, the low resolution (figure 7(c)) enables to detect completely thick rullings but omits all the thin rulings. The medium resolution (figure 7(b)) presents intermediate results but that are not better: only 52.6% recognition.

As a conclusion, the obtained results show that, if we have to deal with only one resolution, the best recognition is obtained with the initial image. Now, we will compare this results with our perceptive method.

Table 2. Comparison of three monoresolution approaches on 4967 rulings extracted from 157 newspaper pages

Method Resolution	Monoresolution High	Monoresolution Medium	Monoresolution Low
Total recognition	69.1%	52.6%	5.6%
Partial recognition	21.2%	8.5%	2.4%
Omission	9.7%	38.8%	92.0 %
Noise	93.9%	16.6%	2.5 %
Time per image	10.9 sec	3.3 sec	2.6 sec

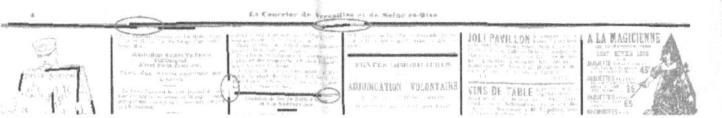

(a) Rulings built with the single high resolution: over-splitting (circled)

(b) Rulings built with the single medium resolution: over-splitting (circled)

(c) Rulings built with the single low resolution: omission of many thin rulings

(d) Rulings built thanks to perceptive approach: good recognition

Fig. 7. Comparison of rulings built with different methods: interest of the perceptive approach

4.4   Using Perceptive Vision

The figure 7(d) shows the good recognition of rulings thanks to the perceptive vision. The table 3 present the results obtained with our perceptive method, faced with the best results obtained in monoresolution, *ie* with high resolution.

The results show the significant interest of our perceptive method for ruling recognition. Thus, with the perceptive approach, 94.4% rulings are entirely recognized, againts 69.1% in monoresolution.

Table 3. Comparison of monoresolution approach with our perceptive method, on 4967 rulings extracted from 157 newspaper pages

Method	Monoresolution (high)	Perceptive vision
Total recognition	69.1%	94.4%
Partial recognition	21.2%	3.0%
Omission	9.7%	2.6%
Noise	93.9%	31.1%
Time per image	10.9 sec	15.4 sec

The computing time is bigger for the perceptive method, due to the need to apply the line segment extractor at three resolutions. However, the large increase of recognition rate compensates for the longer running time.

The weak omission rates with perceptive method (2.6% *vs* 9.7%) is due to the better recognition of thick rulings and speckled rulings that are well perceived thanks to the use of low resolution.

The small partial recognition rate for perceptive approach (3.0% *vs* 21.2%) shows the interest to be guided by hypotheses that are emitted at low resolution, in order to combine line segments at high resolution and to form a single ruling.

The smaller noise rate with perceptive method (31.1% *vs* 93.9%) is due to the prediction/verification strategy that enables to validate the presence of a ruling only if it has been perceived at last on two resolutions. The remaining noise is mainly due to vertical line segments in capital letters of the title. The introduction of more precise knowledge about newspaper pages will enable to take this into account.

5   Application to a Real Problem

The obtained results show that a perceptive approach enable to detect more reliable rulings. These rulings can be used as a base for document structure recognition.

As an example of application, we introduces our perceptive ruling extractor into a generic method for document structure recognition, DMOS-P [1] [6]. Thus, we realized a grammatical description of the structure of newspaper pages, and its decomposition into boxes. The terminals of this grammar are the rulings. Thanks to this description in DMOS-P method, we obtain a system that is able to segment newspaper pages into boxes (figure 8).

We evaluate this method on newspaper pages dated between 1859 and 1944, provided by the *Archives Départementales des Yvelines*. More precisely, we noticed that some of the pages were more complicated than other. Thus, in first pages of the newspapers (figure 8(a)), the rulings have a regular thickness, whereas the last pages (figure 8(c)) contains advertisement with varied kinds of rulings. Consequently, we have created two bases of evaluation: one base of 179 first pages, in which we have built a ground-truth of 4148 boxes, and a second base of 79 last pages with 3480 boxes.

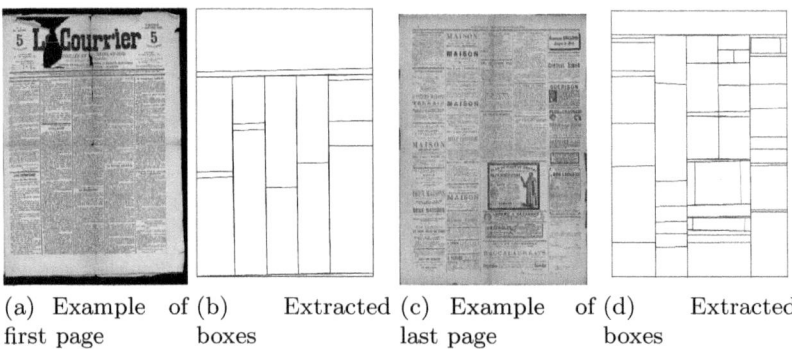

(a) Example of first page   (b) Extracted boxes   (c) Example of last page   (d) Extracted boxes

Fig. 8. Segmentation of a newspaper page using rulings detected with perceptive vision

Table 4. Extracting rulings on 179 first pages (good quality rulings)

Version	Boxes	Over-segmentation	Under-segmentation
Monoresolution	4148	10.5%	10.7%
Perceptive approach	4148	10.2%	7.9%
Improvement		–	**33%**

Table 5. Extracting rulings on 79 last pages (difficult rulings)

Version	Boxes	Over-segmentation	Under-segmentation
Monoresolution	3480	17.1%	11.3%
Perceptive approach	3480	13.7%	6.2%
Improvement		**20%**	**45%**

On these two bases, we propose to compare the segmentation into boxes when the rulings are obtained in monoresolution, and when the rulings are obtained with perceptive vision. The obtained results on the two bases are presented on tables 4 and 5.

The results show that using the perceptive vision for ruling detection improves structure analysis. The improvement are more significant on the last pages. Indeed, on these pages, the rulings are particularly difficult to recognize. On these base, thanks to the rulings extracted with perceptive vision, under-segmentation decreases by 45% while over-segmentation decreases by 20%.

6   Conclusion

In this paper, we have presented a new method for rulings recognition in ancient or damaged documents. This method is generic and does not require specific a priori knowledge on dimensions of the rulings, contrary to the approaches found in the literature.

We have shown that the combination of different points of view is necessary in heterogeneous documents because thick or double rulings are not well perceived at high resolution, whereas thin rulings are not detected at low resolution. Consequently, we proposed a prediction/verification strategy to combine the different perceptions. This works leads to the development of a new positioning tool that enables to set up a correspondence between line segments obtained at different resolutions, and to deal with quantification errors.

The obtained results show the interest of each resolution for a good ruling recognition. The perceptive strategy enables a large increase of rulings recognition. At last, we have also demonstrated that using more reliable detected rulings improve the analysis of more complex structures.

Thanks to the interesting results, this method has led to an industrial transfer to Evodia company, who has treated more than 45,000 newspaper pages thanks to our perceptive method.

References

1. Coüasnon, B.: DMOS, a generic document recognition method: Application to table structure analysis in a general and in a specific way. International Journal on Document Analysis and Recognition, IJDAR 8(2), 111–122 (2006)
2. Gatos, B., Mantzaris, S.L., Chandrinos, K.V., Tsigris, A., Perantonis, S.J.: Integrated algorithms for newspaper page decomposition and article tracking. In: ICDAR 1999: Proceedings of the Fifth International Conference on Document Analysis and Recognition, p. 559. IEEE Computer Society, Washington (1999)
3. Gatos, B., Mantzaris, S.L., Antonacopoulos, A.: First international newspaper segmentation contest. In: International Conference on Document Analysis and Recognition (ICDAR 2001), p. 1190. IEEE Computer Society, Los Alamitos (2001)
4. Hadjar, K., Hitz, O., Ingold, R.: Newspaper page decomposition using a split and merge approach. In: International Conference on Document Analysis and Recognition (ICDAR 2001), p. 1186. IEEE Computer Society, Los Alamitos (2001)
5. Hori, O., Doermann, D.S.: Robust table-form structure analysis based on box-driven reasoning. In: ICDAR, pp. 218–221 (1995)
6. Lemaitre, A., Camillerapp, J., Coüasnon, B.: Multiresolution cooperation improves document structure recognition. International Journal on Document Analysis and Recognition (IJDAR) 11(2), 97–109 (2008)
7. Leplumey, I., Camillerapp, J., Queguiner, C.: Kalman filter contributions towards document segmentation. In: Proceedings of International Conference on Document Analysis and Recognition (ICDAR 1995), pp. 765–769 (1995)
8. Liu, F., Luo, Y., Hu, D., Yoshikawa, M.: A new component based algorithm for newspaper layout analysis. In: International Conference on Document Analysis and Recognition (ICDAR 2001), p. 1176. IEEE Computer Society, Los Alamitos (2001)
9. Xi, D., Lee, S.W.: Extraction of reference lines and items from form document images with complicated background. Pattern Recognition 38(2), 289–305 (2005)

Fuzzy Intervals for Designing Structural Signature: An Application to Graphic Symbol Recognition

Muhammad Muzzamil Luqman[1,2], Mathieu Delalandre[1], Thierry Brouard[1], Jean-Yves Ramel[1], and Josep Lladós[2]

[1] Laboratoire d'Informatique, Université François Rabelais de Tours - France
muhammadmuzzamil.luqman@etu.univ-tours.fr,
{mathieu.delalandre,brouard,ramel}@univ-tours.fr
[2] Computer Vision Center, Universitat Autònoma de Barcelona - Spain
josep@cvc.uab.es

Abstract. The motivation behind our work is to present a new methodology for symbol recognition. The proposed method employs a structural approach for representing visual associations in symbols and a statistical classifier for recognition. We vectorize a graphic symbol, encode its topological and geometrical information by an attributed relational graph and compute a signature from this structural graph. We have addressed the sensitivity of structural representations to noise, by using data adapted fuzzy intervals. The joint probability distribution of signatures is encoded by a Bayesian network, which serves as a mechanism for pruning irrelevant features and choosing a subset of interesting features from *structural signatures of underlying symbol set*. The Bayesian network is deployed in a supervised learning scenario for recognizing query symbols. The method has been evaluated for robustness against degradations & deformations on pre-segmented 2D linear architectural & electronic symbols from GREC databases, and for its recognition abilities on symbols with context noise *i.e. cropped symbols*.

Keywords: symbol recognition, overlapping fuzzy interval, structural signature, Bayesian network.

1   Introduction

Graphics recognition deals with graphic entities in document images and is a subfield of document image analysis. These graphic entities could correspond to symbols, mathematical formulas, musical scores, silhouettes, logos etc., depending on the application domain. Llados & Sanchez [1] have very correctly pointed out that the documents from electronics, engineering, music, architecture and various other fields use domain-dependent graphic notations which are based on particular alphabets of symbols. These industries have a rich heritage of hand-drawn documents and because of high demands of application domains, overtime symbol recognition is becoming core goal of automatic image analysis and understanding systems. Hand-drawn based user interfaces, backward conversion from raster images to CAD, content based retrieval from graphic document databases and browsing of graphic documents are some of the typical applications of symbol recognition. Detailed discussion on the application domains of symbol recognition has been provided by Chhabra [2] and Llados et al. [3].

J.-M. Ogier, W. Liu, and J. Lladós (Eds.): GREC 2009, LNCS 6020, pp. 12–24, 2010.
© Springer-Verlag Berlin Heidelberg 2010

The research surveys by Chhabra [2], Llados et al. [3], Cordella & Vento [4] and Tombre et al. [5] provide a detailed and state of the art historical review of work done in the field of symbol recognition over last two decades. Graphic symbol recognition is generally approached by syntactic, structural, statistical or hybrid methods of pattern recognition. Syntactic approaches involve the use of grammars and syntactical parsing [6] and are usually considered as a special case of structural approaches [3]. Structural and statistical approaches are normally differentiated by the data structures that they employ for pattern representation. Structural approaches use symbolic data structures such as strings, trees and graphs, whereas, the statistical approaches are characterized by the use of feature vectors for representing patterns [6].

Cordella & Vento [4] have provided a detailed listing of methods employing different structural, statistical or hybrid approaches for graphic symbol recognition. The symbolic data structures are very powerful in their representational capabilities. However, the structural approaches lack in the availability of efficient tools for matching and comparison [6]. On the other hand, the use of feature vectors by many statistical approaches limits their representational capabilities but the availability of a much richer repository of mathematical tools in statistical domain [6] and the associated computational advantages, makes them a more favorable choice in certain cases. The use of light weight feature vectors and computationally powerful statistical classifiers allows to design fast and efficient systems which are sufficiently scalable and domain independent.

Several research works have been undertaken to combine structural and statistical approaches, with the aim to utilize their strengths and avoid the weaknesses. Delalandre et al. [7] employ a statistical technique for extracting the components *(that compose the symbol)* and the loops formed by these components. Afterwards, they construct graphs from these loops and deploy an inexact graph matching algorithm for recognition. Hse et al. [8] have used Zernike descriptor with various statistical classifiers for sketched symbol recognition. Barrat et al. [9] have used various shape descriptors with naive Bayes classifier for symbol recognition. Among the hybrid approaches, specially over last decade, the vectorial signatures *(also referred as structural signatures)* (2.1) have gained considerable attention [10]. The vectorial signatures encode the geometric and topologic relations between elementary vectorial primitives. Many recognition and spotting systems have been developed around these signatures [11,12,13,14,15,16].

The rest of paper is organized as follows: §2 gives a general description of our method, §3 is devoted to detailed description of each part of the proposed system, experimental results are presented in §4 and the paper is concluded with some future directions of work in §5.

2   The Proposed Method and Related Works

In this section we outline our proposed method and highlight its placement with past works *(that employ a similar methodology)*. The method is a hybrid of structural and statistical pattern recognition approaches; we exploit representational power of structural approaches and employ computational efficiency of statistical classifiers.

2.1   Vectorial Signatures *(or Structural Signatures)*

Ventura & Schettini [17] were the first to introduce the concept of vectorial signatures for symbol recognition, back in 1994. They extract thin and thick elementary structures *(in terms of geometric constraints between line primitives)* from symbol, describe them by local features and create a signature for symbol. Finally, they deploy a hypothesis-and-test paradigm for detecting occurrences of symbols in line drawings. A recent overview of past works using structural signatures is given by Rusinol [10]. These works include [11,12,13,14,15]. Coustaty et al. [15] have applied structural signature for symbol recognition. They extract segments from the symbol in image by Hough transform, describe their spatial organization by a topological graph and compute a structural signature. They have deployed a Galois Lattice as classifier and have shown the robustness of their method against high levels of degradation. Dosch et al. [14] *(originally proposed for symbol spotting)* and its improvement by Rusinol et al. [13] *(for both recognition and spotting)*, work on a vectorial representation. They extract spatial relationships between pairs of segments and hierarchically organize them into basic shapes. Their signature is comprised of the cardinalities of occurrences of spatial relations between segments in a shape. Zhang et al. [12] work on a vectorial representation, and use circle and arcs as well, in addition to line primitives. They define a structural signature in terms of relations between these primitives and employ a brute force comparison for recognizing a query signature. Qureshi et al. [11] vectorize a graphic symbol, construct its Attributed Relational Graph (ARG) and compute a structural signature for it (the G-Signature). For classification of query symbol they use nearest neighbors rule with Euclidean distance as measure of dissimilarity. Their G-Signature is discriminant in case of hand-drawn deformations and has been shown invariant of rotation and scaling.

These works show the invariance of structural features to transformations and illustrates their representational capabilities. However, we argue that the sensitivity of structural signatures to noise (degradations & deformations) limits these systems to be used for real-life applications, and to scale to large number of symbol models. In this work, we propose to take structural signatures to the domain of fuzzy sets, to enable them to cope with uncertainties, and extend our previous work [16], which in fact takes forward the work of [11]. We have selected Bayesian networks *for dealing with uncertainty in symbol signatures during learning and recognition phases*, and propose to use *(overlapping)* fuzzy intervals instead of rigid boundaries [16] for features in signature. Our motivation behind these choices are the previous works involving Bayesian framework [18] and fuzzy sets [19], that have shown the significance of these methodologies in improving robustness against uncertainties in data. We have increased the scalability capabilities of structural signatures by employing uncertainty-management during signature design, learning and classification phases. The signature is given in Fig.2 and it is discussed in §3.2.

2.2   Bayesian Networks

Bayesian networks are probabilistic graphical models and are represented by their structure and parameters. Structure is given by a directed acyclic graph and it encodes the dependency relationships between domain variables whereas parameters of the network are conditional probability distributions associated with *(each of)* its nodes. A Bayesian network, like other probabilistic graphical models, encodes the joint probability distribution of a set of random variables, and could be used to answer all possible inference queries on these variables. A humble introduction to Bayesian networks is in [20,21].

Bayesian networks have already been applied successfully to a large number of problems in machine learning and pattern recognition and are well known for their power and potential of making valid predictions under uncertain situations. But in our knowledge there are only a few methods which use Bayesian networks for graphic symbol recognition. Recently Barrat et al. [9] have used the naive Bayes classifier in a *'pure'* statistical manner for graphic symbol recognition. Their system uses three shape descriptors: Generic Fourier Descriptor, Zernike descriptor & R-Signature 1D, and applies dimensionality reduction for extracting the most relevant and discriminating features to formulate a feature vector. This reduces the length of their feature vector and eventually the number of variables *(nodes)* in Bayesian network. The naive Bayes classifier is a powerful Bayesian classifier but it assumes a strong independence relationship among attributes given the class variable. We believe that the power of Bayesian networks is not fully explored; as instead of using predefined dependency relationships, if we find dependencies between all variable pairs from underlying data we can obtain a more powerful Bayesian network classifier. This will also help to ignore irrelevant variables and exploit the variables that are interesting for discriminating symbols in underlying symbol set (§3.3 and §3.4).

2.3   Originality of Our Approach

Our method is an original adaptation of Bayesian network learning for the problem of graphic symbol recognition. *For symbol representation*, we use a structural signature. The signature is computed from the ARG of symbol and is composed of geometric & topologic characteristics of the structure of symbol. We use *(overlapping)* fuzzy intervals for computing noise sensitive features in signature. This increases the ability of our signature to resist against irregularities [19] that may be introduced in the shape of symbol by deformations & degradations. *For symbol recognition*, we employ a Bayesian network. This network is learned from underlying training data by using the quite recently proposed genetic algorithms by Delaplace et al. [22]. A query symbol is classified by using Bayesian probabilistic inference *(on encoded joint probability distribution)*. We have selected the features in signature very carefully to best suit them to linear graphic symbols and to restrict their number to minimum; as Bayesian network algorithms are known to perform better for a smaller number of nodes. Our structural signature makes the proposed system robust & independent of application domains and it could be used for all types of 2D linear graphic symbols. Also, relatively basic computations are involved for recognizing a query symbol which enables our system to respond in real time and it could be used, for instance, as a preprocessing step of a traditional symbol recognition method or for indexation & browsing of graphic documents.

3   Detailed Description

In this section we describe the representation, description, learning and classification phases of our system. These phases have been outlined by Cordella & Vento [4] in their research survey on symbol recognition. The authors have remarked that *almost all graphics recognition systems could be looked upon as operating in representation, description and classification phases.*

3.1   Representation Phase

This important & basic phase of our system concerns the formation of an Attributed Relational Graph (ARG) data structure, *as proposed by Qureshi et al. [11]*, and is summarized in Fig.1. The topological and geometric details about structure of symbol are extracted and are represented by an ARG. In first step, the symbol is vectorized and is represented by a set of primitives *(labels 1, 2, 3, 4 in Fig.1)*. In next step, these primitives become nodes and topological relations between them become arcs in ARG. Nodes have *'relative length' (normalized between 0 and 1)* and *'primitive-type' (Vector for filled regions of shape and Quadrilateral for thin regions)* as attributes; whereas arcs of the graph have *'connection-type' (L, X, T, P, S)* and *'relative angle' (normalized between 0° and 90°)* as attributes.

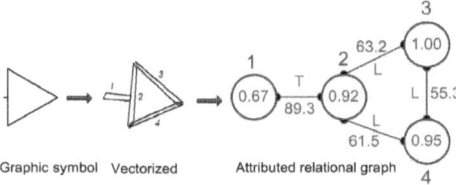

Graphic symbol   Vectorized   Attributed relational graph

Fig. 1. The representation phase

3.2   Description Phase *(Fuzziness of Signature)*

This phase concerns the extraction of features and computation of structural signature, from ARG of an underlying symbol. In order to increase the robustness of our signature and to enable it to resist the irregularities & uncertainties introduced in shape of symbol as result of noise *(degradations & deformations)*, we introduce *(overlapping)* fuzzy intervals to our previous work [16]. Fig.2 presents our proposed structural signature for a symbol. Our motivation behind choosing structural features is to exploit their ability to identify symbols in context [13].

Group-1 & Group-2 *(of features in structural signature)* encode the size of symbol and arrangement of its primitive components, respectively. These features discriminate between symbols of different sizes and also between symbols of same size but with a different arrangement of primitives. Group-3 encodes the density of connections for nodes. This group discriminates between symbols that have similar number of primitives with a similar arrangement but different density of connections at nodes. Group-4 & Group-5 exploits the attributes of primitives and encodes details of length & angle

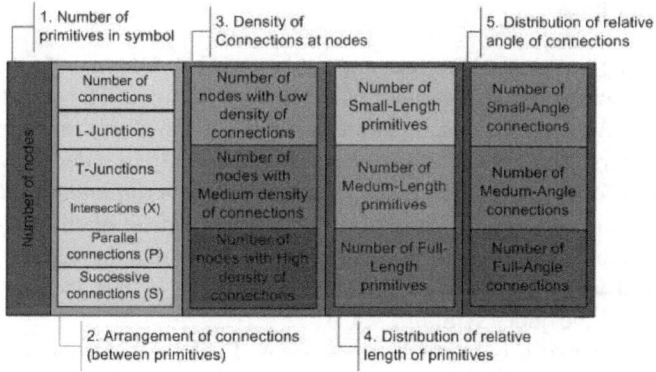

Fig. 2. Structural signature for graphic symbol

attributes. These groups complement the criteria (of Groups-1, Group-2 & Group-3) for outlining boundaries, between symbol classes, in feature space.

The computation of features in Group-1 & Group-2 is straightforward and is achieved by counting the relevant information in ARG of graphic symbol. For features in Group-3, we first compute a list of connection-density counts of all nodes of all ARG *of symbols in underlying symbol set*. And then use this list of connection-density counts for finding connection-density intervals for computing feature in Group-3 of structural signature. We use a histogram based binning technique from [23] for this purpose. The technique is originally proposed for discretization of continuous data and is based on use of Akaike Information Criterion (AIC) [24]. It starts with an initial m-bin histogram of data and finds optimal number of bins for underlying data. Two adjacent bins are merged by using an AIC-based cost function as criterion; until the difference between AIC-before-merge and AIC-after-merge becomes negative. We arrange these bins in overlapping fashion (fuzzy approach) and use them as intervals for computing number of nodes lying in different connection-density intervals. This gives us a distribution of nodes in structural ARG with low, medium and high density of connections, which we use as features of our signature.

Group-4 (and Group-5) is computed by dividing relative length (and relative angle) in three overlapping intervals, as shown in Fig.3 (and Fig.4). The overlapping intervals (fuzzy approach) handle the irregularities caused by distortions and degradations, and ensure that these irregularities do not affect the signature.

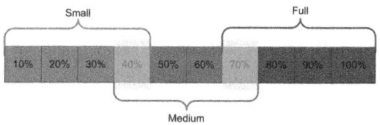

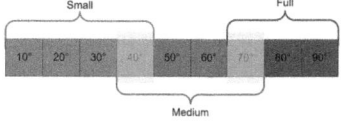

Fig. 3. Intervals for computing number of small, medium and full length primitives

Fig. 4. Intervals for computing number of small, medium and full angle connections

3.3 Learning Phase

After representing the symbols in learning set by ARG and describing them by structural signatures, we proceed to learning of a Bayesian network. The signatures are first discretized [23]. We discretize each feature variable (of signature) separately and independently of others. The class labels are chosen intelligently in order to avoid the need of any discretization for them. The discretization of 'number of nodes' and 'number of arcs' achieves a comparison of similarity of symbols (instead of strict comparison of exact feature values). This discretization step also ensures that the features in signature of query symbol will look for symbols whose number of nodes and arcs lie in same intervals as that of the query symbol.

The Bayesian network is learned in two steps. First we learn the structure of the network by genetic algorithms proposed by Delaplace et al. [22]. These are evolutionary algorithms, but in our case they have provided stable results *(for a given dataset multiple invocations always returned identical network structures)*. Each feature in signature becomes a node of network. The goal of structure learning stage is to find the best network structure from underlying data which contains all possible dependency relationships between all variable pairs. The structure of the learned network depicts the dependency relationships between different features in signature. Fig.5 shows one of the learned structures from our experiments. The second step is learning of parameters of network; which are conditional probability distributions $\Pr(node_i|parents_i)$ associated to nodes of the network and which quantify the dependency relationships between nodes. The network parameters are obtained by maximum likelihood estimation (MLE); which is a robust parameter estimation technique and assigns the most likely parameter values to best describe a given distribution of data. We avoid null probabilities by using Dirichlet priors with MLE. The learned Bayesian network encodes joint probability distribution of the symbol signatures.

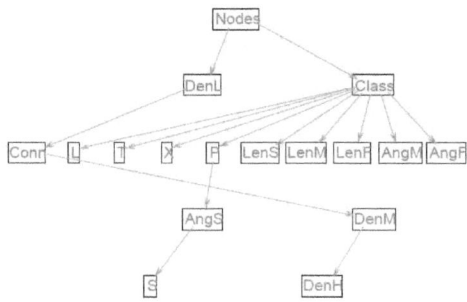

Fig. 5. A Bayesian network structure after learning; each node corresponds to a feature variable

The conditional independence property of Bayesian networks helps us to ignore irrelevant features *in structural signature* for an underlying symbol set. This property states that a node is conditionally independent of its non-descendants given its immediate parents [20]. Conditional independence of a node in Bayesian network is fully exploited during probabilistic inference (see §3.4) and thus helps us to ignore irrelevant features for an underlying symbol set while computing posterior probabilities for different symbol classes (see §3.4).

3.4   Classification Phase *(Graphic Symbol Recognition)*

For recognizing a query symbol we use Bayesian probabilistic inference on the encoded joint probability distribution. This is achieved by using junction tree inference engine which is the most popular exact inference engine for Bayesian probabilistic inference and is implemented in [23]. The inference engine propagates the evidence *(signature of query symbol)* in network and computes posterior probability for each symbol class. Equation 1 gives Bayes rule for our system. It states that posterior probability or probability of a symbol class c_i given a query signature 'evidence e' is computed from likelihood (probability of e given c_i), prior probability of c_i and marginal likelihood (prior probability of e). The marginal likelihood *(Equation 3)* is to normalize the posterior probability; it ensures that the probabilities fall between 0 and 1.

$$\Pr(c_i|e) = \frac{\Pr(e, c_i)}{\Pr(e)} = \frac{\Pr(e|c_i) \times \Pr(c_i)}{\Pr(e)} \qquad (1)$$

where,

$$e = f1, f2, f3, ..., f16 \qquad (2)$$

$$\Pr(e) = \sum_{i=1}^{k} \Pr(e, c_i) = \sum_{i=1}^{k} \Pr(e|c_i) \times \Pr(c_i) \qquad (3)$$

The posterior probabilities are computed for all 'k' symbol classes and the query symbol is then assigned to class which maximizes the posterior probability i.e. which has highest posterior probability for the given query symbol.

4   Experimentation

The organization of four international symbol recognition contests over last decade [25,26,27,28], has provided our community an important test bed for evaluation of methods over a standard dataset. These contests were organized to evaluate and test the symbol recognition methods for their scalability and robustness against binary degradation and vectorial deformations. The contests were run on pre-segmented linear symbols from architectural and electronic drawings, as these symbols are representative of a wide range of shapes [26]. GREC2005 [27] & GREC2007 [28] databases are composed of the same set of models, whereas GREC2003 [26] database is a subset of GREC2005.

4.1   Symbols with Vectorial and Binary Noise

We experimented with synthetically generated 2D symbols of models collected from database of GREC2005 [27,31]. In order to get a true picture of the performance of our

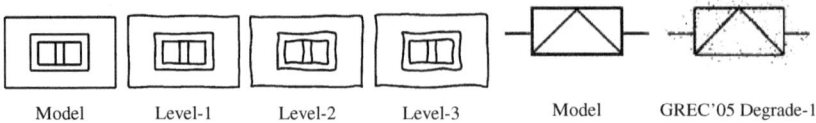

| Model | Level-1 | Level-2 | Level-3 | Model | GREC'05 Degrade-1 |

Fig. 6. Model symbol with deformations; used for simulating hand-drawn symbols and applied using an application from project Epeires [27]

Fig. 7. Model symbol with degraded example; used to simulate photocopying / printing / scanning and applied using ImageMagick [29] & QGar package [30]

proposed method on this database, we have experimented with 20, 50, 75, 100, 125 & 150 symbol classes. We generated our own learning & test sets *(based on deformations & degradations of GREC2005)* for our experiments. For each class the perfect symbol *(the model)* along with its 36 rotated and 12 scaled examples was used for learning; as the features have already been shown invariant to scaling & rotation [11,16] and because of the fact that generally Bayesian network learning algorithms perform better on datasets with large number of examples. The system has been tested for its scalability on clean symbols (rotated & scaled), various levels of vectorial deformations and for binary degradations of GREC symbol recognition contest (Fig.6 and Fig.7). Each test dataset was composed of 10 query symbols for each class.

Table 1. Results of symbol recognition experiments

Number of classes (models)		20	50	75	100	125	150
Clean symbols (rotated & scaled)		100%	100%	100%	100%	100%	99%
Hand-drawn deformation	Level-1	99%	96%	93%	92%	90%	89%
	Level-2	98%	95%	92%	90%	89%	87%
	Level-3	95%	77%	73%	70%	69%	67%
Binary degrade		98%	96%	93%	92%	89%	89%

Table 1 summarizes the experimental results. A 100% recognition rate for clean symbols illustrates the invariance of our method to rotation & scaling. Our method outperforms all GREC participants *(available results from GREC2003 [26] and GREC2005 [27] competetions)* in scalability tests and is comparable to contest participants for low levels of deformation & degradations. The recognition rates decrease with level of deformation and drop drastically for high binary degradations. This is an expected behaviour and is a result of the irregularities produced in symbol signature; which is a direct outcome of the noise sensitivity of vectorization step, as also pointed out by [3]. We used only clean symbols for learning and (thus) the recognition rates truely illustrate the robustness of our system against vectorial and binary noise. Fig.8 compares our results with [11] *(The system proposed in [11] presents recognition rates only for 20 models)*.

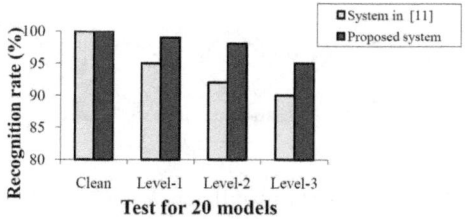

Fig. 8. Comparison of recognition rates

4.2 Symbols with Contextual Noise

A second set of experimentation was performed on a synthetically generated corpus, of symbols cropped from complete documents [32]. These experiments focused on evaluating the robustness of the proposed system against context noise i.e. the structural noise introduced in symbols when they are cropped from documents. We believe that this type of noise gets very important when we are dealing with symbols in context in complete documents and to the best of our knowledge; no results have yet been published for this type of noise. We have performed these experiments on two subsets of symbols: consisting of 16 models from floor plans and 21 models from electronic diagrams. The models are derived from GREC2005 database [27,31] and are given in Fig.9 and Fig.10. For each class the perfect symbol (model), along with its 36 rotated and 12 scaled examples was used for learning. The examples of models, for learning, were generated using ImageMagick [29] and the test sets were generated synthetically [32] with different levels of context-noise (Fig.11) in order to simulate the cropping of symbols from documents. Test symbols were randomly rotated & scaled and multiple query symbols were included for each class. The test datasets are available at [33].

Table 2 summarizes the results of experiments for context noise. We have not used any sophisticated de-noising or pretreatment and our method derives its ability to resist against context noise, directly from underlying vectorization technique, the fuzzy approach used for computing structural signature and the capabilities of Bayesian

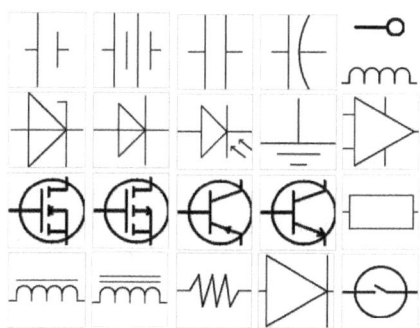

Fig. 9. Model symbols from electronic drawings

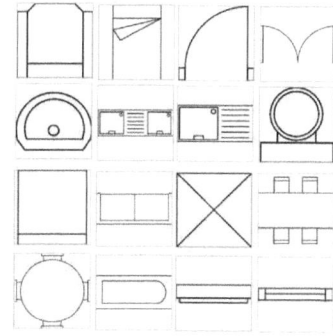

Fig. 10. Model symbols from floor plans

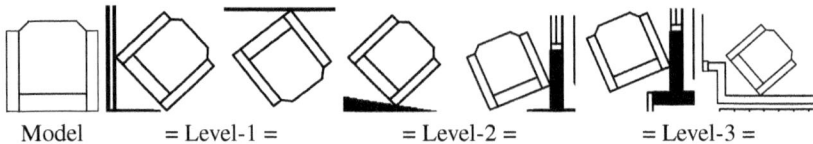

| Model | = Level-1 = | = Level-2 = | = Level-3 = |

Fig. 11. An arm chair with different levels of context noise

networks to cope with uncertainties. The models for electronic diagrams contain symbols consisting of complex arrangement of lines & arcs, which affects the features in structural signature as the employed vectorization technique is not able to cope with arcs & circles; as is depicted by the recognition rates for these symbols. But keeping in view the fact that we have used only clean symbols for learning and noisy symbols for testing, we believe that the results show the ability of our signature to exploit the sufficient structural details of symbols and it could be used to discriminate and recognize symbols with context noise.

Table 2. Results of symbol recognition experiments for context noise

	Noise	Model symbol (classes)	Query symbol (each class)	Recognition rate (match with topmost result)	Recognition rate (a match in top-3 results)
Floor plans	Level-1	16	100	84%	95 %
	Level-2	16	100	79%	90 %
	Level-3	16	100	76%	87 %
Average recog. rate				80%	91%
Electronic diagrams	Level-1	21	100	69%	89%
	Level-2	21	100	66%	88%
	Level-3	21	100	61%	85%
Average recog. rate				65%	87%

5   Conclusion

Structural methods are the strongest methods for graphics representation and statistical classifiers provide efficient recognition techniques. By designing a mechanism to convert a structural representation to feature vector, the whole range of statistical tools *(classifiers)* are opened for that structural representation. First, we have presented an *overlapping* fuzzy interval based methodology to convert an ARG based representation of graphic symbol to a feature vector. Our signature exploits the structural details of symbols. And second, an original adaptation of Bayesian network learning for the problem of graphic symbol recognition, has been presented. We represent symbols by signatures and encode their joint probability distribution by a Bayesian network. We then use Bayesian probabilistic inference on this network to classify query symbols. Experimental results show an improvement in recognition rates and scalability of the old system.

Our system does not use any sophisticated de-noising or pretreatment and it drives its power to resist against deformations and degradations, directly from representation, description, learning and classification phases. We have addressed the issue of sensitivity of structural representations to noise and deformations; by introducing *overlapping* fuzzy intervals for computing structural signature. The features in signature are affected by the small quadrilaterals that are produced during vectorization (in case of noisy symbols), which produce irregularities in signature. The use of fuzzy approach for computing structural signature and probabilistic inference of Bayesian networks gives our system a certain level of resistance against these irregularities.

We believe that the recognition rates will be improved for real learning sets which include deformed and degraded examples as well. The system is extensible to new models. The signature is invariant to rotation & scaling and robust against deformations & degradations. It is adapted to underlying symbol set and has a resistance against context noise. The proposed system has the capability to generate its learning set from models and could be used for 2D linear symbols from a wide range of application domains. The use of lightweight signature and statistical classifier makes our method efficient *(could be used for real time queries)* and scalable to a large number of symbol classes. In future we plan to use this method, as quick graphic symbol discrimination technique, for designing a system for symbol spotting and indexation of line drawing documents.

References

1. Llados, J., Sanchez, G.: Symbol recognition using graphs. In: ICIP, pp. 49–52 (2003)
2. Chhabra, A.K.: Graphic symbol recognition: An overview. In: Chhabra, A.K., Tombre, K. (eds.) GREC 1997. LNCS, vol. 1389, pp. 68–79. Springer, Heidelberg (1998)
3. Llados, J., Valveny, E., Sanchez, G., Marti, E.: Symbol recognition: Current advances and perspectives. In: Blostein, D., Kwon, Y.-B. (eds.) GREC 2001. LNCS, vol. 2390, pp. 104–128. Springer, Heidelberg (2002)
4. Cordella, L.P., Vento, M.: Symbol recognition in documents: a collection of techniques? IJDAR 3(2), 73–88 (2000)
5. Tombre, K., Tabbone, S., Dosch, P.: Musings on symbol recognition. In: Liu, W., Lladós, J. (eds.) GREC 2005. LNCS, vol. 3926, pp. 23–34. Springer, Heidelberg (2006)
6. Bunke, H., Gunter, S., Jiang, X.: Towards bridging the gap between statistical and structural pattern recognition: Two new concepts in graph matching. In: Singh, S., Murshed, N., Kropatsch, W.G. (eds.) ICAPR 2001. LNCS, vol. 2013, pp. 1–11. Springer, Heidelberg (2001)
7. Delalandre, M., Trupin, É., Ogier, J.M.: Symbols recognition system for graphic documents combining global structural approaches and using a XML representation of data. In: Fred, A.L.N., Caelli, T., Duin, R.P.W., Campilho, A.C., de Ridder, D. (eds.) SSPR&SPR 2004. LNCS, vol. 3138, pp. 425–433. Springer, Heidelberg (2004)
8. Hse, H., Newton, A.R.: Sketched symbol recognition using zernike moments. In: ICPR, pp. 367–370 (2004)
9. Barrat, S., Tabbone, S., Nourrissier, P.: A bayesian classifier for symbol recognition. In: GREC, October 24, vol. 7 (2007)
10. Rusinol, M.: Geometric and Structural-based Symbol Spotting. Application to Focused Retrieval in Graphic Document Collections. PhD thesis, Universitat Autonoma de Barcelona (2009)

11. Qureshi, R.J., Ramel, J.Y., Cardot, H., Mukherji, P.: Combination of symbolic and statistical features for symbols recognition. In: ICSCN, pp. 477–482 (2007)
12. Zhang, W., Liu, W.: A new vectorial signature for quick symbol indexing, filtering and recognition. In: ICDAR, vol. 9, pp. 536–540 (2007)
13. Rusiñol, M., Lladós, J.: Symbol spotting in technical drawings using vectorial signatures. In: Liu, W., Lladós, J. (eds.) GREC 2005. LNCS, vol. 3926, pp. 35–46. Springer, Heidelberg (2006)
14. Dosch, P., Llados, J.: Vectorial signatures for symbol discrimination. In: Lladós, J., Kwon, Y.-B. (eds.) GREC 2003. LNCS, vol. 3088, pp. 154–165. Springer, Heidelberg (2004)
15. Coustaty, M., Guillas, S., Visani, M., Bertet, K., Ogier, J.M.: On the joint use of a structural signature and a galois lattice classifier for symbol recognition. In: Liu, W., Lladós, J., Ogier, J.-M. (eds.) GREC 2007. LNCS, vol. 5046, pp. 61–70. Springer, Heidelberg (2008)
16. Luqman, M.M., Brouard, T., Ramel, J.Y.: Graphic symbol recognition using graph based signature and bayesian network classifier. In: ICDAR, vol. 10, pp. 1325–1329 (2009)
17. Ventura, A., Schettini, R.: Graphic symbol recognition using a signature technique. In: ICPR, vol. 2, pp. 533–535 (1994)
18. Valveny, E., Marti, E.: Deformable template matching within a bayesian framework for handwritten graphic symbol recognition. In: Chhabra, A.K., Dori, D. (eds.) GREC 1999. LNCS, vol. 1941, pp. 193–208. Springer, Heidelberg (2000)
19. Mitra, S., Pal, S.K.: Fuzzy sets in pattern recognition and machine intelligence. FSS 156(3), 381–386 (2005)
20. Charniak, E.: Bayesian networks without tears. AI Magazine 12(4), 50–63 (1991)
21. Heckerman, D.: A tutorial on learning with bayesian networks. In: Innovations in Bayesian Networks. SCI, vol. 156, pp. 33–82 (2008)
22. Delaplace, A., Brouard, T., Cardot, H.: Two evolutionary methods for learning bayesian network structures. In: Wang, Y., Cheung, Y.-m., Liu, H. (eds.) CIS 2006. LNCS (LNAI), vol. 4456, pp. 288–297. Springer, Heidelberg (2007)
23. Leray, P., François, O.: BNT structure learning package: Documentation and experiments. Technical report, Laboratoire PSI - INSA Rouen- FRE CNRS 2645 (November 2004)
24. Colot, O., Olivier, C., Courtellemont, P., El Matouat, A.: Information criteria and abrupt changes in probability laws. In: Signal Processing VII: Theories and Applications, pp. 1855–1858 (1994)
25. Aksoy, S., Ye, M., Schauf, M., Song, M., Wang, Y., Haralick, R.M., Parker, J.R., Pivovarov, J., Royko, D., Sun, C., Farneback, G.: Algorithm performance contest. In: ICPR, vol. IV, pp. 870–876 (2000)
26. Valveny, E., Dosch, P.: Symbol recognition contest: A synthesis. In: Lladós, J., Kwon, Y.-B. (eds.) GREC 2003. LNCS, vol. 3088, pp. 368–385. Springer, Heidelberg (2004)
27. Dosch, P., Valveny, E.: Report on the second symbol recognition contest. In: Liu, W., Lladós, J. (eds.) GREC 2005. LNCS, vol. 3926, pp. 381–397. Springer, Heidelberg (2006)
28. Valveny, E., Dosch, P., Fornes, A., Escalera, S.: Report on the third contest on symbol recognition. In: Liu, W., Lladós, J., Ogier, J.-M. (eds.) GREC 2007. LNCS, vol. 5046, pp. 321–328. Springer, Heidelberg (2008)
29. http://www.imagemagick.org/ (As on March 20, 2010)
30. http://www.qgar.org/ (As on March 20, 2010)
31. http://symbcontestgrec05.loria.fr/symboldescription.php (As on March 20, 2010)
32. Delalandre, M., Pridmore, T., Valveny, E., Locteau, H., Trupin, E.: Building synthetic graphical documents for performance evaluation. In: Liu, W., Lladós, J., Ogier, J.-M. (eds.) GREC 2007. LNCS, vol. 5046, pp. 288–298. Springer, Heidelberg (2008)
33. http://mathieu.delalandre.free.fr/projects/sesyd/queries.html (As on March 20, 2010)

Interactive Conversion of Web Tables

Raghav Krishna Padmanabhan[1], Ramana Chakradhar Jandhyala[1],
Mukkai Krishnamoorthy[1], George Nagy[1], Sharad Seth[2], and William Silversmith[1]

[1] ECSE, DocLab, Rensselaer Polytechnic Institute, Troy, NY USA 12180
nagy@ecse.rpi.edu
[2] CSE, University of Nebraska-Lincoln, Lincoln, NE USA 68502
seth@cse.unl.edu

Abstract. Two hundred web tables from ten sites were imported into Excel. The tables were edited as needed, then converted into layout independent Wang Notation using the Table Abstraction Tool (TAT). The output generated by TAT consists of XML files to be used for constructing narrow-domain ontologies. On an average each table required 104 seconds for editing. Augmentations like aggregates, footnotes, table titles, captions, units and notes were also extracted in an average time of 93 seconds. Every user intervention was logged and audited. The logged interactions were analyzed to determine the relative influence of factors like table size, number of categories and various types of augmentations on the processing time. The analysis suggests which aspects of interactive table processing can be automated in the near term, and how much time such automation would save. The correlation coefficient between predicted and actual processing time was 0.66.

Keywords: Document Understanding, Interactive Table Interpretation, Performance Evaluation, Ontology Construction, Table Abstraction Tool.

1 Introduction

Our objective is to harvest web tables in order to assist our parent project, TANGO, to construct, with as little human intervention as possible, an ontology in the relatively narrow domain of geopolitics [1]. Since web tables can be readily imported into spreadsheet programs like MS-Excel which provide a natural coordinate system for tables, we developed the Table Abstraction Tool (TAT) to convert Excel tables into Wang Notation [2]. TAT was coded in Visual Basic for Applications (VBA) for ease of access to internal Excel formatting variables. If a table exhibits features that cannot be handled by TAT, then the operator uses Excel commands to change the table into a TAT- admissible format. After verifying the validity of the edited table, TAT creates a category notation which preserves the relationship of the header hierarchies to the content cells. TAT also processes augmentations like aggregates and footnotes tagged by the operator. The Augmented Wang Notation (AWN), which contains both the category information and the augmentations, is embedded into an XML file for portability. The edits can be visually verified by highlighting the relationship between designated headers and content cells. This proofing method was developed in an earlier tool, WNT, which imported web tables into MATLAB rather than Excel [3].

J.-M. Ogier, W. Liu, and J. Lladós (Eds.): GREC 2009, LNCS 6020, pp. 25–36, 2010.

Although we have conducted (and reported elsewhere) experiments on partial automation of data extraction from web tables, we believe that interactive processing will be necessary for some tables for the foreseeable future and that the properties of tables that preclude complete automation are worthy of careful study.

Comprehensive reviews of two decades of research on table processing appear in [4, 5]. Algorithms were first developed for specifying cell location in terms of rulings or, in the case of unruled tables, the geometric alignment and typographic similarity of cell content (e.g., [6,7,8,9]). A recent proposal for an end-to-end system divides the task into table detection, segmentation, function analysis, structural analysis and interpretation, but was not implemented and does not define which tables can and cannot be processed [10]. None of the methods that address web tables (e.g. [11]) carry the analysis to the layout-independent multi-category level. Some of the reasons why we do not expect table recognition to be fully automated in the near future were presented at GREC 1999 [12]. Our model of table processing consists of six interrelated tasks:

Task 1. Table Recognition: Detection of tables within a larger document or corpus, and determination of their exact locations and extents. This is not trivial with unruled Web tables [13].

Task 2. Geometric Structure Extraction: Recognition of the geometric grid structure that characterizes all tables and associated text within the table frame from grid coordinates. Most classical table processing research, especially on scanned tables, addressed this task (e.g. [14]).

Task 3. Table Interpretation: Associating content cells with the heading structure and describing their relationship independently of the geometric layout of the table. This step targets the underlying *logical table.* We have recently developed a formalism to link Task 3 with Task 2 [15].

Task 4. Table Understanding: Determining the conceptual relationships (*is-a, part-of, owns, quantifies, describes*) of the table entries to the contents of other tables, databases, or ontologies. This step, which we call *table understanding* [16], calls for external knowledge from either the vicinity of the table or extraneous sources. It is necessary for conflating tabular data from diverse sources.

Task 5. Metadata Extraction: Extracting and encoding table attributes that do not cleanly fit into either the geometric or the logical views but appear within or adjacent to the table. Examples are *table title, caption, aggregates, footnotes,* and *units.* They have been largely ignored in the table processing literature.

Task 6. Adaptation: Recalling and exploiting the errors and interventions recorded in processing earlier tables to modify the automated aspects of processing the current table. The objective is to develop a system that improves with use, i.e., an evolutionary system that decreases the need for human intervention. Some researchers call this task *learning.*

Here, we present an experimental investigation focused on Tasks 3 and 5. In this experiment 200 tables were randomly chosen from ten large web sites and were processed by one operator.

In Section 2, we list the novel aspects of our interactive procedure. In Section 3, we describe an experimental protocol designed to evaluate the various factors that affect interactive table processing. Section 4 presents the analysis of operator interaction time throughout the processing of the 200 web tables. Section 5 summarizes our observations and offers some projections about what aspects of table processing could be automated in the short term.

2 Novel Aspects

Our work differs from earlier work with respect to

1. Focusing on end-to-end processing of tables from large web sites;
2. Making use of commercial software to import web tables into a spreadsheet and using familiar spreadsheet operations to edit the tables as necessary;
3. Facilitating content analysis by extracting the relationship of headers to content cells rather than only the geometric cell structure;
4. Making provisions for augmentations.
5. Timing, logging, and analyzing all operator interactions.

2.1 Excel Tables

Although several algorithms have been published for finding the cell structure of web tables, with the passage of time this has become a non-issue in research. Excel has built-in provisions for parsing the hypertext and allocating its content to cells. For most sites, it is sufficient to *select* the table, *copy* it, and *paste* into a worksheet. Alternatively, after selection one may use the Excel import menu command. This process is not foolproof. Sometimes the contents of a multi-line table cell are distributed over several worksheet cells, or separate table cells are merged into one worksheet cell. Excel also tries to interpret the data, for instance turning hyphenated numerals into a calendar date. Gratuitous data conversions can be prevented by pre-formatting the target worksheet as *text*. Any errors in conversion must be corrected by the operator. These corrections can be interleaved with the edits necessary to render the table admissible for algorithmic processing by TAT. In the experiments reported below, the interaction time is included under *editing*. In spite of the occasional conversion problems, letting Excel do the heavy lifting has allowed us to concentrate on the more subtle issues.

2.2 Wang Notation

Xinxin Wang in her 1996 dissertation [17] proposed an abstract "table" data type where each logical dimension is defined by a category tree of *labeled domains*. Consider the tables of Fig. 1. The data cell containing "5.0" (a *delta cell* in Wang terminology), is specified by a path through each of the three category trees: DEMOGRAPHICS→IMMIGRANT, YEAR→1990, and COUNTRY→CANADA.

POPULATION IN MILLIONS		DEMOGRAPHICS			
		NATIVE		IMMIGRANT	
		YEAR			
		1990	2000	1990	2000
	CANADA	22.7	25.4	5.0	5.6
COUNTRY	USA	221	249.9	27.4	31.5

(a)

		DEMOGRAPHICS	
		NATIVE	IMMIGRANT
COUNTRY	YEAR		
	1990	22.7	5.0
CANADA	2000	25.4	5.6
	1990	221.3	27.4
USA	2000	249.9	31.5

(b)

Fig. 1. (a) A three-category table; (b) another table with the same Wang Notation

There are several conventions for laying out hierarchical table headings. As row and column headers are conceptually similar, geometric symmetry would suggest that the roles of horizontal and vertical orientations are interchangeable in the layout of table headers (Wang Notation does not distinguish them). However, row headings *above* row subheadings are common in English tables, as in Fig. 1b.

		DEMOGRAPHICS	
		NATIVE	IMMIGRANT
COUNTRY	YEAR		
	1990	22.7	5
CANADA	2000	25.4	5.6
	1980	221.3	27.4
USA	2000	249.9	31.5

(a)

		DEMOGRAPHICS	
		NATIVE	IMMIGRANT
COUNTRY	YEAR		
	1990	22.7	5
CANADA	2000	25.4	5.6
	2000	221.3	27.4
USA	2000	249.9	31.5

(b)

Fig. 2. (a) A well-formed table (WFT) with two categories; (b) not a WFT

A table is *well formed* if it can be represented in Wang Notation. A necessary condition for a well formed table (WFT) is that any combination of paths, one through each category tree, must specify a unique delta cell. Equivalently, the cardinality of the Cartesian product of the unique paths through the category trees must be equal to the number of delta cells (which is eight in all of the above tables).The table on top in Fig. 2 is not a 3-D table, because there is no delta cell that can be specified by the path COUNTRY→CANADA,DEMOGRAPHICS→IMMIGRANT, and YEAR→1980. It is, however, a 2-D WFT, with the category trees of Fig. 3. The table below is not a WFT because the paths COUNTRY→USA, DEMOGRAPHICS→IMMIGRANT, and YEAR→2000 lead to either "27.4" or "31.5".

TAT checks whether a table is *TAT-admissible* before it extracts its Wang category notation. Tables are TAT-admissible even if the roots of some category trees are missing. If the header DEMOGRAPHICS were missing in the table of Fig. 1, TAT would simply generate a unique virtual header *VH xxxxxxxxx* (Virtual Header with the x's indicating the date-time at which the header was generated in the format: *mmdd hhmmss*) as the parent of subcategories NATIVE and IMMIGRANT.

Category 1 (four unique paths)   Category 2 (two unique paths)

```
Category 1 (four unique paths)            Category 2 (two unique paths)
     COUNTRY
          CANADA                                       DEMOGRAPHICS
                    YEAR
                                                              NATIVE
                         1990
                                                              IMMIGRANT
                         2000
          USA
                    YEAR
                         1980
                         2000
```

Fig. 3. Category trees of the table of Fig. 2a, shown in TAT's internal indented notation

2.3 Augmentations

An *augmentation* is information appearing in a table that is not part of the header-to-content cell mappings. An augmentation may apply to the entire table (e.g., *Table Title*, *Table Caption*, *Notes*), to one or more rows or columns (*Unit*), or to a single cell of the table (*Footnote*). The most interesting augmentation is the *aggregate*. For instance, NORTH AMERICA could appear in Fig. 1 instead of COUNTRY (Fig. 4). If no population is listed for NORTH AMERICA, then it is just a header. But if the totals for Canada and USA are listed in that row, then the corresponding paths will be NORTH AMERICA→NORTH AMERICA, NORTH AMERICA→CANADA, NORTH AMERICA→USA. Therefore this aggregate functions both as a category root and as a category leaf cell. Aggregates must be annotated by the operator because TAT cannot yet identify them automatically.

POPULATION IN MILLIONS	DEMOGRAPHICS			
	NATIVE		IMMIGRANT	
	YEAR			
	1990	2000	1990	2000
NORTH AMERICA	244.0	275.3	32.4	37.1
CANADA	22.7	25.4	5.0	5.6
USA	221.3	249.9	27.4	31.5

Fig. 4. The header NORTH AMERICA is both a category root and an aggregate, and must be so tagged in the XML output file

3   Experimental Protocol

We sought to determine the main factors that affect the conversion time of web tables using TAT. We used the collection of rare and unusual tables from Lopresti and Nagy [18] as a guide for selecting tables to evaluate TAT. In particular, we excluded tables that were not well-formed or had any of the following characteristics:

1.   Non-rectilinear structure.
2.   Text in languages other than English.
3.   Cells containing graphic symbols or figures.
4.   Recursive structure, i.e., a table with a table as one of its content cells.
5.   Concatenation (tables formed by concatenating two or more tables).
6.   Sources other than the World Wide Web and formats other than HTML, Microsoft Excel or CSV.
7.   Domains other than Geopolitical or Scientific Research data.
8.   For convenience, we also excluded tables that span more than one HTML page or Excel sheet.

The experimental protocol was developed in a pilot study. The pilot study was used to determine the final format of the *analysis table,* which would contain all of the experimental data to be collected. A bug in TAT that limited the size of the tables to 100 rows was also found and fixed. The tables used in the pilot study were excluded from the evaluation reported below.

We collected and processed 200 Excel and HTML tables from ten non-profit websites (Table 1). Importing an HTML file into Excel takes negligible time and hence both Excel and HTML tables were treated as one. By "collect", we mean:

1.   Browse the websites and save the selected files of tables
2.   Store separately the set of 12 tables that we looked at but rejected.
3.   Number serially all the accepted tables for subsequent reference and pseudo-randomization.

Table 1. URLs of table sources

Site #	Table Source
1	http://www.statcan.gc.ca/
2	http://www.sciencedirect.com/
3	http://www.worldbank.org/
4	http://www.ssb.no/english/
5	http://www.ojp.usdoj.gov
6	http://www.geohive.com/
7	http://www1.lanic.utexas.edu/la/region/aid/aid98/
8	http://eia.doe.gov/
9	http://ies.ed.gov/
10	http://www.census.gov/population/www/socdemo/voting/cps2006.html

4 Experimental Results and Discussion

The main experiment was conducted in 15 sessions. Processing a table required three consecutive steps:

1. *Editing* the table, i.e., transforming it using Excel operations into TAT-admissible form;

2. *Annotating* the table: this requires clicking on the corner cells of the header and delta cell regions, and on cells containing the table title, caption, aggregates, footnote citations, footnotes, units, and other notes.

3. *Post-processing*, which consists of checking TAT's category assignments by highlighting selected header and delta cells, and either initiating a correction cycle or starting the XML generation algorithm.

Seven tables in the list could not be processed either because they were poorly constructed or because Excel could not interpret their content correctly. Two of these actually failed to match our criteria: they were collected by mistake. One of them was too large (~85000 cells).

The total processing time increases with the number of cells (Table 2) for two reasons: Larger tables typically have more augmentations and require scrolling to edit them into TAT-compatible format. The total processing time includes checking the category assignments by highlighting header and delta cells to display their relationship. It also includes generating the XML output file, which is typically a few seconds.

Table 2. Effect of table size (# of cells in the table) on Total Processing Time

Number of Cells (RxC)	Number of Tables	Avg. Total Processing Time (sec)
< 201	78	134.2
201-400	44	212.3
401-600	25	242.6
601-800	13	430.3
>800	33	394.8
All tables	**193**	**230.5**

On an average, 3-D tables take just 27 seconds more than 2-D tables to process (Table 3). Our data set had too few tables of lower or higher dimensionality for reliable estimates of processing times.

Table 3. Effect of Wang Dimensionality on Total Processing Time

Wang Dimensionality	Number of Tables	Avg. Total Processing Time (sec)
1	2	150.5
2	140	224.3
3	49	251.0
4	2	247.0
All tables	**193**	**230.5**

Table 4 shows that the editing time more than doubles for tables with more than two aggregates compared to tables without aggregates. This does not mean that all the tables with aggregates are not TAT-admissible. But to derive the correct Wang notation/XML, we must transform the table into a form which preserves the category trees (Fig. 4). Tables with aggregates also take much longer to annotate than tables without aggregates. The maximum number of aggregates was 43 in a single table. As illustrated in Fig. 4, aggregates often also serve as top-level row headers. Detecting them requires lexical as well as structural analysis.

Table 5 shows that the presence of footnotes also significantly increases annotation time, but has relatively little effect on editing time. The highest number of footnote cells encountered in a single table was 214. The current implementation requires the user to select each of those cells and annotate it. However, the format of the footnotes below the tables and the corresponding footnote references within the table is uniform enough to allow hope for automated footnote annotation.

The number of cells, the Wang dimensionality, and the prevalence of aggregates and footnotes provide a measure of the amount of operator interaction required to process the table. We predicted the total processing time and the global correlation coefficient by multilinear regression on these four "features." The correlation coefficients between the actual and predicted processing times are shown in Table 6. Sources 6 and 7 contained some poorly constructed/unconventional tables. This resulted in large processing times compared to well-constructed tables with similar features. The global correlation coefficient for all tables without regard to their source was 0.66. The weighted source correlation coefficient was 0.72

Table 4. Effect of aggregates on Editing and Annotation times

Number of Aggregates	Number of Tables	Avg. Editing Time (sec)	Avg. Annotation Time (sec)
0	106	77.8	60.3
1	44	120.2	118.9
2	15	106.5	100.7
>2	28	177.2	171.6
All tables	**193**	**104.1**	**93.0**

Table 5. Effect of footnotes on Editing and Annotation times

Number of Footnotes	Number of Tables	Avg. Editing Time (sec)	Avg. Annotation Time (sec)
0	120	99.9	71.4
1	21	89.8	74.8
2	17	124.3	138.1
>2	35	117.4	155.9
All tables	**193**	**104.1**	**93.0**

Table 6. Source-specific correlation coefficients

Source	Number of Tables	Source-specific correlation coefficient
1	20	0.78
2	15	0.85
3	18	0.97
4	21	0.62
5	24	0.79
6	26	0.40
7	15	0.42
8	24	0.87
9	23	0.72
10	7	0.99
Total	193	0.72

Table 7. Preparation and Action times for each user intervention

Action	Preparation or Idle Time (T_p)	Action Time (T_a)	Ratio (T_p/T_a)
Editing into TAT-admissible form	**6.7**	**97.4**	**0.1**
Annotation	**39.6**	**52.4**	**0.8**
Select Title	3.4	2.5	1.4
Select Caption	0.7	0.9	0.8
Augmentations	5.0	2.6	1.9
Footnotes	3.5	14.6	0.2
Notes	1.8	3.7	0.5
Aggregates	3.2	7.0	0.5
Units	0.7	0.6	1.2
Delta Cell Selection & WFT check	10.4	11.8	0.9
Category Selection	10.9	8.7	1.3
Post-processing	**16.1**	**19.0**	**0.8**
Highlighting (category check)	14.7	11.1	1.3
Generate XML	1.4	7.9	0.2
TOTAL	**62.4**	**168.8**	**0.4**

The time elapsed between the completion of an action and the initiation of the next action was interpreted as the preparation time for the next action. The preparation and action times are shown for each activity in Table 7. As seen earlier, the presence of many aggregates and footnotes significantly increases processing time, and most tables have some of these augmentations. Overall they account for almost 15% of the total processing time, but still 30% less than the fundamental operations of marking category headers and delta cells. Checking the categories assigned by TAT takes significant time (~26 seconds on average). The action time for XML file generation is actually machine time.

5 Summary

TAT was evaluated by a single operator in 15 sessions that took a total of 24.7 hours. The samples were collected from ten web sites which contain thousands of tables relevant to the geopolitical domain. Two hundred tables according to prescribed criteria were processed in a pseudo-random order using TAT. Each selected sample was edited if necessary, and every editing operation was time-stamped and recorded. The time required for editing the table into the desired format along with the interaction to process title, caption, footnotes, units, and aggregates was logged. The Wang Notation for seven of the two hundred tables could not be determined. After processing a table, the operator verified its Wang Notation visually through the TAT functionality which highlights the categories and subcategories associated with the selected delta cell.

Tables with Wang dimensionality 3, which is where layout-independence becomes really significant, took approximately 27 seconds more than tables with Wang dimensionality 2. As expected, there was a strong positive correlation between the processing time and table features (size, dimensionality, aggregates and footnotes). The time distributions have significant positive skew because of a few difficult tables.

Tables with aggregates took much more time than tables without them. Aggregates often result in repeated cells in a header column, which is not TAT-admissible. This requires that the table be modified using Excel commands. The current implementation of selecting and annotating the aggregate cells and footnotes in TAT becomes cumbersome in tables with many aggregates. In the geopolitical domain, it is common to have hundreds of cells with footnote references. Manually selecting these cells is a human intensive, time consuming and error prone task. Thus, there is a great need to automate the identification and annotation of aggregates and footnotes, a task that appears quite feasible. Spanning cells containing *units* should also be relatively easy to detect automatically.

Only a few of the sample tables were processed by TAT without some preliminary editing. The greatest potential savings in time is to make TAT accept a larger variety of table formats. More specifically, it should save the edit sequences applied by operator, generalize them to an arbitrary number of rows and columns, and apply them to new tables in previously seen formats. We are currently working on algorithms to accomplish this [19].

We are also developing methods to automatically determine the delta-cell and header regions, which would save by itself over 15% of the interaction time. We are exploring the problem of table segmentation using visual cues in the table. The proposed method relies on visual distinctions (typeface, type size, capitalization, alignment) between cells of a table, many of which have been explored in earlier studies by others. The cell's features can be captured in a feature vector with both numerical and categorical attributes. By comparing the feature vectors of adjacent cells using a comparison function, a difference table can be formed and used to perform orientation analysis, category-delta space segmentation and identification of aggregates and footnotes. Automating this phase would pave the way for faster processing and conversion to a layout independent format that would complement the "learning" approach outlined in the previous paragraph.

To determine inter-operator variability in processing time, we are currently planning another experiment with multiple operators on the same corpus of tables. If there is little variability between operators then we can construct an operator-independent formula for predicting processing time as a function of table features.

Conversion of multiple tables from large web sites to Augmented Wang Notation is only the first step towards extracting the intra- and inter-table relationships that are the essential constituents of a domain-specific ontology of semi-structured data.

Acknowledgments. This work was supported by the National Science Foundation under Grants# 041414854 and 0414644 and the Rensselaer Center for Open Source Software. We acknowledge the help of Professor David Embley of BYU in developing the XML formats and the treatment of annotations.

References

1. Tijerino, Y.A., Embley, D.W., Lonsdale, D.W., Ding, Y., Nagy, G.: Toward Ontology Generation from Tables. World Wide Web: Internet and Web Information Systems 8(3), 261–285 (2005)
2. Padmanabhan, R.: Table Abstraction Tool, RPI DocLab, Master's Thesis, May 16 (2009)
3. Jha, P., Nagy, G.: Wang Notation Tool: Layout Independent Representation of Tables. In: Proceedings of the Nineteenth International Conference on Pattern Recognition (ICPR 2008), Tampa (April 2008)
4. Zanibbi, R., Blostein, D., Cordy, J.R.: A survey of table recognition: Models, observations, transformations, and inferences. International Journal of Document Analysis and Recognition 7(1), 1–16 (2004)
5. Lopresti, D., Embley, D.W., Hurst, M., Nagy, G.: Table Processing Paradigms: A Research Survey. International Journal of Document Analysis and Recognition 8(2-3), 66–86 (2006)
6. Sobue, T., Watanabe, T.: Identification of Item Fields in Table-form Documents with/without Line Segments. In: Proceedings of IAPR Workshop on Machine Vision Applications, Tokyo, Japan, November 12-14, pp. 522–525 (1996)
7. Klink, S., Kieninger, T.: Rule-based document structure understanding with a fuzzy combination of layout and textual features. International Journal of Document Analysis and Recognition 4(1), 18–26 (2001)
8. Laurentini, A., Viada, P.: Identifying and understanding tabular material in compound documents. In: Proceedings of the Eleventh International Conference on Pattern Recognition (ICPR 1992), The Hague, pp. 405–409 (1992)

9. Itonori, K.: A table structure recognition based on textblock arrangement and ruled line position. In: Proceedings of the Second International Conference on Document Analysis and Recognition (ICDAR 1993), Tsukuba Science City, Japan, pp. 765–768 (1993)
10. Silva, E.C., Jorge, A.M., Torgo, L.: Design of an end-to-end method to extract information from tables. International Journal of Document Analysis and Recognition 8(2), 144–171 (2006)
11. Krüpl, B., Herzog, M., Gatterbauer, W.: Using visual cues for extraction of tabular data from arbitrary HTML documents. In: Proceedings of the 14th Int'l. Conf. on World Wide Web, pp. 1000–1001 (2005)
12. Lopresti, D., Nagy, G.: Automated Table Processing: An (Opinionated) Survey. In: Proceedings of the Third IAPR International Workshop on Graphics Recognition, Jaipur, India, pp. 109–134 (September 1999)
13. Wang, Y., Hu, J.: Automatic Table Detection in HTML Documents. In: Web Document Analysis: Challenges and Opportunities, October 2003, pp. 135–154 (2003)
14. Handley, J.C.: Table analysis for multiline cell identification. In: Proceedings of Document Recognition and Retrieval VIII (IS\&T/SPIE Electronic Imaging), San Jose, CA, vol. 4307, pp. 44–55 (2001)
15. Jandhyala, R.C., Nagy, G., Seth, S., Silversmith, W., Krishnamoorthy, M., Padmanabhan, R.: From tessellations to table interpretation. In: Carette, J., Dixon, L., Coen, C.S., Watt, S.M. (eds.) Calculemus 2009. LNCS, vol. 5625, pp. 422–437. Springer, Heidelberg (2009)
16. Embley, D.W., Lopresti, D., Nagy, G.: Notes on Contemporary Table Recognition Workshop on Document Analysis Systems. In: Bunke, H., Spitz, A.L. (eds.) DAS 2006. LNCS, vol. 3872, pp. 164–175. Springer, Heidelberg (2006)
17. Wang, X.: Tabular Abstraction, Editing, and Formatting, Ph.D Dissertation, University of Waterloo, Waterloo, ON, Canada (1996)
18. Lopresti, D., Nagy, G.: A Tabular Survey of Automated Table Processing, Graphics Recognition: Recent Advances. In: Chhabra, A.K., Dori, D. (eds.) GREC 1999. LNCS, vol. 1941, pp. 93–120. Springer, Heidelberg (2000)
19. Seth, S., Jandhyala, R., Krishnamoorthy, M., Nagy, G.: Analysis and Taxonomy of Column Header Categories for Web Tables. To appear in Proceedings of the Document Analysis Systems, Boston (June 2010)

Comparing Graph Similarity Measures for Graphical Recognition*

Salim Jouili[1], Salvatore Tabbone[1], and Ernest Valveny[2]

[1] LORIA UMR 7503 - University of Nancy 2
BP 239, 54506 Vandoeuvre-lès-Nancy Cedex, France
{salim.jouili,tabbone}@loria.fr
[2] Centre de Visió per Computador, Dep. Ciències de la Computació
Universitat Autònoma de Barcelona, Edifici O, Campus UAB, 08193 Bellaterra, Spain
ernest@cvc.uab.cat

Abstract. In this paper we evaluate four graph distance measures. The analysis is performed for document retrieval tasks. For this aim, different kind of documents are used including line drawings (symbols), ancient documents (ornamental letters), shapes and trademark-logos. The experimental results show that the performance of each graph distance measure depends on the kind of data and the graph representation technique.

Keywords: Graph matching, Graph retrieval, Structural representation, Performance evaluation.

1 Introduction

In document retrieval applications, it is necessary to define some description of the document based on a set of features. These descriptions are then used to search and to determine which documents satisfy the query selection criteria. The effectiveness of a document retrieval system ultimately depends on the type of representation used to describe it. In pattern recognition, the document representation can be broadly divided into statistical and structural methods [6]. In the former, the document is represented by a feature vector, and in the latter, a data structure (e.g. graphs or trees) is used to describe objects and their relationships in the document. The classical retrieval systems are often limited to work with a statistical representation due to the need of computing distances between documents (feature vectors) or finding a representative cluster of documents. However, when a numerical feature vector is used to represent the document, all structural information is discarded although the structural representation is more powerful in terms of its representational abilities [6]. In the last decades, many structural approaches have been proposed. These approaches deal, especially, with graph-based representations. Nevertheless, dealing with graphs suffers, on the one hand from the high complexity of the graph matching problem

---

* This work is partially supported by the French National Research Agency project NAVIDOMASS referenced under ANR-06-MCDA-012 and Lorraine region.

J.-M. Ogier, W. Liu, and J. Lladós (Eds.): GREC 2009, LNCS 6020, pp. 37–48, 2010.

which is a problem of computing distances between graphs, and on the other hand from the robustness to structural noise which is a problem related to the capability to cope with structural variations and differences in the size of the graph. In order to overcome this problem, several approximate graph matching methods have been proposed [11,15,19,21]. In this paper, our attention is focused on the comparison of different graph similarity measures in the context of document retrieval.

Graph similarity measures use different techniques to minimize the complexity and to optimize the robustness to structural noise. Robles-Kelly and al. [21] propose a spectral seriation approach to reduce the graph matching to a string edit distance in a probabilistic framework. Jouili and al. [11] simplify the problem to a bipartite graph matching by making use of node signatures. Lopresti and al. [15] use a probe technique to reduce the graph matching to distance between vectors. Papadopoulos and al. [19] introduce an histogram-based technique.

In this paper, we present an evaluation of these four graph distance measures on four different document data sets. We use the well-known GREC [20] data base which consists of graphs representing symbols from architectural and electronic drawings. Here the ending points (ie corners, intersections and circles) are represented by nodes which are connected by undirected edges and labeled as lines or arcs. We have also performed a retrieval evaluation on an ornamental letters data set which contains lettrine (graphical object) extracted from digitized ancient document [1]. Since one lettrine contains a lot of information (i.e. texture, decorated background, letters), the graphs are extracted from a region-based segmentation [9] of the lettrine with a user-based parameterization technique. The nodes of the graph are represented by the regions and the edges describe their adjacency relationships. We have also evaluated the graph similarity measures on a shape data set [23] in which the graph is extracted by making use of a skeletonizing algorithm and a delaunay triangulation of detected endpoints. Finally, the graph similarity measures are evaluated on a set of trademark-logos in which the graph is extracted by making use of an interest points detector [10] and the delaunay triangulation.

The performance evaluation is performed using the Precision-Recall curves. Through this evaluation, we will examine the robustness of each graph similarity distance and this will allow us to investigate the applicability of each measure to the problem of retrieval for different kinds of documents.

2   Graph-Based Representations

Region-Based approaches have been one of the most important research issues in content-based image retrieval. Representing images at the region level captures not only the local variations of regions but also their spatial organizations. Graph-based representations are widely used in region-based segmentation. To

---

[1] Provided by the CESR - University of Tours on the context of the ANR Navidomass project http://l3iexp.univ-lr.fr/navidomass/

incorporate both region attributes and adjacent relationship an image is usually represented as an attributed graph. Classical image representations such as colors histograms, texture descriptors, or shape descriptions do not take into account the regions localization in the document.

Graph-based representations are used in many applications, for instance, to represent circuit diagrams [4], for shape recognition [8], image matching [14,2], or old document analysis [12]. Other works on graph-based representation [1,3,18,17], use different methods to incorporate features of the document image. The methods vary according to the characteristics of the data and the aims of the representation (i.e. matching or retrieval). Bunke [4] illustrates an example of converting a circuit diagram to a graph by representing the lines in the circuit diagram; each graph node represents a line endpoint, corner or intersection point, and node attributes record the image coordinates (x,y) of this feature. In [12] the authors manipulate initial letters from old documents. They proceed by segmenting the initial letter into different information layers to obtain "Information layers of homogeneous zones". Then, each homogeneous zone of the initial letter is converted to a node of graph with two attributes: size and shape descriptions, and each edge contains two attributes: angle and distance. Baeza-Yates and al. [2] also represent images as attributed graphs and adopt the graph edit distance to calculate the image distance. In another way using graphs in image analysis, Pan and al. [18] introduce a graph-based automatic image annotation. The authors propose a graph-based method to assign automatically keywords to an image. The main idea of this work is to represent all the images, as well as their attributes (caption words and regions) as nodes and link them according to their known association into a graph. For the task of image annotation, they use a "3-layer" graph, with one layer for image nodes, one layer for annotation term nodes, and one layer for the image regions.

In this section, we have seen that graphs can be widely used as a data structure-model in the pattern recognition domain. Moreover, most of the previous graph-based representations aim to measure some similarity between objects for recognition or retrieval tasks. This leads to the development of several similarity measures for graphs.

3   Graph Matching Measures

An important step in structural pattern recognition is the representation of documents by a graph data structure. This structural representation should provide a description of the characteristics of the images efficiently for the task under consideration (e.g. retrieval). The retrieval problem can then be addressed in the corresponding graph space without addressing the original images. The process of comparing graphs is generally referred as graph matching. Generally, given two graphs $G_1=(V_1,E_1)$ and $G_2=(V_2,E_2)$, the graph matching methods are divided into two broad categories: the first one contains exact matching methods called graph isomorphism that requires to find a one-to-one mapping $f:V_1 \mapsto V_2$ such that $(u,v) \in E_1$ if $(f(u),f(v)) \in E_2$ with $|V_1|=|V_2|$. The second category contains inexact matching methods, where a strict correspondence among the nodes or the edges of the two graphs can not be found. Therefore, in these cases no iso-morphism can be expected between both graphs, and the graph matching does not consist in finding the *exact* matching but the *best* matching between them. To perform such a structural matching, various formalisms have been proposed, using error-tolerant methods based on continuous optimization [16], quadratic programming, and spectral decomposition of graph matrices [21]. Other methods try to characterize the properties of graphs using a vector-based representation in order to profit from the existing vector measures[15,11,19]. Most of the inexact graph matching measures are based on some sort of edit operations. The basic idea is to define the similarity of graphs based on the effort needed to make the graphs identical. This is an extension of the well known string edit distance [13] to the graph edit distance (GED) [22]. For a review of graph similarity measures we refer the readers to [7,5].

The matching methods selected for our evaluation belong to different for-malisms. The spectral technique proposed by the Robles-Kelly's method [21] has proven to obtain good performance results. The graph matching based on node signature [11] uses a local decomposition of graphs and an assignment method to carry out an optimum node-to-node correspondence. Papadopoulos and al. [19] provide a histogram-based representation for graphs to compute the edit distance between graphs as a sequence of three different primitive opera-tions. Finally, using the new concept of probe, Lopresti [15] introduces the graph probing which is characterized by its rapidity.

3.1   Graph Edit Distance from Spectral Seriation

Robles-Kelly and al. [21] use a spectral method to represent graphs by strings, and then the similarity of graphs is measured according to the edit distance of strings in a probabilistic framework. The graph edit distance is the cost of the shortest edit path in an edit lattice for transforming the data graph into the model. The rows and columns of edit lattice are indexed by two strings $Y=\{y_1,y_2,,y_{|V_D|}\}$ for data graph $G_D=(V_D,E_D)$ and $X=\{x_1,x_2,,x_{|V_M|}\}$ for the model graph $G_M=(V_M,E_M)$, with null symbol ε, and V_D and E_D being the point set and the edge set of the data graph. The problem of computing the edit dis-tance is posed as that of finding the least expensive path $\Gamma^* = \langle \gamma_1, \gamma_2, ..., \gamma_k,, \gamma_L \rangle$ from (y_1,x_1) to $(y_{|V_D|},x_{|V_M|})$ through the edit lattice based on the Levenshtein distance. Each state $\gamma_k \in (V_D \cup \varepsilon) \times (V_M \cup \varepsilon)$ of the edit path is a Cartesian pair. Then cost functions are defined for elementary matches, according to the cost edit path Γ^* (i.e., graph edit distance) computed using the following equation:

$$d(X,Y) = C(\Gamma^*) = \sum_{\gamma_k \in \Gamma} \eta(\gamma_k \rightarrow \gamma_{k+1}) \qquad (1)$$

where $\eta(\gamma_k \rightarrow \gamma_{k+1}) = -(lnP(\gamma_k|\phi_X^*(x_i), \phi_Y^*(y_j)) + lnP(\gamma_{k+1}|\phi_X^*(x_{i+1}), \phi_Y^*(y_{j+1})) + lnR_{k,k+1})$, and the edge compatibility coefficient $R_{k,k+1}$ is

$R_{k,k+1}=$

$$\frac{P(\gamma_k|\gamma_{k+1})}{P(\gamma_k)P(\gamma_{k+1})} = \begin{cases} \rho_M\rho_D \\ \quad \text{if } \gamma_k \rightarrow \gamma_{k+1} \text{ is a diagonal transition on the edit lattice} \\ \rho_M \\ \quad \text{if } \gamma_k \rightarrow \gamma_{k+1} \text{ is a vertical transition on the edit lattice} \\ \rho_D \\ \quad \text{if } \gamma_k \rightarrow \gamma_{k+1} \text{ is a horizontal transition on the edit lattice} \\ 1 \\ \quad \text{if } y_j = \varepsilon \text{ or } x_i = \varepsilon \text{ and } y_{j+1} = \varepsilon \text{ or } x_{i+1} = \varepsilon \end{cases}$$

$$P(\gamma_k|\phi_X^*(x_i),\phi_Y^*(y_j)) = \begin{cases} \frac{1}{\sqrt{2\pi}\sigma}exp\{-\frac{1}{2\sigma^2}(\phi_X^*(x_i)-\phi_Y^*(y_j))^2\} \\ \quad \text{if } y_j \neq \varepsilon \text{ and } x_i \neq \varepsilon \\ \alpha \\ \quad \text{if } y_j = \varepsilon \text{ and } x_i \neq \varepsilon \end{cases}$$

where ρ_M and ρ_D are respectively the edge densities of the graphs G_M and G_D ($\rho_M = \frac{|V_M|^2}{E_M}$) and ϕ_X^* and ϕ_Y^* are, respectively, the leading eigenvectors of the adjacency matrices for the graph G_M and G_D. In the remainder, we denote this graph matching technique by GEDSS.

3.2 Graph Matching Based on Node Signatures

Jouili and al. [11] propose a new algorithm for matching and computing the distance between graphs. This approach is based on node signatures notion. In order to construct a signature for a node in an attributed graph, all available information into the graph and related to this node is used. The collection of these informations should be refined into an adequate structure which can provides distances between different node signatures. In this perspective, the node signature is defined as a set composed by four subsets which represent the node attribute, the node degree and the attributes of its adjacent edges and the degrees of the nodes on which these edges are connected. Given a graph $G=(V,E,\alpha,\beta)$, the node signature of $n_i \in V$ is defined as follows:

$$\gamma(n_i) = \Big\{ \ \alpha_i, \ \theta(n_i), \ \{\theta(n_j)\}_{\forall ij \in E}, \ \{\beta_{ij}\}_{\forall ij \in E} \Big\}$$

where

- α_i the attribute of the node n_i.
- $\theta(n_i)$ the degree of n_i.
- $\{\theta(n_j)\}_{\forall ij \in E}$ the degrees set of the nodes adjacent to n_i.
- $\{\beta_{ij}\}_{\forall ij \in E}$ the attributes set of the incident edges to n_i.

Then, to compute a distance between node signatures, the *Heterogeneous Euclidean Overlap Metric* (HEOM) is used. The HEOM uses the *overlap* metric for symbolic attributes and the normalized Euclidean distance for numeric attributes. Next the similarities between the graphs is computed: firstly, a definition of the distance between two sets of node signatures is given. Subsequently, a matching distance between two graphs is defined based on the node signatures sets. Let S_γ be a collection of local descriptions, the set of node signatures S_γ of a graph $g=(V,E,\alpha,\beta)$ is defined as :

$$S_\gamma(\, g\,) = \left\{\, \gamma(n_i) \mid \forall n_i \in V \,\right\}$$

Let $A=(V_a,E_a)$ and $B=(V_b,E_b)$ be two graphs. And assume that $\phi : S_\gamma(A) \to S_\gamma(B)$ is a function. The distance d between A and B is given by φ which is the distance between $S_\gamma(A)$ and $S_\gamma(B)$

$$d(A,B) = \varphi(S_\gamma(A), S_\gamma(B)) = \min_{\phi} \sum_{\gamma(n_i) \in S_\gamma(A)} d_{nd}(\gamma(n_i), \phi(\gamma(n_i)))$$

The calculation of the function $\varphi(S_\gamma(A), S_\gamma(B))$ is equivalent to solve an assignment problem, which is one of the fundamental combinatorial optimization problems. It consists of finding a maximum weight matching in a weighted bipartite graph. This assignment problem can be solved by the Hungarian method. The permutation matrix P, obtained by applying the Hungarian method to the cost matrix, defines the optimum matching between two given graphs. In the remainder, we denote this graph matching technique by GMNS.

3.3   Graph Probing Approach

Lopresti and al. [15] introduce the paradigm of graph probing. This technique consist on using a probe into the graphs to determine some particular information. The measure of similarity between two graphs is an L1 norm distance of the two corresponding vectors. For the construction of vectors, Lopresti present three classes of construction each one led by a question, Class 0: "*How many vertices with degree n are present in graph G = (V,E)?*", Class 1: "*How many vertices with in-degree m and out-degree n are present in G?*", Class2 : "*How many vertices labeled as **att** are present in G?*". The use of such class depends on the type of graph.

Therefore, for each graph, a representative vector is computed and the corresponding graph distance. Concretely, let G = (V,E) be an undirected graph, the vector associated to G is: $PR(G) \equiv (n_0, n_1, n_2, ...)$ where $n_i = |\{\, v \text{ in } V \mid \deg(v)=i\}|$. So the distance between two graphs is $L_1(PR1, PR2)$. In the remainder, we denote the graph probing technique by GP.

3.4   Graph Histogram Approach

Papadopoulos and al. [19] present a similarity measure for graphs, which is based on the concept of edit operations. They propose three different primitive operations, which are vertex insertion, vertex deletion and vertex update. While vertex insertions or deletions have a trivial meaning, the update operation is needed to insert or delete edges incident to a vertex. Additionally they introduce the degree sequence of a graph, i.e. the non-increasing sequence of the degrees of vertices in a graph. The similarity distance between two graphs is defined as the minimum number of primitive operations which are required so that the two graphs have the same degree sequence. To calculate the similarity measure, the sorted graph histogram is introduced, which is a histogram of the degrees of the vertices in a graph. Papadopoulos and al. show also that the L_1-distance between two sorted graph histograms defines their similarity distance. Additionally it is proven that the similarity distance satisfies the metric properties. In some cases, the sorted degree histograms of the graphs in a database are of different dimensionality if not all graphs are of the same order. To allow the use of index structures for vector spaces, the authors introduce a histogram folding technique to achieve a constant dimensionality of the histograms for all graphs. In the remainder, we denote this method by GH.

Table 1 provides a description of the selected graph similarity measures. In this table, we show the capability of each graph similarity measure to deal with labeled or unlabeled graphs and it provides a node-to-node matching.

Table 1. A summarization of the selected graph similarity measures

	Handle labeled graphs	Handle unlabeled graphs	Explicit node-to-node matching
Robles-Kelly [21]	no	yes	no
Jouili [11]	yes	yes	yes
Lopresti [15]	yes	yes	no
Papadopoulos [19]	no	yes	no

4 Experimental Results

In this section, we provide a performance comparison of the graph similarity measures described above. In addition, even if the statistical/structural approaches comparison is not the aim of this paper, we also use the generic Fourier descriptor (GFD) well known for its good performance [24] to show the general behavior of structural approaches vs a statistical one. In GFD, feature vectors are created from images by extracting information in the frequency domain. This statistical descriptor is invariant to rotation and scaling. In our experiments, we use the Euclidean distance to compute the distance between images represented by GFD feature vectors. Four graphic databases have been used to perform the comparison study. The precision/recall curves are used to measure retrieval performances. The *leave-one-out* protocol is used to provide precision/recall curves.

4.1 Data Sets

The graph retrieval tasks considered in this paper includes the retrieval of line drawings (symbols), ancient documents (ornamental letters), the set of

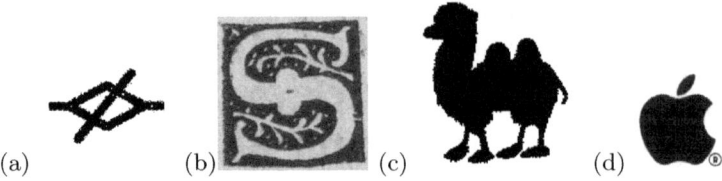

(a)   (b)   (c)   (d)

Fig. 1. Samples from: (a) GREC database, (b) Ornamental letters database, (c) Shape database, (d) Logo database

trademark-logos and LEMS shape database. Figure 1 represents samples for each database used.

- **GREC database**: The GREC database [20] (see figure 1(a)) consists of graphs representing symbols from architectural and electronic drawings. Here, the ending points (ie corners, intersections and circles) are represented by nodes which are connected by undirected edges and labeled as lines or arcs. The graph database used in our experiments has of 528 graphs, 24 classes and 22 graphs per class.
- **Ornamental letters database**: The ornamental letters database (see figure 1(b)) contains lettrine (graphical objects) extracted from digitized ancient document [2]. Since one lettrine contains a lot of information (i.e. texture, decorated background, letters), the graphs are extracted from a region-based segmentation [9] of the lettrine with a user-based parameterization technique. The nodes of the graph are represented by the regions and the edges describe their adjacency relationships. The graph database used in our experiments consists of 280 graphs, 4 classes and 70 graphs per class.
- **Shape database**: We use the shapes provided by the LEMS laboratory of the Brown University [23] (see figure 1(c)). The graphs are extracted from the shapes by skeletonizing and applying a polygonal approximation to the skeleton to obtain straight line segments. For each line segment, we locate endpoints and the graphs are based on the Delaunay triangulations of these endpoints. The graph database used in our experiments has 216 graphs, 18 classes and 12 graphs per class.
- **Logo database**: This database (see figure 1(d)) consists of graphs representing binary images of trademark-logos. Here, graphs are extracted by the delaunay triangulations on the detecting points of interest using Harris algorithm [10]. The graph database used in our experiments consists of 80 graphs, with 10 classes and 8 graphs per class.

Table 2 provides a description of the characteristics of the used graph data sets. Each graph data set is described by the maximum number and the average of nodes and edges (max nodes/edges and ϕ nodes/edges). Let us recall that all the graphs used in this paper are weighted graphs. That is, there is no labels attached to the nodes, but each edge is weighted by a unique numeric label.

---

[2] Provided by the CESR - University of Tours on the context of the ANR Navidomass project http://l3iexp.univ-lr.fr/navidomass/

Table 2. Description of the four graph data sets

	max nodes	max edges	ϕ nodes	ϕ edges
GREC	24	29	11.54	11.6
Ornamental Letter	178	4314	97.6	1779.6
Shape	73	2592	29.1	471.8
Logo	292	802	102.5	264.9

4.2   Results Analysis

The results of our experiments for these four databases with the four graph matching measures are presented in figure 2.

From the precision-recall curves, we can remark that the performance of the graph matching methods depend on the databases and more particularly the description we put in the graph. For the GREC database, the matching measures (GP, GH and GMNS) that use simple structural modification perform similarly and better than the GEDSS method which use a string representation for graphs. We realize that for graphs with low edge and node densities (as the case of the GREC database) the string-based representation is not discriminant. In addition, the GMNS method provides a performance peak for low recall values, and it joins the performance of the GP and GH methods for high recall values. The discrimination of the node signatures provides a good robustness for this kind of database.

From the results provided on the Lettrine database, we see that all the distance measures provide similar results with a little less performance for the GP technique. This may be explained by the fact that the different methods produce a quite similar response to the structural errors between the graphs used to represent the ornamental letters. In the other way, one can conclude that this kind of graph representation (region adjacency graph) of the ornamental letter is more or less robust to different graph matching methods.

In the case of the shape database, the performance of the graph probing fails clearly in comparison with other distance measures. It seems that the probe of the node degree is not a good discriminating feature for this database which presents important structural errors between graphs in different classes. Further, the GEDSS method which has shown previously good results for similar databases (see [21]), provides the better retrieval results.

For the logo database, all the distance measures provide similar behaviors. Here, the graph probing keeps the leader position among the other distances. In addition, the provided results of all the distance measures are particularly better in comparison with the other databases. This may be due to the suitable graph representation used for this database. We can think that the graph representation approaches used for other databases is not necessary the most suitable. In addition, different distance measures provide quite similar results for a given graph representation as the case of the Ornamental letters database. From all these results, we can remark that the GP and GEDSS methods are more sensitive to the representation we put in the graph.

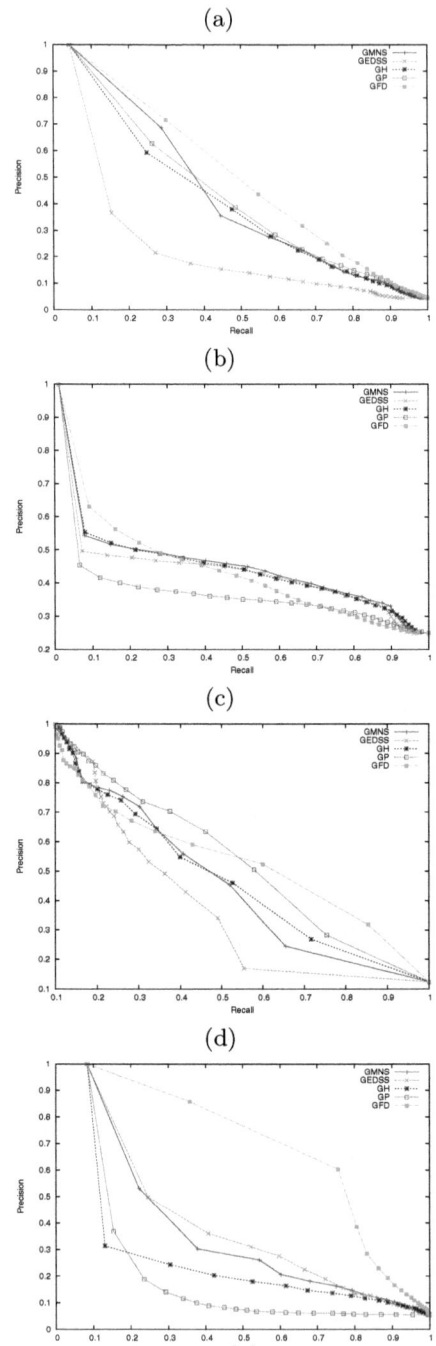

Fig. 2. Precision-Recall curves on: (a) GREC database, (b) Ornamental letters database, (c) Logo database, (d) Shape database

In addition, the GFD descriptor outperforms the results of all the graph similarity measures for the Shape database. However, for the Ornamental letters and the Logo databases in which the structural information is more important we remark that the structural approaches provide better results.

5 Conclusion and Perspectives

In this paper we have compared the performance of four graph matching methods for graph retrieval with different kind of document databases. The evaluation is performed using Precision rate against Recall rate. Our experimental results show that the performance of each graph distance measure depends on the databases and the approaches are also more and less robust to the variability of the representation. That is to say, a given graph distance can provide a good performance for one database and poor performance for an other database. Moreover, for a good graph representation we can remark that the performances of different graph matching methods are quite similar. In future works we want to study the behavior of these methods against the representation we put in the graph and the type of database.

References

1. Ambauen, R., Fisher, S., Bunke, H.: Graph edit distance with node splitting and merging, and its application to diatom identification. In: Hancock, E.R., Vento, M. (eds.) GbRPR 2003. LNCS, vol. 2726, pp. 95–106. Springer, Heidelberg (2003)
2. Baeza-Yates, R., Valiente, G.: An image similarity measure based on graph matching. In: Proc. International Symposium on String Processing Information Retrieval, pp. 28–38 (2000)
3. Barrow, H.G., Popplestone, R.J.: Relational descriptions in picture processing. Machine Intelligence 4, 377–396 (1971)
4. Bunke, H.: Attributed of programmed graph grammars and their application to schematic diagram interpretation. IEEE Transactions on Pattern Analysis and Machine Intelligence 4(6), 574–582 (1982)
5. Bunke, H.: Recent developments in graph matching. In: Proceedings of the 15th International Conference on Pattern Recognition, vol. 2, pp. 117–124 (2000)
6. Bunke, H., Günter, S., Jiang, X.: Towards bridging the gap between statistical and structural pattern recognition: Two new concepts in graph matching. In: Singh, S., Murshed, N.A., Kropatsch, W.G. (eds.) ICAPR 2001. LNCS, vol. 2013, pp. 1–11. Springer, Heidelberg (2001)
7. Conte, D., Foggia, P., Sansone, C., Vento, M.: Thirty years of graph matching in pattern recognition. International Journal of Pattern Recognition and Artificial Intelligence 18(3), 265–298 (2004)
8. Di Ruberto, C., Rodriguez, G., Casta, L.: Recognition of shapes by morphological attributed relational graphs. In: AIIA (2002)
9. Felzenszwalb, P.F., Huttenlocher, D.P.: Efficient graph-based image segmentation. International Journal of Computer Vision 59(2), 167–181 (2004)
10. Harris, C., Stephens, M.: A combined corner and edge detection. In: Proceedings of The Fourth Alvey Vision Conference, pp. 147–151 (1988)

11. Jouili, S., Tabbone, S.: Graph matching using node signatures. In: Torsello, A., Escolano, F., Brun, L. (eds.) GbRPR 2009. LNCS, vol. 5534, pp. 154–163. Springer, Heidelberg (2009)

12. Karray, A., Ogier, J.-M., Kanoun, S., Alimi, M.A.: An ancient graphic documents indexing method based on spatial similarity. In: Liu, W., Lladós, J., Ogier, J.-M. (eds.) GREC 2007. LNCS, vol. 5046, pp. 126–134. Springer, Heidelberg (2008)

13. Levenshtein, V.I.: Binary codes capable of correcting deletions, insertions, and reversals. Soviet Physics Doklady 10(8), 707–710 (1966)

14. Li, C.-Y., Hsu, C.-T.: Region correspondence for image retrieval using graph-theoretic approach and maximum likelihood estimation. In: International Conference on Image Processing, pp. 421–424 (2004)

15. Lopresti, D.P., Wilfong, G.T.: A fast technique for comparing graph representations with applications to performance evaluation. IJDAR 6(4), 219–229 (2003)

16. Neuhaus, M.: Edit distance-based kernel functions for structural pattern classification. Pattern Recognition (2006)

17. Ounis, I., Pasca, M.: Modeling, indexing and retrieving images using conceptual graphs. In: Quirchmayr, G., Bench-Capon, T.J.M., Schweighofer, E. (eds.) DEXA 1998. LNCS, vol. 1460, pp. 226–239. Springer, Heidelberg (1998)

18. Pan, J., Yang, H., Faloutsos, C., Duygulu, P.: Gcap: Graph-based automatic image captioning. In: Proceedings of the 4th International Workshop on Multimedia Data and Document Engineering (2004)

19. Papadopoulos, A.N., Manolopoulos, Y.: Structure-based similarity search with graph histograms. In: Proceedings of International Workshop on Similarity Search (DEXA IWOSS 1999), September 1999, pp. 174–178 (1999)

20. Riesen, K., Bunke, H.: IAM graph database repository for graph based pattern recognition and machine learning. In: da Vitoria Lobo, N., Kasparis, T., Roli, F., Kwok, J.T., Georgiopoulos, M., Anagnostopoulos, G.C., Loog, M. (eds.) S+SSPR 2008. LNCS, vol. 5342, pp. 287–297. Springer, Heidelberg (2008)

21. Robles-Kelly, A., Hancock, E.R.: Graph edit distance from spectral seriation. IEEE Trans. Pattern Anal. Mach. Intell. 27(3), 365–378 (2005)

22. Sanfeliu, A., Fu, K.: A distance measure between attributed relational graphs for pattern recognition. IEEE Trans. on Systems, Man and Cybernetics 13(3), 353–362 (1983)

23. Sharvit, D., Chan, J., Tek, H., Kimia, B.B.: Symmetry-based indexing of image databases. J. Visual Communication and Image Representation 9, 366–380 (1998)

24. Zhang, D., Lu, G.: Shape-based image retrieval using generic fourier descriptor. Signal Processing: Image Communication 17, 825–848 (2002)

Robust and Precise Circular Arc Detection

Bart Lamiroy and Yassine Guebbas

Nancy Université – INPL – LORIA
Équipe Qgar – Bât. B
Campus Scientifique – BP 239
54506 Vandoeuvre-lès-Nancy Cedex – France
Bart.Lamiroy@loria.fr

Abstract. In this paper we present a method to robustly detect circular arcs in a line drawing image. The method is fast, robust and very reliable, and is capable of assessing the quality of its detection. It is based on Random Sample Consensus minimization, and uses techniques that are inspired from object tracking in image sequences. It is based on simple initial guesses, either based on connected line segments, or on elementary mainstream arc detection algorithms. Our method consists of gradually deforming these circular arc candidates as to precisely fit onto the image strokes, or to reject them if the fitting is not possible, this virtually eliminates spurious detections on the one hand, and avoiding non-detections on the other hand.

1 Introduction

Finding circular arcs is one the recurring problems in graphical document interpretation or symbol recognition. The main difficulty with the existing approaches is that they often are of considerable complexity (e.g. Hough-like [1] or feature grouping approaches [2]) sensitive to image quality, line thickness, or rely on a number of user defined parameters or thresholds that make them extremely difficult to apply to generic problems or on heterogeneous document sets.

The approach developed in this paper reduces the set of needed parameters to a minimal set of very elementary and visually significant values and can be applied without prior knowledge of the document set, regardless of line widths, connectedness or complexity. It relies on elementary (3,4)-distance transform skeletonization [3] and segment detection [4]. Unlike extremely efficient methods like [5], ours does not require reasonable segmentation of arcs. This work is tightly related to [6].

The following section establishes how to determine if a single circular arc is present, provided we have a rough initial guess of its position, and how to robustly detect and locate it using RANSAC (Random Sample Consensus [7]). Section 3 then explains how to generalize to detecting and localizing any number of circles, without *a priori* knowledge of their position. The last two sections conclude by eliminating spurious detections and by establishing the limits of the approach.

J.-M. Ogier, W. Liu, and J. Lladós (Eds.): GREC 2009, LNCS 6020, pp. 49–60, 2010.
© Springer-Verlag Berlin Heidelberg 2010

2   Determining the Presence of a Circular Arc

In this section we address the problem of detecting a circular arc, given an initial estimate of its center (x_c, y_c), its radius σ, and its two endpoints p_l and p_r[1]. This estimate, as we shall see further, can be very approximate. The main goal, in this first stage, is to detect whether or not, an arc is present in the image, near the vicinity of the given parameters.

2.1   General Algorithm

We are mainly exploiting the algorithm described in [6], with one major adjunction. The cited method has been developed to identify and locate full circles, and therefore only needs to consider adapting to two variables: the center (x_c, y_c), and the radius σ. This is not the case anymore for detecting arcs, since two parameters are added: p_l and p_r, the left and right endpoints.

The general approach we develop consists of taking the set $\mathcal{P} = \{p_i\}$ of all pixels p_i lying on the discrete circular arc $\mathcal{A}^0$ defined by (x_c, y_c), σ, p_l and p_r. As in, [6], we define, for each of these pixels p_i, the discrete line Δ_i, starting at (x_c, y_c), and passing through p_i. Let q_i be the pixel on Δ_i that is the closest black pixel to p_i. Let $\mathcal{Q}_a^0 = \{q_i\}$. $\mathcal{Q}_a^0$ therefore is the set of all black pixels closest to the initial estimate $\mathcal{A}^0$ in the direction of the circle radius.

Figure 1 gives an illustration of this estimation. Initial guesses are drawn in blue. For each conjectured circle, green pixels are those found at the correct distance from the center, while red ones lie on the radius and are closest to the circle.

Now, let $\mathcal{C}^1$ be the best fitting circle over $\mathcal{Q}_a^0$ (any criterion can be used, but we are using the Least Median of Squares – *cf.* section 2.2), and let us generalize the previous step, such that $\mathcal{Q}_c^t$ contains the set of all black pixels closest to the theoretical circle $\mathcal{C}^t$ in the direction of the circle radius (and similarly for $\mathcal{Q}_a^t$).

By construction, $\mathcal{Q}_a^t \subset \mathcal{Q}_c^t$, and while this new set of points allows for a re-estimation of (x_c, y_c), σ, the other parameters p_l and p_r need to be re-evaluated as well. The approach is the following:

Let τ^t be the error measure between $\mathcal{A}^t$ and $\mathcal{Q}_a^t$. *i.e.* τ^t represents the fitness between the model $\mathcal{A}^t$ and its corresponding data $\mathcal{Q}_a^t$. Let $\mathcal{A}_<^t \subset \mathcal{A}^t$ a smaller circular arc[2] than $\mathcal{A}^t$ such that $\tau_<^t > \tau^t$ and that

$$\forall \mathcal{A}^{t\star} | \mathcal{A}_<^t \subset \mathcal{A}^{t\star} \subset \mathcal{A}^t : \tau^{t\star} < \tau^t. \tag{1}$$

---

[1] In this document we shall conveniently ignore the fact that there is a small ambiguity with defining an arc by the center of its corresponding circle, the radius and the endpoints: one also has at least orientation of the arc to consider as to know what part of the circle between the two endpoints is belonging to the arc, and which part isn't.

[2] "smaller" meaning having the same center and radius, but having a smaller aperture while being fully included in the "larger" one.

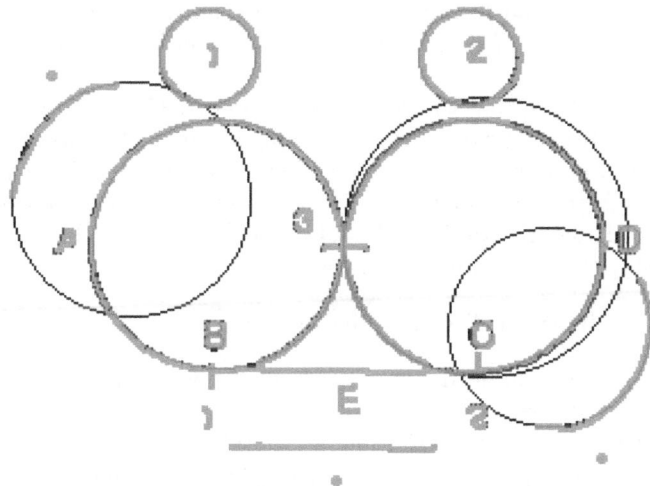

Fig. 1. Example of circle hypotheses: in blue, the initial guess; in green, points correctly lying on the conjectured circle; in red, point closest to the circle

In other terms, $\mathcal{A}_<^t$ is a smaller arc that fits the dataset better than $\mathcal{A}^t$ and all intermediate arcs fit less. This means that $\mathcal{A}_<^t$ is the largest sub-arc fitting the data better than $\mathcal{A}^t$.

We do a similar search by increasing the arc size thus obtaining $\mathcal{A}^t \subset \mathcal{A}_>^t$ a larger circular arc than $\mathcal{A}^t$ such that $\tau_>^t > \tau^t$ and that

$$\forall \mathcal{A}^{t\star} | \mathcal{A}^t \subset \mathcal{A}^{t\star} \subset \mathcal{A}_<^t : \tau^{t\star} < \tau^t, \tag{2}$$

$\mathcal{A}_>^t$ thus being the smallest super-arc fitting the data better than $\mathcal{A}^t$.

We can then define $\mathcal{A}^{t+1}$ as being the $\mathrm{argmax}_\tau \{\mathcal{A}_<^t, \mathcal{A}^t, \mathcal{A}_>^t\}$. Continuing this iteration until $\mathcal{A}^t = \mathcal{A}^{t+1}$ will yield the best estimate of the arc (if any) closest to the initial $\mathcal{A}^0$.

In the following sections we detail the different steps of this general approach.

2.2   Using RANSAC and LMedS

Since there is no guarantee that any $\mathcal{A}^t$ or $\mathcal{Q}^t$ may effectively contain points that form a circle, it may be extremely hazardous to use global minimization approaches (like Least Squares, for instance) [8]. It is known that these estimators are very sensitive to outliers or spurious data that does not conform to the required model [9]. Using these functions would invariably lead to degenerate convergence.

RANSAC [7] is much better suited for fitting very noisy data – especially data containing measures that do not belong to the model that is to be estimated – The approach consists of selecting the strict minimum of data points required for estimating an instance of the model (*e.g.* three points for estimating a circle) and then computing the residual error of the other data points to this model. This is done a number of times, and the final model is the one with the lowest residual error.

More formally: let $\mathcal{Q}^t$ be the set of model points. $\mathcal{Q}^t$ supposedly, and in the worst case, contains a ratio of τ outliers. Let q_n, q_n' and q_n'' be three random points belonging to $\mathcal{Q}^t$, and let $\mathcal{C}_n$ be the circle defined by and passing through q_n, q_n' and q_n''. Let $\delta\left(\mathcal{C},p\right)$ be the distance of a point p to a circle $\mathcal{C}$, and let $\mathrm{Med}_\tau\left(\mathcal{S}\right)$ be the τ-quantile median value of the set $\mathcal{S}$. We then define the residual error of a set of model points $\mathcal{Q}^t$ to a circle $\mathcal{C}_n$ as

$$\mathrm{RsdErr}\left(\mathcal{Q}^t,\mathcal{C}_n\right) = \mathrm{Med}_\tau\left(\left\{\delta\left(\mathcal{C}_n,p\right) \,|\, p \in \mathcal{Q}^t\right\}\right). \tag{3}$$

RsdErr gives the maximum distance of a set of points to a circle, discarding a proportion of τ outliers.

With RANSAC we choose R random subsets of 3 points within $\mathcal{Q}^t$, each giving rise to the computation of a circle $\mathcal{C}_n$. For each subset, we compute the corresponding $\mathrm{RsdErr}\left(\mathcal{Q}^t,\mathcal{C}_n\right)$, thus obtaining

$$\mathcal{C}^{t+1} = \underset{\mathcal{C}_n,n\in[1...R]}{\mathrm{argmin}}\left(\mathrm{RsdErr}\left(\mathcal{Q}^t,\mathcal{C}_n\right)\right). \tag{4}$$

The number of required subsets can be formally deduced from both the quality of the data (expected rate of outliers τ), the dimensionality of the problem (here 6, since we need three points for estimating a circle, each point having two dimensions) and the required confidence in the result [7].

3   Robust Arc Detection

The previously presented method does a very good job of robustly determining whether there is a circular arc close to a given center and radius (x_c, y_c) and σ. However, it needs some initial guess on where to search. The method we are developing here proceeds in three main phases:

1. Generate a high number of possible arc candidates, without consideration of uniqueness, overlapping or exact localization.
2. Verify the quality of each candidate using the approach described in section 2. The output of this verification is a list $\mathcal{A}$ of genuine arcs, correctly fitted on the image data.
3. Detect and merge multiple and/or partial detections of the same curves as to obtain a set of unique, disjoint arcs.

3.1   Arc Candidate Generation

In order to obtain the largest possible set of arc candidates, we automatically segment the image using a basic Rosin & West line segment vectorization [4]. We then simply enumerate all connected pairs of segments. Each pair gives us three points, which is exactly the amount of data that allows for getting an initial guess for a circular arc: (p_1, p_2, p_3). These points define a unique circle on the one hand, and furthermore, since they are ordered – p_2 being in the middle – they define the left and right extrema for the definition of an arc.

This approach is combined with direct arc detection from [4] as to produce the largest possible set of arc candidates to bootstrap our localization method (*cf.* section 2).

3.2   Merging of Multiple Detections

Since the method is based on unfiltered hypothesis generation, it has a clear tendency toward over-segmentation, as shown in Figure 2. The main idea behind being tolerant towards this over-segmentation is to be confident that (almost) all image pixels belonging to an arc are covered by at least one initial arc candidate. Merging arcs should therefore result in a full coverage of each arc of the image by one unique, genuine arc. Merging arc candidates representing the same circular arc in the image requires two distinct operations: merging estimates covering the same pixels and merging arc candidates not sharing the same pixels but being partial estimates of a same wider arc. These two operations can be performed by first increasing the aperture of the arcs (*cf.* section 3.2.1), thus making hypothetical arcs share pixels, and, secondly, merging the arc candidates sharing

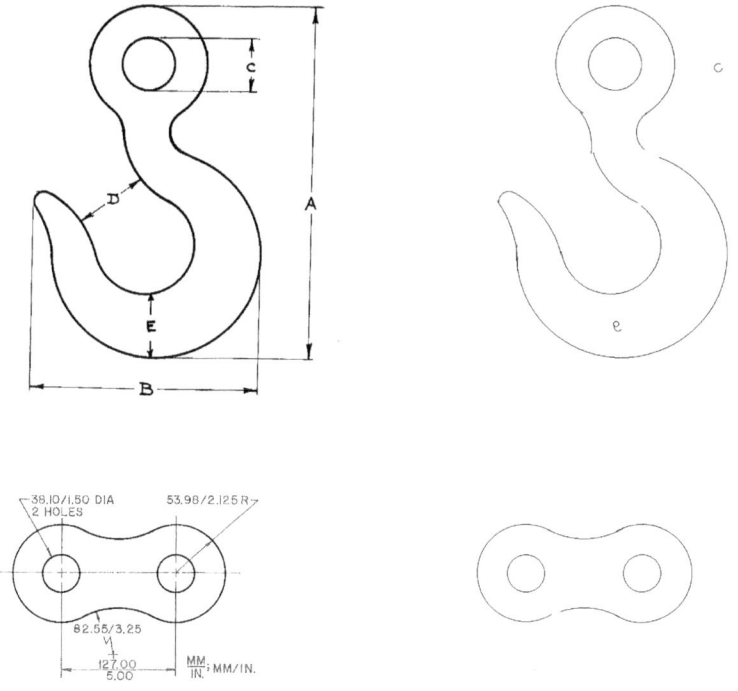

Fig. 2. GREC 2007 contest images: original image (left) – final segmentation (right)

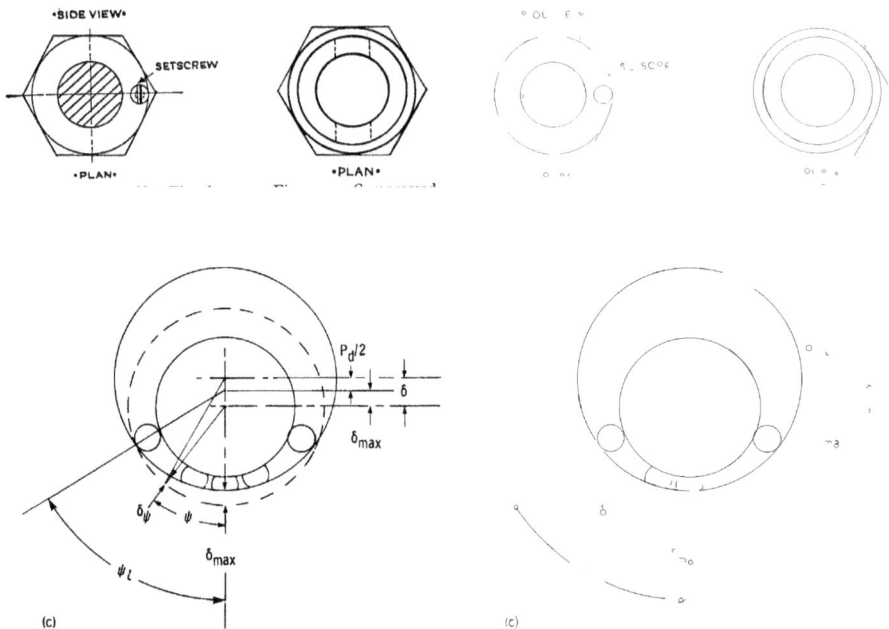

Fig. 3. GREC 2009 contest images: original image (left) – final segmentation (right)

pixels (*cf.* section 3.2.3). For merging arcs, we do not use the full circle image, but the image skeleton [6].

3.2.1   Increasing Aperture

To increase the aperture of an arc, we first set a threshold to the maximum distance between a point from the discrete hypothetical arc and the closest pixel, as shown in Figure 4, where the distance is measured on the line going through the center of the hypothetical arc and a point on the hypothetical arc.

The increase of the aperture of an arc is done by starting from the endpoints of the candidate arc p_l and p_r and then increasing the aperture pixel-wise, as long as the distance to the closest pixel remains below the threshold.

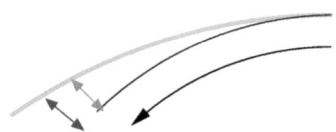

Fig. 4. Distance between a point of the hypothetical arc and the closest pixel: in blue the hypothetical arc; in Grey, image pixels; in green, distance between a point on the hypothetical arc and the closest pixel; in red, threshold

3.2.2   Finding Which Arcs to Merge

Once the set of maximal arc candidates obtained, overlapping ones or those lying on the same image curve need to respond to the following criteria in order to be merged:

1. A non-empty intersection between two arcs means that these two arcs are likely part of a same covering arc. However two arcs having common closest pixels are not necessarily sub-arcs of a same arc as shown in Figure 5.

2. Arcs having comparable radii are merge candidates. This is checked through a radius ratio with the formula:

$$\frac{|r_1 - r_2|}{\max(r_1, r_2)} < \texttt{RatioRadiusError}. \tag{5}$$

   Checking the center of the circle is less robust since small changes in curvature may be visually insignificant, but generate large differences in the center position.

3. Arcs having opposed normal vectors (*cf.* Figure 5) are not eligible for merging, even though they may overlap. This criterion is verified by choosing a point I from the overlapping part of two arcs, and constructing a vector $\overrightarrow{IO_1}$ that originates from I and finishes at O_1 the center of the first arc, and defining similarly a vector $\overrightarrow{IO_2}$. We then compute the scalar product of these vectors:

$$\overrightarrow{IO_1} \cdot \overrightarrow{IO_2} = |\overrightarrow{IO_1}||\overrightarrow{IO_2}| \cos\theta \tag{6}$$

   If the sign of the product is positive, we consider that the two arcs stem from the same covering one.

Once these criteria are verified, three different configurations may occur for merging. They are depicted in Figure 6. In configuration A the two arcs are "adjacent" sharing some pixels, in configuration B one arc includes another and in configuration C the two arcs are "explementary". The covering arc is formed by considering that the two arcs belong to the same circle. Therefore the resulting arc is the union of the two arcs as if they had the same center and the same radius, in other words, the computation of the arc's angle and aperture is based on the angles and apertures of the two arcs.

Fig. 5. Intersection of arcs with opposed curvature signs or with significantly different radii

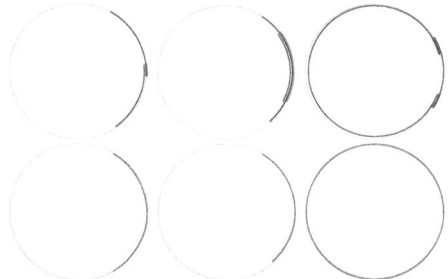

Fig. 6. Intersection configurations (top) and corresponding covering arcs (bottom). Configurations are labeled from left to right: A,B and C.

3.2.3   Merging Arcs

The last phase consists of creating the final, genuine arcs by merging the selected candidates corresponding to the previously described criteria. The method developed here tries to find the three points of the equilateral triangle which is circumscribed by the merged arc circle. This increases the odds of having the best fitting circle as with this method we avoid choosing either noisy or numerically instable points. This procedure begins by choosing either the arc 1 or 2, and then computes the edge length of the circumscribed equilateral triangle

$$\text{edgeLength} = 2r_i \sin \frac{\pi}{3}, \tag{7}$$

where r_i is the radius of the circle.

Now, let $\mathcal{R}_i$ be the subset of image curve points $(\mathcal{Q})$ which are close to the arc candidate $\mathcal{C}_i$

$$\mathcal{R} = \{p \in \mathcal{Q} | \delta\left(\mathcal{C}, p\right) < \text{RsdErr}\left(\mathcal{Q}, \mathcal{C}\right)\}. \tag{8}$$

We can then define a partitioning of the points $(\mathcal{D}_1$ and $\mathcal{D}_2)$ belonging to the two arc candidates, as well as their intersection $\mathcal{I}$

$$\mathcal{I} = \mathcal{R}_1 \cap \mathcal{R}_2, \mathcal{D}_1 = \mathcal{R}_1 \backslash \mathcal{R}_2, \mathcal{D}_2 = \mathcal{R}_2 \backslash \mathcal{R}_1. \tag{9}$$

We then define p_3 and p_1 such that

$$p_3 p_1 = \min_{p_i \in \mathcal{I}, p_j \in \mathcal{D}_2} |\text{edgeLength} - p_i p_j| \tag{10}$$

and find p_2 such that

$$p_2 p_3 + p_2 p_1 = \max_{p_i \in \mathcal{D}_2} \left(p_i p_3 + p_i p_1\right). \tag{11}$$

If we consider that the initial arcs belong effectively to the same circle, this method constructs the equilateral triangle and gives three points of a same circle. The more the three points are distant from each other, the more accurate the construction of the new circle is.

In fact, we are likely to find a better distributed set of three points over a circle if instead of using $\mathcal{D}_1$ and $\mathcal{D}_2$ we use $\mathcal{R}_1$ and $\mathcal{R}_2$.

4 Experiments

In collaboration with E. Barney Smith [10] we have been conducting an exhaustive survey of the influence of all possible internal parameters and external noise variations on the quality of the arc detection. Full analysis and report of this work is beyond the scope of this paper and will be published separately. Fig 7 shows some samples of used synthetic data for assessing the quality and precision of our approach.

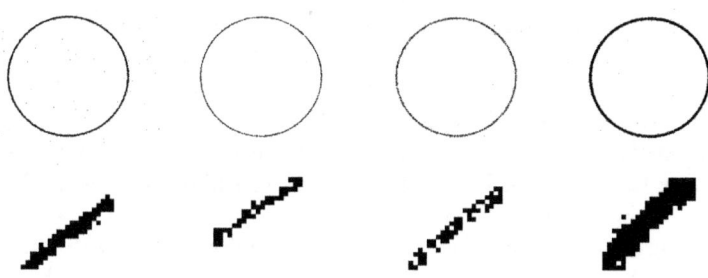

Fig. 7. Various Degraded Synthetic Circles and Selected Zooms

The overall precision in circle detection is extremely high. Precision was measured using three different metrics:

Circle Center Precision is obtained by measuring the euclidian distance of the detected arc circle to the theoretical circle center.

Circle Radius Precision is obtained by measuring the absolute difference between the detected radius an the theoretical radius.

Overlapping is a metric correlating the two previous ones, and expresses the overlapping surface ratio of both circles. It is normalized to $[0, 1]$, where 1 signifies perfectly identical circles, and 0 perfectly disjoint circles.

Tested over a wide range of parameters (τ outlier quantile, required coverage rates – *cf.* next section – ...) our method gives the following results:

	Worst Case	Best Case
Avg. Center Precision Error (pixels)	0.67	0.38
St.Dev. Center Precision Error	0.71	0.52
Avg. Radius Precision Error (pixels)	0.11	0.01
St.Dev. Radius Precision Error	0.37	0.12
Avg. Overlapping (%)	98.2	99.2
St.Dev. Overlapping (%)	6.0	2.3

This translates into estimation errors upto a pixel for center and radius. Coverage standard deviation might seem high for the announced detection precisions, but comes from situations where we tested on small circles (radius 8 pixels) where a single pixel shift accounts for a significant proportion of non overlapping.

Figures 2 to 3 show results on the GREC 2007 and 2009 contest images. The initial images are in black, while detected arcs are in Grey (right column).

4.1 Parameters and Their Influence

All parameters mentioned here are either direct transpositions of the algorithm described in this paper, or are direct call parameters of the software available for download (*cf.* note below).

One parameter that has an influence on determining whether two arcs are partial estimates of a same global arc is `RatioRadiusError` (*cf.* section 3.2.2). To compare the radii of two arcs, experiments show that a value of 64% for `RatioRadiusError` makes the merge possible for most arcs having common black pixels and avoids merging arcs with significantly different radii as shown in Figure 8. For the image in Figure 8 65% was too high and resulted in a loss of precision as shown within the rectangle.

The `filterCoverage` parameter is used to keep only those arc candidates that have a sufficient percentage of pixels effectively lying on pixels of the image. This process uses the original image instead of the skeleton image. The coverage percentage of `filterCoverage` is set to 89% to ensure keeping only accurate estimates.

`preFilterCoverage` checks if the percentage of pixels of the discrete estimate of the given arc lying effectively on black pixels of the original image, oversteps a cover percentage and returns a boolean. This process uses the original image instead of the skeleton image. The cover percentage of `preFilterCoverage` can be lower than the cover percentage defined for `filterCoverage`, as some intermediate arcs that do not withstand the cover percentage of `filterCoverage` might be merged with another arc resulting in an arc that does withstand the cover percentage of `filterCoverage`. The percentage 89% proved to be a good value for the cover percentage of `preFilterCoverage`, while 90% removed some good circles. This is often due to the fact that in reality, the images have slight deformations, an percieved circles are actually ellipses.

The merging algorithm can be improved by increasing the aperture of the arcs "virtually'. In other words, each arc has two angles and two apertures. The

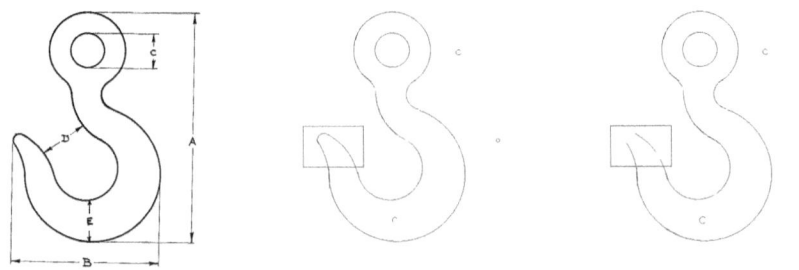

Fig. 8. RatioRadiusError tuning;the original image;64% filter; 65% filter

"virtual" angle and aperture are those which are increased and used to check if two arcs have to be merged. The actual angle and aperture are not changed. In fact, while increasing the actual aperture, the discrete arc of an estimate might no longer withstand the `preFilterCoverage` processing, as it is likely to have more pixels that are not black. Thus by keeping the initial angle and aperture unchanged we maintain arcs that were not merged. Moreover when performing the merge, the points used to construct the new arc are from the closest points of the actual arc, which is more accurate especially if we choose the points from those who belong as well to the black pixels of the image. The "virtual" increase uses the threshold defined in 3.2.1, namely a value of 3.

5   Conclusion and Further Work

In this paper we have presented a highly efficient and complete arc detection algorithm that needs extremely few parameters or contextual knowledge to operate. We have validated it on quite difficult images, coming from the GREC 2007 and 2009 contest. Further work will include stroke width integration in order to obtain a more precise localization of the arcs, as well as a more quantitative assessment of the positioning and localization of the detected arcs.

Acknowledgments

Authors acknowledge funding from the PROCORE-FRANCE/HONG KONG JOINT RESEARCH SCHEME (F-HK04/05T, 9050187): Knowledge representation issues for the performance evaluation of graphic symbol recognition methods. B. Lamiroy more particularly thanks Prof. Liu Wenyin for having received him at the City University of Hong Kong for finalizing this work. Authors aknowledge reporting snippet of yet unpublished work in collaboration with E. Barney Smith on evaluation of noise influence in section 4. Furthermore, the circle overlapping metric described in section 4, is the result of fruitful, informal discussions with E. Magagnin. B. Lamiroy was a visiting scientist at Lehigh University at the time of publication of this article.

Access to Source Code

The source code of this work is available for download and evaluation under LGPL at `http://gforge.inria.fr/projects/visuvocab/`

References

1. Olson, C.F.: Constrained Hough Transforms for Curve Detection. Computer Vision and Image Understanding 73(1), 329–345 (1999)
2. Chen, T.C., Chung, K.L.: An Efficient Randomized Algorithm for Detecting Circles. Computer Vision and Image Understanding 83(2), 172–191 (2001)

3. di Baja, G.S.: Well-Shaped, Stable, and Reversible Skeletons from the (3,4)-Distance Transform. Journal of Visual Communication and Image Representation 5(1), 107–115 (1994)
4. Rosin, P.L., West, G.A.: Segmentation of Edges into Lines and Arcs. Image and Vision Computing 7(2), 109–114 (1989)
5. Hilaire, X., Tombre, K.: Robust and Accurate Vectorization of Line Drawings. IEEE Transactions on PAMI 28(6), 890–904 (2006)
6. Lamiroy, B., Gaucher, O., Fritz, L.: Robust Circle Detection. In: Proceedings of 9th International Conference on Document Analysis and Recognition, Curitiba (Brazil), pp. 526–530 (2007)
7. Fischler, M.A., Bolles, R.C.: Random Sample Consensus: A Paradigm for Model Fitting with Applications to Image Analysis and Automated Cartography. Communications of the ACM 24(6), 381–395 (1981)
8. Rousseeuw, P.J., Leroy, A.M.: Robust Regression and Outlier Detection. Wiley Series in Probability and Mathematical Statistics. John Wiley and Sons, Chichester (1987)
9. Berman, M.: Large Sample Bias in Least Squares Estimators of a Circular Arc Center and Its Radius. Computer Vision, Graphics and Image Processing 45, 126–128 (1989)
10. Smith, E.H.B.: Characterization of image degradation caused by scanning. Pattern Recognition Letters 19(13), 1191–1197 (1998)

Automatic Palette Identification of Colored Graphics

Vinciane Lacroix*

CISS Department, Royal Military Academy, 1000 Brussels, Belgium

Abstract. The median-shift, a new clustering algorithm, is proposed to automatically identify the palette of colored graphics, a pre-requisite for graphics vectorization. The median-shift is an iterative process which shifts each data point to the "median" point of its neighborhood defined thanks to a distance measure and a maximum radius, the only parameter of the method. The process is viewed as a graph transformation which converges to a set of clusters made of one or several connected vertices. As the palette identification depends on color perception, the clustering is performed in the L*a*b* feature space. As pixels located on edges are made of mixed colors not expected to be part of the palette, they are removed from the initial data set by an automatic pre-processing. Results are shown on scanned maps and on the Macbeth color chart and compared to well established methods.

Keywords: palette extraction, clustering, mean-shift.

1 Introduction

The first step of color graphic vectorization is the identification of its palette. According to a recent study [1], approximately 13.5 million images are vectorized in the United States every years, consuming more than 7 million man hours. These images are made of photos, artworks, logos, etc. Commercial vectorisation software [1] exist but do not provide satisfying results in a full automated mode [2].

Despite this demand, there has been limited research done on colored image vectorisation except from a specific application: scanned maps [3], [4], [5] for which the vectorisation is performed on each colored layer, thus after the color extraction process.

Color palette reduction — when the final number of classes is a power of two the process is also known as "color image quantization" — such as kmeans, fuzzy-kmeans, median-cut, and octrees or any clustering method (also called unsupervised classification) could be considered for palette identification. For a comparison of color image quantization methods see [6], for a list and discussion on unsupervised classification see [7].

---

* This study is funded by the Belgian Ministry of Defense.
[1] see http://en.wikipedia.org/wiki/
Comparison_of_raster_to_vector_conversion_software

J.-M. Ogier, W. Liu, and J. Lladós (Eds.): GREC 2009, LNCS 6020, pp. 61–68, 2010.

In this paper we propose the "median-shift", a new clustering method, so called by analogy with the mean-shift [2] [8], an iterative procedure that shifts each data point towards the "median" of data points of its neighborhood. The method is used to identify the palette of some scanned maps and other graphics.

The aim of any clustering method — and in particular of palette extraction — is to find a small set of representative of the whole data set. Many authors have suggested that clustering methods applied to multi-variate images should make use of the spatial information as neighboring pixels are likely to belong to the same cluster (see [9]). The spatial information is introduced either before, during or after the clustering. For example, a common pre-processing is made by regularization or pixel grouping. Markov Random Fields [10] are examples of methods including spatial constraints into their process. The authors of the mean-shift, by adding the spatial coordinates to the feature space, also introduce spatial constraints into the clustering process. Finally, voting schemes are examples of post-processing methods. In this article, the spatial information is first used in pre-processing by filtering out edges of the luminance image, as they are most probably mixed pixels.

The full pre-processing is described in section 2. Section 3 presents the median-shift while section 4 explains how the algorithm is implemented for palette extraction and addresses the case of scanned maps. Section 5 shows the results of the procedure on several maps and on the Macbeth color chart in comparison to well established methods. Section 6 provides summary, discussion and conclusions.

2   Pre-processing

The aim of the pre-processing is twofold: removing potential mixed pixels and finding a space providing a better cluster separation.

Apart from the noise generating unexpected colors, problems for automatic palette identification come from the superposition of colors and from edge pixels generating disturbing colors. In order to remove part of these outliers, an automatic thresholding on the norm of the gradient of the luminance is performed, as according to Koschan [11] 90% of all edges are in the intensity image. This operation may however delete pixels located on lines; the latter are then brought back by a fully automatic ridge extraction process [12] on the luminance image.

A colored pixel is represented in a 3D space. As our aim is to identify some "target colors" such that each present color could be replaced by its closest target, the space and the distance are of prime importance. The *uniform* L*a*b* space (noted "Lab" in this paper), in which the Euclidean distance reflects the perceived distance, has thus been chosen for a better color separation. The data set is then made of filtered pixels described by their Lab coordinates.

---

[2] There exist many variants of the mean-shift, but according to the author's knowledge, none is using the median instead of the mean.

3   Clustering

Let the data be a set of points $\mathbf{x}$ embedded in a n-dimensional feature space: $\mathbf{x} = (x_1, ..., x_n)$ and let $d(\mathbf{x}, \mathbf{y})$ be a distance defined in this space. The neighborhood of $\mathbf{x}$, $V_R(\mathbf{x})$, is defined as the set of $\mathbf{y}$'s such that $d(\mathbf{x}, \mathbf{y}) < R$. In this framework, a point is *isolated* if the cardinal of its neighborhood is one, and *connected* otherwise. The "median point" $\bar{\mathbf{x}}$ of $V_R(\mathbf{x})$ is defined as the point $\bar{\mathbf{x}} = (\bar{x}_1, ..., \bar{x}_n)$ such that $\bar{x}_i$ is the median of the ith component of all points in $V_R(\mathbf{x})$.

The median-shift algorithm is an iterative process which shifts each data point $\mathbf{x}$ at time t to $\bar{\mathbf{x}}$; it can be seen as a graph transformation. Each vertex $v(\mathbf{x}, w)$ of the graph G is characterized by a vector $\mathbf{x} = (x_1, ..., x_n)$, corresponding to a point of the data set, and a weight w initially set to 1. The vertices of G are connected if the points are neighbors. A cluster $C_R(\mathbf{x})$ is defined as the connected component of G containing the vertex $\mathbf{x}$.

At time t the graph G is transformed into G' (initially empty) according to the following rule: for each vertex $v(\mathbf{x}, w)$ of G the median point $\bar{\mathbf{x}}$ is computed; if the corresponding vertex already exists in G', its weight is incremented by w, otherwise it is created with a weight equal to w. The edges of G' are then updated before the operation is repeated at $t = t + 1$ with $G = G'$.

At $t = 0$, the graph G is a set of one or several clusters. The convergence of G is thus related to the convergence of each connected component. Except in very few cases depending on the distance definition and the distance between clusters, disconnected clusters will remain disconnected.

These exceptions set aside, any cluster made of isolated vertex will remain stable. If a cluster is made of two or more points all connected to each other, the cluster will collapse in one point, would it be a new or an existing one, thus converging.

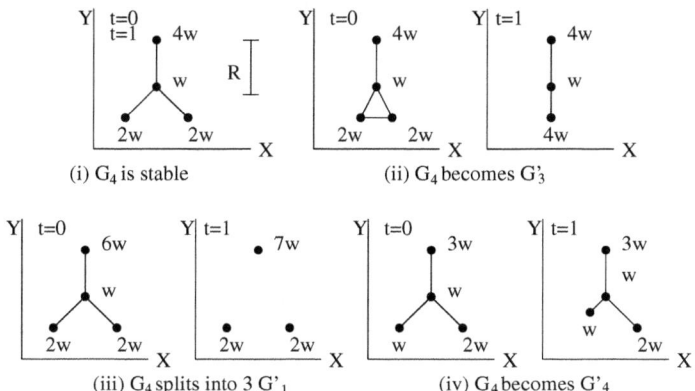

(i) G_4 is stable

(ii) G_4 becomes G'_3

(iii) G_4 splits into 3 G'_1

(iv) G_4 becomes G'_4

Fig. 1. Graph transformation scenarios

For a more complex cluster G_n characterized by n vertices, several scenarios may take place: (i) it is stable (ii) some vertices vanish thus leading to a graph G'_{n-i} (iii) it splits into several clusters $G^1_{n-i}, ..., G^m_{n-k}$, with a total number of vertices lower or equal to n (iv) it is transformed into G'_n. In this list of scenarios, only the last one could be problematic with respect to convergence, as in all other non stable cases, the total number of vertices is decreasing. However, though the number of vertices remains the same in (iv), the distances between them decrease, and at some distance below a threshold, the points could be considered as being at the same location, leading to G'_{n-1}. Figure 1 shows several scenarios for a 4 vertices graph in a 2D feature space. In practice convergence to complex clusters (i.e. made of several vertices) occurs when the radius is too big compared to the variations of the density.

The final result is thus a graph made of clusters containing one or several vertices, each cluster being separated from each other by at least a distance R. Note that the mean-shift may also be viewed as a graph transformation and its converging graph has the same property.

4   Implementation

The authors of the mean-shift [8] suggest to label a data point according to the cluster it converges to. In the palette extraction process however, this might result in assigning a point to a very different color. The following strategy is thus suggested. The median-shift algorithm is used to find the most important colors. The too small clusters ($< T$) are ignored. For each remaining cluster, the pixels are put aside if their distance to the cluster vertices is larger than R/2. If the set of all these outsiders is significant ($> PC\%$ of the initial set), the set is used again in a median-shift procedure providing additional clusters. The most important clusters are accepted until the number of ignored pixels is below the threshold.

Several distances can be used but for the current application the euclidean distance in the Lab color space is convenient. Several radii could also be used, but a radius between 14 and 18 seems optimal to separate colors of most graphics. As the radius represents the maximum distance between two clusters, it means that colors of a given cluster are not further than a distance between 7 and 9 from their center; knowing that under the unity (in the Lab space) human cannot discriminates colors, this range seems reasonable.

A typical map is about 64 cm by 40 cm. Recommended scan resolution varies between 300 to 600 samples per inch, so that a scan map can be as large as 10078 pixels by 6299 pixels. A 512 by 512 sample could be too small to have the chance to get all pixel colors, while a 1024 by 1024 would seem reasonable. The following strategy is thus proposed for extracting the palette of large images.

A medium-size image (1024 by 1024) is extracted, pre-processed and divided in small images (512 by 512) on which the median-shifted is run. A new median-shift may be used for combining all clusters assigning to each cluster vertex a weight equal to the number of pixels no more distant than $R/2$.

5   Results

The procedure has been used to identify the colors of six maps and a photographic chart. The pre-processing involves a Gaussian gradient computation ($\sigma = 0.7$) used for the edge and ridge outputs. The mean of all non-zero edges m_e and non-zero bright and dark ridges, m_b and m_d respectively, are used as threshold to derive the mask of selected pixels. So far an image cut of size 512×512 has been considered in each map. Two distances have been used: Euclidean and maximum component difference ("box distance") with $R = 18$ (Euclidean distance) and $R = 15$ (box distance), $T = 100$, $PC = 4$. The computation time is highly dependent on the image content: 32, 36, 47, 64, 83 and 208 sec (box distance) and 30, 35, 43, 62, 78, and 210 sec (Euclidean distance) is needed for the median-shift computation. Results on four maps are shown on Figure 2. In order to better judge the quality of the palette extraction a labeling (i.e. assigning a color palette to all pixels) is performed. The palette is compared to the ones extracted by Vector-magic (http://vectormagic.com/home), and several color reduction algorithms provided by the VPmap-Pro software: median cut, kmeans, minimum distance and octree. For the latter, no pre-processing could be performed. Due to the bad quality of the results obtained with the octree method, these results are not reported. The first column shows the original images and in their lower right corner, a partial zoom. In order to judge the quality of the clusters extracted, the second column shows images labeled thanks to the median-shift algorithm (box distance) in which each pixel is assigned to the nearest vertex or to "undefined" if the box distance is larger than 3R/2. The last column shows all extracted palettes; from left to right: median-shift box distance $R = 15$, median-shift Euclidian distance $R = 18$, minimum distance (VPmap-Pro), Vector Magic, median-cut(VPmap-Pro), kmeans (VPmap-Pro).

All extracted palettes are displayed in the last column. For Vector-Magic the best number of classes has been chosen manually. VPmap fully automatic color reduction requires a minimum color classes of 12, value which has been chosen for all images.

The algorithm has also been tested on the photo of the Macbeth color chart used in [13] and compared to other algorithms: the implementation of the kmeans proposed in [13] and the mixture of Gaussians [14]. Figure 3 shows the Macbeth color chart in (a), the results of the median-shift in Lab space using Euclidian distance with $R = 14$ in (b), of the kmeans in RGB in (c) and in Lab space with 25 classes in (d), and finally of a mixture of Gaussians in RGB with 25 classes in (f). An "x" inside a square means that the square received a wrong label, while a "y" means the class has been correctly identified but the cluster is not a good visual representative of the class.

The median-shift parameter has been initially set to 18, as in previous experiments but at $R = 18$, two complex clusters are generated, suggesting to use a smaller R. At $R = 17$, all colors are separated but the three light grays are assigned to the same cluster and would thus have a "4x" score. A further separation occurs at $R = 14$, still keeping the two light grays in one class and the two dark grays with the background (3x); this is somehow expected as the

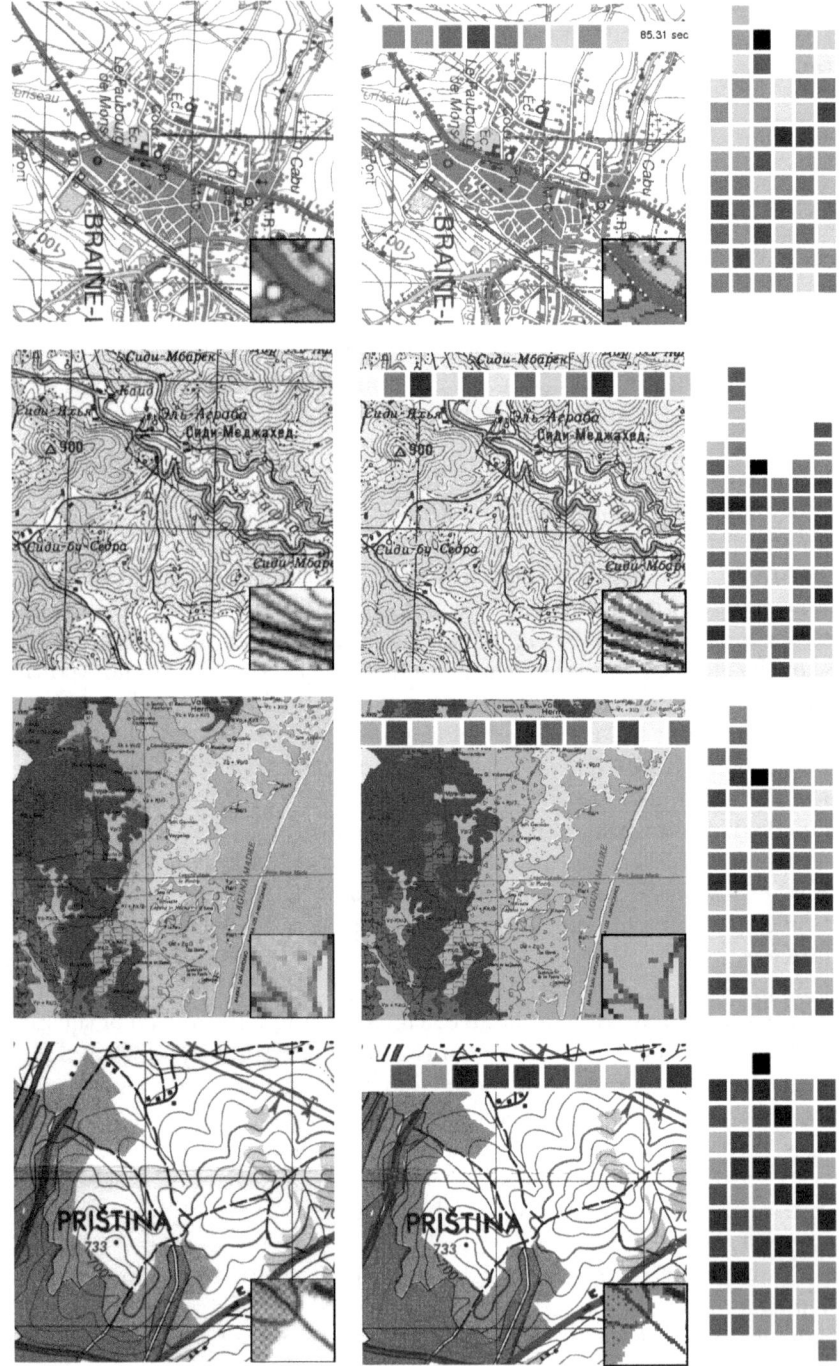

Fig. 2. Palette extraction on various maps (see text)

distance between the two light grays is about 7, and the background lies between the two darkest grays at a distance of about 9, both distances being lower than R. Note that the *original* Macbeth color chart has slightly different values; in particular each grey is separated from its neighbour by a distance of 15 (or 14 for the darkest).

The median-shift applied in Lab space is thus excellent for colors (no "x" and no "y" in colored squares); in particular, it is the only algorithm able to discriminate the deep blues (distance about 24) and the yellows (distance about 21) but is less good in discriminating shades of gray (3x). Other distances [15] like CIE1994, CIE2000, or CMC may provide better results.

The results of kmean depends on its initialization. Palus [13] proposed two initialization schemes; in this experiment both provided the same number of "x" and "y". In the RGB space three yellows and one green-yellow are merged into a unique yellow, two pairs of blues and one pair of pinks are merged, making a total of 6x. The algorithms is slightly better in the Lab space (5x, 1y).

The mixture of Gaussians gives the worse results in this experiment (8x,3y). This algorithm may also discover the best number of classes, 40 in this case, resulting in over-segmentation without resolving the difficult pairs of blues, yellows and grays.

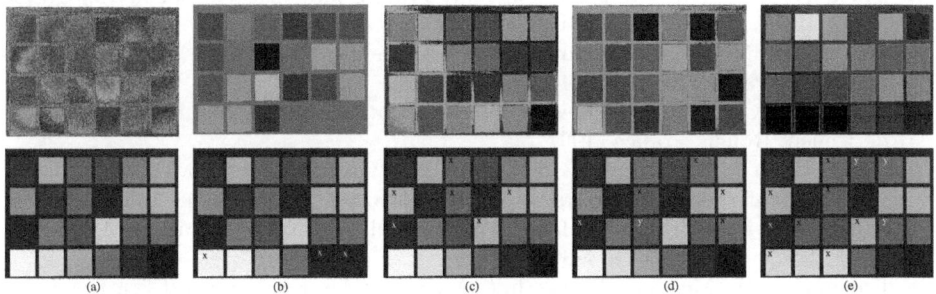

Fig. 3. Comparison of palette extraction on the Macbeth color chart. Upper row: pseudo colors; lower row: true colors. (a) Original Macbeth chart; (b) Median-shift ($R = 14$); (c) kmeans in RGB (25 classes); (d) kmeans in Lab (25 classes); (e) mixture of Gaussians in RGB space (25 classes); "x" and "y" denotes wrong class assignment, and visually not acceptable cluster value respectively.

6   Summary and Conclusions

A strategy to identify the palette of scanned graphics has been proposed. A first pre-processing transforms the data into the Lab space and removes pixels with a less well defined color, for a better cluster separation.

The remaining pixels are clustered thanks to the median-shift, a new clustering algorithm which requires one parameter R related to the expected cluster radius in the feature space associated to a distance. As far as palette extraction is concerned, the Lab feature space with Euclidian distance and R between 14 and

18 (or smaller in case of complex clusters) seems suitable for many applications. The procedure is applied on scanned graphics such as maps and a photographic chart showing improvement compared to well-established methods. In particular, the algorithm shows excellent discriminative power on saturated colors, but is slightly less efficient when dealing with low saturated ones.

Acknowledgments

The author wishes to thank Pasquale Nardone for the graph transformation suggestion, Henrick Palus and Dirk Borghys for the provision of results on the Macbeth color chart.

References

1. Diebel, J.R.: Bayesian Image Vectorization: The Probabilistic Inversion of Vector Image Rasterization. Phd thesis, Standford University (2008)
2. Hilaire, X.: RANVEC and the Arc Segmentation Contest: Second Evaluation. In: Liu, W., Lladós, J. (eds.) GREC 2005. LNCS, vol. 3926, pp. 362–368. Springer, Heidelberg (2006)
3. Chen, Y., et al.: Extracting Contour Lines From Common-Conditioned Topograpic Maps. IEEE TGARS 44(4), 1048–1057 (2006)
4. Deseilligny, M.P.: Lecture automatique de cartes, Phd thesis, Université René Descartes, Paris, France (1994)
5. Robert, R.: Contribution à la lecture automatique de cartes, Phd thesis, Université de Rouen, Rouen, France (1997)
6. Braquelaire, J.-P., Brun, L.: Comparison and Optimization of Methods of Color Image Quantization. IEEE Trans. on Image Processing 6(7), 1048–1052 (1997)
7. Jain, A.K., Duin, R.P.W., Mao, J.: Statistical Pattern Recognition: A Review. IEEE Trans. PAMI 22(1), 4–37 (2000)
8. Comaniciu, D., Meer, P.: Mean Shift: A Robust Approach Toward Feature Space Analysis. IEEE Trans. PAMI 24(5), 603–619 (2002)
9. Tran, T.N., Wehrens, H.R.M.J., Buydens, L.M.C.: Clustering multispectral images: a tutorial. Chemometrics and Int. Lab. Syst. 77, 3–17 (2005)
10. Price, K.: Computer Vision Biography, 8.8.3 MRF Models for Segmentation, http://www.visionbib.com/bibliography/segment369.html (accessed 05-09)
11. Koschan, A., Abidi, M.: Detection and Classification of Edges in Color Images. Sig. Proc. Mag., Spec. Issue on Color Img. Proc. 22(1), 64–73 (2005)
12. Lacroix, V., Acheroy, M.: Feature-Extraction Using the Constrained Gradient. IS-PRS J. of Photogram. and RS 53(2), 85–94 (1998)
13. Palus, H.: On color image quantization by the k-means algorithm. 10. In: Droege, D., Paulus, D. (eds.) Workshop Farbbildverarbeitung, Universität Koblenz-Landau, Tönning, Der Andere Verlag (2004)
14. Bouman, C.A.: Cluster: An unsupervised algorithm for modeling Gaussian mixtures (1997), http://www.ece.purdue.edu/~bouman
15. Ohta, N., Robertson, A.R.: Colorimetry Fundamentals and Applications. John Wiley & Sons, Ltd., Chichester (2005)

Detection of Circular Arcs in a Digital Image Using Chord and Sagitta Properties

Sahadev Bera[1], Partha Bhowmick[2], and Bhargab B. Bhattacharya[1]

[1] Advanced Computing and Microelectronics Unit
Indian Statistical Institute, Kolkata - 700108, India
sahadev_r@isical.ac.in, bhargab@isical.ac.in
[2] Department of Computer Science and Engineering
Indian Institute of Technology, Kharagpur - 721302, India
bhowmick@gmail.com

Abstract. This paper presents a new technique for detection of digital circles and circular arcs using *chord property* and *sagitta property*. It is shown how a variant of the *chord property* of an Euclidean circle can be used to detect a digital circle or a circular arc. Based on this property, digital circular arcs are first extracted and then using the *sagitta property*, their centers and radii are computed. Several arcs are merged together to form a complete digital circle or a larger arc. Finally, a technique based on Hough transform is used to improve the accuracy of computing the centers and radii. Experimental results have been furnished to demonstrate the efficiency of the proposed method.

Keywords: Circle detection; Chord property; Digital geometry; Sagitta property; Hough transform.

1 Introduction

Fast and accurate recognition of circles or circular arcs in a digital image is a challenging problem with practical relevance. There exist several algorithms, most of which are based on Hough transform or its variants [2,3,4,5,8,13,9,17]. In particular, circle detection is important in computer vision applications as well as in medical imaging. Hough transform (HT) has been widely used to extract digital primitives, such as straight lines, circles, etc. Though HT is robust against noises, clutters, object defects, and shape distortions, it often requires intensive computation and a large amount of memory.

Several researchers have developed modified HT methods for detecting circles in digital images. In one such method, the parameter space is decomposed into several lower dimension parameter spaces [18]. Gradient information of each edge pixel is used in another method to reduce the computing time or the memory requirement [5,10]. A third variety uses the geometric properties of circles to improve the performance [7]. However, these methods mainly focus on the robustness and accuracy of detection.

Kim *et al.* [9] have proposed a two-step circle detection algorithm, given a pair of intersecting chords, in which the first step is to compute the center of the circle. In the second step, the radius histogram is used to identify the circle. Xu *et al.* [17] presented a randomized HT, which reduces the storage requirement and computational time significantly compared to other methods based on the conventional HT. Chen *et al.* [2]

J.-M. Ogier, W. Liu, and J. Lladós (Eds.): GREC 2009, LNCS 6020, pp. 69–80, 2010.

proposed an efficient randomized algorithm (RCD) for detecting circles that does not use HT. The underlying concept in RCD is to first select four edge pixels randomly in the image and then to use a distance criterion to determine whether there might exist a possible circle, and finally to collect further evidence for determining whether or not, it is indeed a circle. Chiu *et al.* [3] proposed an effective voting method for circle detection, which also does not use the HT.

In this work, we propose a novel technique of recognizing digital circles as well as circular arcs, based on the *chord property* and the *sagitta property* [16]. First, all the arc segments are extracted and then using the *chord property*, the circularity of each arc segment is verified and the circular segments are identified. Then the *sagitta property* is applied to determine the radii of the circular arc segments, and in turn, the corresponding centers. Finally, two arc segments with closest radii and centers are merged iteratively to obtain a complete circle or a larger circular arc segment. To improve the accuracy of computing the centers and radii, a technique based on restricted Hough transform (RHT) is used.

2   Proposed Work

Most of the earlier methods are suitable for identifying only an isolated digital circle or digital circular arc. If a digital circle or a circular arc intersects other circular arcs or digital straight line segments, then these methods may not be applicable. To handle such cases, first we need to detect the digital circular segments separately and then merge them efficiently to form a complete digital circle or a larger circular arc segment. This will enhance the scope of circle detection algorithms.

Since in a digital image, the contour may be given as thick curve segments, we use thinning [6] as preprocessing before applying the algorithm. The subsequent steps may be briefed as follows.

2.1   Finding the Intersection Points

In order to detect each segment separately, first of all we detect all the points of intersection (among the digital curve segments) and end points (for open digital curves), and store them in a list $\mathcal{P}$. As we detect circular segments first and then merge them to form a complete circle or a larger circular segment, we have to do some special treatment for a free/isolated closed curve. Consider S to be a free and closed digital curve segment. We put two virtual points of intersection, say, $p_1 \in S$ and $p_2 \in S$, such that, if S_1 and S_2 be the two resultant segments ($S_1 \cup S_2 = S$) whose (virtual) end points are p_1 and p_2, then the lengths of S_1 and S_2 differ by at most unity.

2.2   Storing the Curve Segments

A *thin digital curve segment* is a set of pixels having two end pixels and a minimal list of pixels that establish connectivity between the end pixels using 8-neighbor rule [6]. For each point $p_i \in \mathcal{P}$, the corresponding segment(s) incident at p_i is/are extracted. If p_i is an end point of a digital curve segment S, then there is only one segment, i.e., S,

End point 1	End point 2	Center	Radius	Pointer to link list of curve points

Fig. 1. The data structure $\mathcal{L}$ that stores the information for each detected curve segment

incident at p_i. If p_i is a virtual point of intersection created for a free and closed digital curve, namely $S = S_1 \cup S_2$, then there are two segments, i.e., S_1 and S_2, incident at p_i. And number of digital curve segments incident at p_i is three or more if and only if p_i is an actual point of intersection between two or more digital curve segments.

To discard the spurious segments, we consider the length of (i.e., number of digital points constituting) each segment incident at each $p_i \in \mathcal{P}$. If a segment is negligibly small (10 pixels or less), then it is discarded; otherwise, it is stored in a list of segments, $\mathcal{L}$, whose node structure is shown in Figure 1. To identify the circular arc segments, we need to first remove the digital straight line segments from the list $\mathcal{L}$. It may be mentioned here that there are several techniques to determine digital straightness [1,11,12,14]. We have used the concept of area deviation [15], which is realizable in purely integer domain using a few primitive operations only. The method is as follows.

Let $S := \langle a = c_1, c_2, \dots, c_k = b \rangle$ be a digital curve segment with end points a and b. Let c_i $(2 \le i \le k - 1)$ be any point on the segment S other than a and b. Let h_i be the distance of the point c_i from the real straight line segment $\overline{ab}$. Then S is considered to be a single digital straight line segment starting from a and ending at b, provided

$$\max_{2 \le i \le k-1} |\triangle (a, c_i, b)| \le \tau_h d_T (a, b). \tag{1}$$

Here, $|\triangle (a, c_i, b)|$ denotes twice the magnitude of area of the triangle with vertices $a :=$ (x_1, y_1), $c_i := (x_i, y_i)$, and $b := (x_k, y_k)$, and $d_T(a, b) := \max(|x_1 - x_k|, |y_1 - y_k|)$ is the maximum isothetic distance between a and b, and $\tau_h = 2$ in our experiments.

2.3 Deviations of Chord Property

Let θ_m be the angle subtended by (chord) $\overline{ab}$ at the midpoint m of a segment $S \in \mathcal{L}$ with end points a and b. Let θ_c be the angle subtended by $\overline{ab}$ at an arbitrary point $c \in S \setminus \{a, b\}$. Then, according to the *chord property*, $\theta_c = \theta_m$ if the segment S is a part of the Euclidean (real) circle. But this is not exactly true for a digital circle. Hence, we use a variant of this useful chord property (Fig. 2). If $\theta_c \in [\theta_m - \epsilon, \theta_m + \epsilon]$ for all $c \in S$ excepting a few points near its two ends, ϵ being a small positive quantity (Fig. 2), then the digital curve S is circular.

Let $\mathcal{C}^{\mathbb{R}}(o, r)$ be the real circle centered at $o(0, 0)$ and having radius $r \in \mathbb{Z}^+$. Let $\mathcal{A}^{\mathbb{R}}(\alpha, \beta)$ be an arc of $\mathcal{C}^{\mathbb{R}}(o, r)$ having end points $\alpha(x_\alpha, y_\alpha) \in \mathcal{C}^{\mathbb{R}}(o, r)$ and $\beta(x_\beta, y_\beta) \in \mathcal{C}^{\mathbb{R}}(o, r)$, such that $x_\alpha, x_\beta \in \mathbb{Z}$. Let $\gamma(x_\gamma, y_\gamma) \in \mathbb{R}^2$ be an arbitrary point in $\mathcal{A}^{\mathbb{R}}(\alpha, \beta) \setminus \{\alpha, \beta\}$. Let the chord $\overline{\alpha\beta}$ subtends an angle ϕ_γ at γ. Let $a, b, c \in \mathbb{Z}^2$ be the respective points in the digital circle $\mathcal{C}^{\mathbb{Z}}(o, r)$ corresponding to α, β, γ, and the angle subtended by the line segment $\overline{ab}$ at c be ϕ_c.

Now, consider that $\mathcal{A}^{\mathbb{R}}(\alpha, \beta)$ is an arc of $\mathcal{C}^{\mathbb{R}}(o, r)$ in Octant 1, which corresponds to the digital (circular) arc $\mathcal{A}^{\mathbb{Z}}(a, b)$. Since $a(x_a, y_a)$, $b(x_b, y_b)$, and $c(x_c, y_c)$ are the

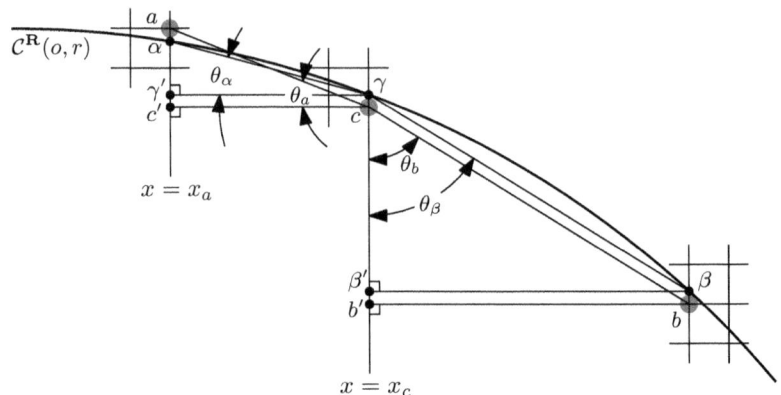

Fig. 2. Deviation of the chord property. Points in $\mathbb{R}^2$ $(\alpha, \beta, \dots)$ or in the real circle $\mathcal{C}^{\mathbb{R}}(o, r)$ are shown in black, and points shown as larger gray blobs (a, b, c) belong to the digital circle, $\mathcal{C}^{\mathbb{Z}}(o, r)$. $\mathcal{A}^{\mathbb{R}}(\alpha, \beta)$ is an arc of $\mathcal{C}^{\mathbb{R}}(o, r)$ in Octant 1, which corresponds to the given digital (circular) arc $\mathcal{A}^{\mathbb{Z}}(a, b)$. As c changes its place along $\mathcal{A}^{\mathbb{Z}}(a, b)$ such that $y_\gamma - \frac{1}{2} < y_c < y_\gamma + \frac{1}{2}$, the angle $\phi_c (= \theta_a + \theta_b + \pi/2)$ gets deviated by $\pm\epsilon$.

respective points of $\mathcal{C}^{\mathbb{Z}}(o, r)$ corresponding to $\alpha(x_\alpha, y_\alpha)$, $\beta(x_\beta, y_\beta)$, and $\gamma(x_\gamma, y_\gamma)$ of $\mathcal{C}^{\mathbb{R}}(o, r)$, we have $x_\alpha = x_a \in \mathbb{Z}$, $x_\beta = x_b \in \mathbb{Z}$, and $x_\gamma = x_c \in \mathbb{Z}$ (Fig. 2). Further, owing to the digitization scheme, we have

$$y_\alpha - \tfrac{1}{2} < y_a < y_\alpha + \tfrac{1}{2}, \ y_\beta - \tfrac{1}{2} < y_b < y_\beta + \tfrac{1}{2}, \ y_\gamma - \tfrac{1}{2} < y_c < y_\gamma + \tfrac{1}{2}. \tag{2}$$

Let c' and γ' be the respective feet of perpendiculars dropped from c and γ to the vertical line $x = x_a$, and b' and β' be those from b and β to the vertical line $x = x_c$. Let θ_a, θ_α, θ_b, and θ_β be the acute angles subtended at c and γ by the corresponding perpendiculars. Then, $\phi_c = \theta_a + \theta_b + \pi/2$ and $\phi_\gamma = \theta_\alpha + \theta_\beta + \pi/2$. Clearly, for $c \in \mathcal{A}^{\mathbb{Z}}(a, b)$, we have

$$\max(\phi_c) = \max(\theta_a + \theta_b) + \pi/2 \le \max(\theta_a) + \max(\theta_b) + \pi/2, \tag{3}$$
$$\min(\phi_c) = \min(\theta_a + \theta_b) + \pi/2 \ge \min(\theta_a) + \min(\theta_b) + \pi/2. \tag{4}$$
$$\text{or,} \quad \phi_c \in [\min(\theta_a) + \min(\theta_b) + \pi/2, \max(\theta_a) + \max(\theta_b) + \pi/2]. \tag{5}$$

Let $\delta_{y_{ac}} = y_a - y_c$, $\delta_{x_{ca}} = x_c - x_a$, and $\delta_{y_{\alpha\gamma}} = y_\alpha - y_\gamma$. Then the *deviation of θ_a from θ_α* is given by

$$\delta_{\theta_{a\alpha}} = \theta_a - \theta_\alpha = \tan^{-1}\frac{y_a - y_c}{x_c - x_a} - \tan^{-1}\frac{y_\alpha - y_\gamma}{x_\gamma - x_\alpha} \tag{6}$$

$$= \tan^{-1}\frac{\delta_{y_{ac}}}{\delta_{x_{ca}}} - \tan^{-1}\frac{\delta_{y_{\alpha\gamma}}}{\delta_{x_{ca}}} \quad (\text{since } x_a = x_\alpha \text{ and } x_c = x_\gamma)$$

$$= \tan^{-1}\left(\frac{(\delta_{y_{ac}} - \delta_{y_{\alpha\gamma}})\delta_{x_{ca}}}{\delta_{x_{ca}}^2 + \delta_{y_{ac}}\delta_{y_{\alpha\gamma}}}\right) < \tan^{-1}\left(\frac{\delta_{x_{ca}}}{\delta_{x_{ca}}^2 + \delta_{y_{ac}}\delta_{y_{\alpha\gamma}}}\right) \quad (\text{Eqn. 2}) \tag{7}$$

$$\approx \tan^{-1}\left(\frac{\delta_{x_{ca}}}{\delta_{x_{ca}}^2 + \delta_{y_{ac}}^2}\right) = \tan^{-1}\frac{1}{|\overline{ca}|} \quad (\text{with } \delta_{y_{ac}} \approx \delta_{y_{\alpha\gamma}}). \tag{8}$$

Thus, the deviation of θ_a from θ_α is at most $\tan^{-1}(1/\overline{ca})$, and higher the distance of a from c, lesser is the deviation. Similar deviation, namely $\delta_{\theta_{b_\beta}}$, also comes into play while considering θ_b, and as the distance of c from b increases, the deviation becomes insignificant. Hence, if m be the middle pixel (one of two, if there are two such) of $\mathcal{A}^{\mathbb{Z}}(a, b)$, then the maximum possible deviation of ϕ_m from ϕ_γ is given by

$$\tau_\phi = 2\tan^{-1}\frac{1}{|\overline{am}|} = 2\tan^{-1}\frac{2}{|\overline{ab}|}. \tag{9}$$

For practical cases, the distance of c or m from a or b is quite low, and hence such deviations have to be considered for proper results. Our algorithm possesses this feature, resulting to its satisfactory performance in terms of both precision and robustness.

2.4 Verifying the Circularity

The list $\mathcal{L}$ contains segments which are not digitally straight, as explained in Sec. 2.2. That is, each segment S in $\mathcal{L}$ is made of at least one circular segment with or without one or many intervening straight pieces. So, for each segment S, we check its circularity using the *chord property* as explained in Sec. 2.3. If the segment S consists of both circular and straight components, then we extract its circular part(s) only from S, store these circular segment(s) in the list $\mathcal{L}$ with necessary updates, and remove the original segment S from $\mathcal{L}$.

We start checking the circularity of $S := \langle a = c_1, c_2, \ldots, c_k = b\rangle$ starting from the end point, a. We consider an *appropriately small prefix of S*, namely $S_j := \langle a = c_1, c_2, \ldots, c_j\rangle$, where $j = \min(\tau_s, k)$, and verify the circularity for one-third of the points lying in the central region of S_j, namely

$$S_j^{(m)} := \left\langle c_{\lfloor j/3\rfloor}, c_{\lfloor j/3\rfloor+1}, \ldots, c_{m-1}, c_m, c_{m+1}, \ldots, c_{2\lfloor j/3\rfloor-1}, c_{2\lfloor j/3\rfloor}\right\rangle.$$

In our experiments, we have considered $\tau_s = 15$. Two-third points (one-third from either end) of S are discounted from circularity test as they are prone to excessive deviations of chord property, as explained in Sec. 2.3. Hence, if c_m $(m = \lfloor j/2\rfloor)$ be the midpoint of S_j and the angle subtended by the chord $\overline{ac_j}$ at m is estimated to be ϕ_m, then S_j is considered to be satisfying the chord property, provided the angles ϕ_c subtended by $\overline{ac_j}$ at all points $c \in S_j^{(m)} \setminus \{a, c_j\}$ satisfy the following equation.

$$\max_{c \in S_j^{(m)}} \{|\phi_c - \phi_m|\} \leq \tau_\phi \tag{10}$$

If S_j is found to be circular, then we augment it to $S_{j'} := \langle a = c_1, c_2, \ldots, c_{j'}\rangle$, where $j' = \min(j + \lfloor\frac{1}{3}\tau_s\rfloor, k)$, in order to include the next $\lfloor\frac{1}{3}\tau_s\rfloor$ pixels from S, and again verify the chord property for $S_{j'}^{(m')}$ with $m' = \lfloor j'/2\rfloor$. The process is continued until all points in S are verified or the chord property fails for some prefix S_j (or $S_{j'}$) of S.

2.5 Combining the Arcs

If S is a circular arc with end points a and b, then its *sagitta* is the straight line segment drawn perpendicular to the chord $\overline{ab}$, which connects the midpoint μ of ab with S. The

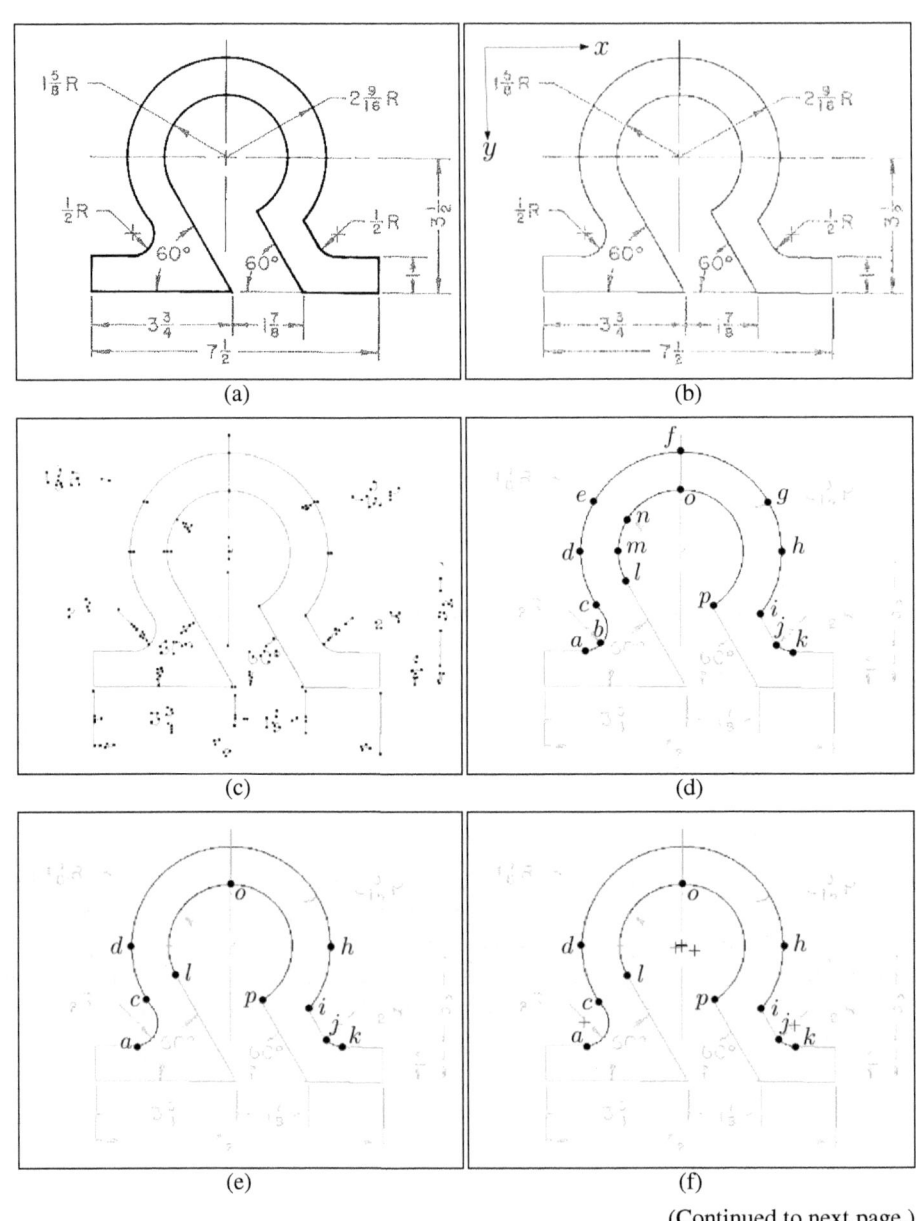

(Continued to next page.)

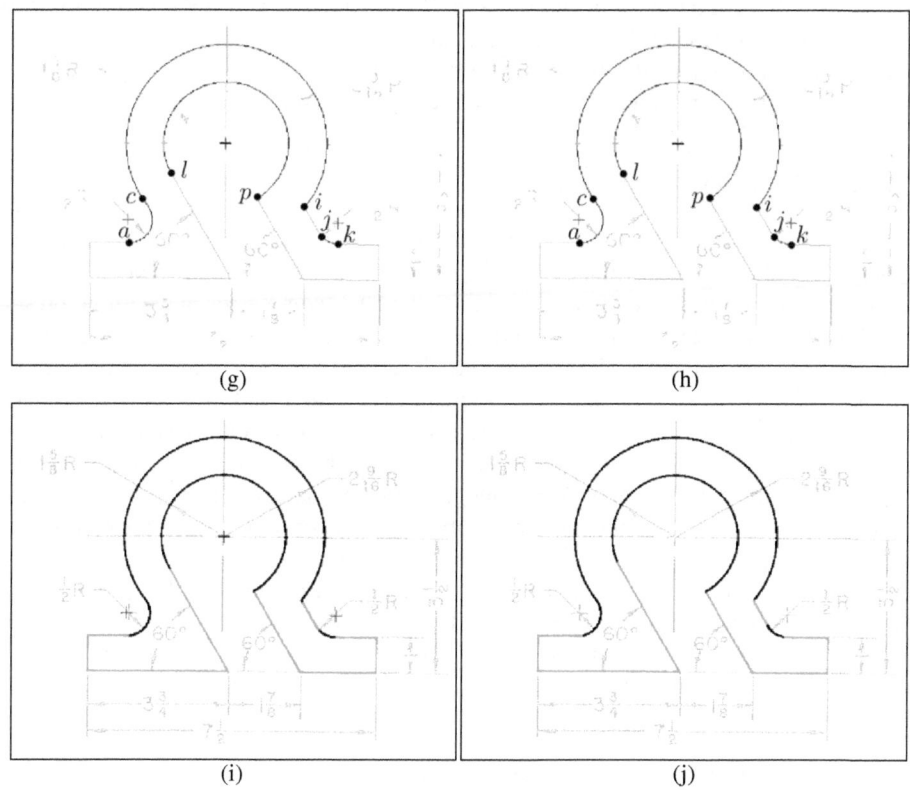

Fig. 3. Step-wise snapshots of the algorithm on 6.pbm: (a) input; (b) after thinning; (c) intersection points and end points in L, removing small arcs; (d) detected circular arcs by *chord property*; (e) after combining adjacent arcs; (f) centers detected by *sagitta property*; (g) after merging circular arcs; (h) after applying RHT; (i) final result; (j) ground-truth;

sagitta property is as follows: If the perpendicular to ab at μ intersects S at s, then the radius of the circle whose arc is S, is given by

$$r = \frac{d^2(a,b)}{8d(\mu,s)} + \frac{d(\mu,s)}{2} \qquad (11)$$

where, $d(a,b)$ denotes the Euclidean distance between the points a and b.

The radius, and hence the center, of each circular arc $S \in \mathcal{L}$ are computed using the *sagitta property*, and stored in the node corresponding to S (Fig. 1). While combining the circular arcs, necessary care has to be taken for the inevitable error that creeps in owing to the usage of sagitta property, which is a property of real circles only. Since we deal with digital curve segments, the *cumulative error* of the effective radius computed for a combined/growing circular arc using the aforesaid sagitta property is very likely to increase with an increase in the number of segments constituting that arc. Hence, to enhance accuracy, we merge two digital circular segments S and S' into $S'' := S \cup S'$, if (i) S and S' have a common end point in $\mathcal{P}$ and (ii) S'' satisfies the *chord property*. Since the node corresponding to each segment in the list $\mathcal{L}$ contains end points, center, radius, and a pointer to the list of curve points, the attributes of the segment S are updated by those of S'', and the data structure $\mathcal{P}$ is updated accordingly.

2.6   Finalizing the Centers and Radii

In spite of the treatments to reduce discretizations error while employing chord property to detect circular arcs and while employing sagitta property to combine two or more circular arcs and compute the effective radius and center, some error may still be present in the estimated values of the radii and the center. To remove such errors, we apply a restricted Hough transform (RHT) on each circular arc $S \in \mathcal{L}$ with a small parameter space [2]. Let $q(x_q, y_q)$ and r be the respective center and radius of S estimated using the sagitta property (Sec. 2.5). Then the restricted parameter space is taken as $[x_q - \delta, x_q + \delta] \times [y_q - \delta, y_q + \delta] \times [r - \delta, r + \delta]$, such that $\delta = \tau_H r$, where $0 < \tau_H < 1$ ($\tau_H = \frac{1}{8}$ in our experiments). A 3D integer array, H, is taken corresponding to this parameter space, each of whose entry is initialized to zero. For every three distinct and non-collinear points c, c', and c'' from S, we estimate the center $q'(x', y')$ and the radius r' of the (real) circle passing through c, c', and c''. If each of $rd(x')$, $rd(y')$, and $rd(r')$ lies within the corresponding bounds of H, then the entry in H corresponding to $rd(x', y', r')$ is incremented ($rd(x') = \lfloor x' + \frac{1}{2} \rfloor$, etc.). Finally, the entry in H corresponding to the maximum frequency provides the final center and radius of S.

3   Demonstration of the Proposed Method

A demonstration of the proposed method on a sample image (6.pbm) is shown in Fig. 3. All the digital curve segments in the image are extracted and stored in the list $\mathcal{L}$. At first, the straight line segments are removed from $\mathcal{L}$ with a small computation time. For example, $L(f(304, 45), o(304, 94))$, $L((304, 94), (303, 155))$, etc. are some of the straight line segments that are removed[1]. Then using the *chord property*, the circular segments are detected with necessary updates in the list $\mathcal{L}$. For example, the digital curve segment from $d(174, 174)$ to $b(199, 293)$ consists of two circular segments. Those two circular segments are extracted; one segment from $d(174, 174)$ to $c(199, 249)$ is stored in the node of the original segment and the other one from $c(199, 250)$ to $b(199, 293)$ is stored in a newly created node and inserted in the list $\mathcal{L}$. Similarly, the segment from $m(223, 174)$ to $(307, 349)$ consists of a circular arc and a straight line segment. From this segment, we extract the circular part and remove the straight part. Necessary updates in $\mathcal{L}$ and $\mathcal{P}$ are made. The circular segments in the list L are shown in Fig. 3(d) by different colors and are enumerated in Table 1.

Two or more smaller adjacent arcs are combined if they jointly satisfy the *chord property* in order to get larger arcs for reducing the computational error in the next step of applying the *sagitta property* (Sec. 2.5). After such combining/merging, the number of

---

[1] Here we use $L((x, y), (x', y'))$ to denote the digital straight line segment joining the points (x, y) and (x', y').

Table 1. Changes in $\mathcal{L}$ after successive stages on the image of Fig. 3

Seg. No.	End point 1	End point 2	# curve points	Center	Radius	Curve points
(a) After detecting the circular arcs						
1	f(303, 45)	e(193, 107)	118	—	—	(303, 45), (302, 45), (301, 45), ... , (193, 107)
2	f(305, 45)	g(414, 109)	118	—	—	(305, 45), (306, 45), (307, 45), ... , (414, 109)
3	o(304, 93)	p(344, 243)	188	—	—	(304, 93), (305, 94), (306, 94), ... , (344, 243)
4	e(192, 108)	d(174, 172)	65	—	—	(192, 108), (191, 109), (191, 110), ... , (174, 172)
5	g(415, 110)	h(432, 174)	65	—	—	(415, 110), (416, 111), (417, 112), ... , (432, 174)
6	n(236, 131)	o(301, 94)	82	—	—	(236, 131), (236, 130), (236, 129) ... , (304, 94)
7	n(235, 132)	m(223, 172)	42	—	—	(235, 132), (234, 132), (233, 133), ... , (223, 172)
8	d(174, 174)	c(199, 249)	76	—	—	(174, 174), (174, 175), (174, 176), ... , (199, 249)
9	m(223, 174)	l(233, 213)	40	—	—	(223, 174), (223, 175), (223, 176), ... , (233, 213)
10	h(432, 175)	i(404, 255)	81	—	—	(432, 175), (432, 176), (432, 177), ... , (404, 225)
11	c(199, 250)	b(200, 293)	45	—	—	(199, 250), (200, 251), (201, 252), ... , (200, 293)
12	b(199, 294)	a(181, 303)	20	—	—	(199, 294), (199, 295), (198, 296), ... , (181, 303)
13	j(426, 295)	k(445, 304)	20	—	—	(426, 295), (427, 296), (428, 296), ... , (445, 304)
(b) After combining adjacent arcs and detection of centers by *sagitta property*						
1	o(304, 93)	p(344, 243)	188	(302, 174)	81	(304, 93), (305, 94), (306, 94), ... , (344, 243)
2	h(432, 174)	d(174, 172)	366	(303, 174)	129	(432, 174), (432, 173), (432, 172), ... , (174, 172)
3	d(174, 174)	c(199, 249)	76	(294, 176)	120	(174, 174), (174, 175), (174, 176), ... , (199, 249)
4	l(233, 213)	o(301, 94)	164	(302, 173)	79	(233, 213), (233, 212), (232, 211), ... , (301, 94)
5	h(432, 175)	i(404, 255)	81	(319, 180)	113	(432, 175), (432, 176), (432, 177), ... , (404, 255)
6	c(199, 250)	a(181, 303)	65	(177, 272)	31	(199, 250), (200, 251), (201, 252), ... , (181, 303)
7	j(426, 295)	k(445, 304)	20	(446, 277)	27	(426, 295), (427, 296), (428, 296), ... , (445, 304)
(c) After merging						
1	l(233, 213)	p(344, 243)	352	(302, 173)	80	(233, 213), (233, 212), (232, 211), ... , (344, 243)
2	c(199, 249)	i(404, 255)	523	(303, 174)	124	(199, 249), (198, 248), (197, 247), ... , (404, 255)
3	c(199, 250)	a(181, 303)	65	(177, 272)	31	(199, 250), (200, 251), (201, 252), ... , (181, 303)
4	j(426, 295)	k(445, 304)	20	(446, 277)	27	(426, 295), (427, 296), (428, 296), ... , (445, 304)
(d) After RHT						
1	l(233, 213)	p(344, 243)	352	(303, 173)	80	(233, 213), (233, 212), (232, 211), ... , (344, 243)
2	c(199, 249)	i(404, 255)	523	(303, 174)	129	(199, 249), (198, 248), (197, 247), ... , (404, 255)
3	c(199, 250)	a(181, 303)	65	(177, 272)	31	(199, 250), (200, 251), (201, 252), ... , (181, 303)
4	j(426, 295)	k(445, 304)	20	(447, 276)	28	(426, 295), (427, 296), (428, 296), ... , (445, 304)

Table 2. Results for some samples images

Image	size in pixels	N_c	N_g	N_p	N_{fa}	N_{fr}	D_r	T
6.pbm	905 × 562	8321	2639	2673	82 (3.11%)	48 (1.82%)	98.18	0.075
7.pbm	907 × 779	14817	11435	11629	390 (3.41%)	196 (1.71%)	98.29	0.243
2007-1.tif	368 × 460	10215	–	5495	–	–	–	0.092
2007-2.tif	368 × 460	8569	–	3954	–	–	–	0.073

N_c: # curve pixels in the original image. N_g: # pixels on circular arcs in the ground-truth image. N_p: # pixels on circular arcs detected by the proposed algorithm. N_{fa}: false-acceptance $\left(\frac{N_{fa}}{N_g} \times 100\%\right)$. N_{fr}: false-rejection $\left(\frac{N_{fj}}{N_g} \times 100\%\right)$. D_r: Detection rate $= \frac{(N_p - N_{fa})}{N_g} \times 100\%$. T: Total execution time (seconds).

circular segments gets reduced to almost 50%, as reflected in Table 1 and Fig. 3(e). Next, the radius and the center of each arc in $\mathcal{L}$ are computed using the *sagitta property* and stored in the node of the corresponding arc. Figure 3(f) shows the center of each circular arc as '+'. The detailed information of the circular segments stored in the list $\mathcal{L}$ is given in Table 1. The radius and center of the combined arc are estimated as the weighted arithmetic means of the radii and centers of the constituent arcs, respectively; the weight is

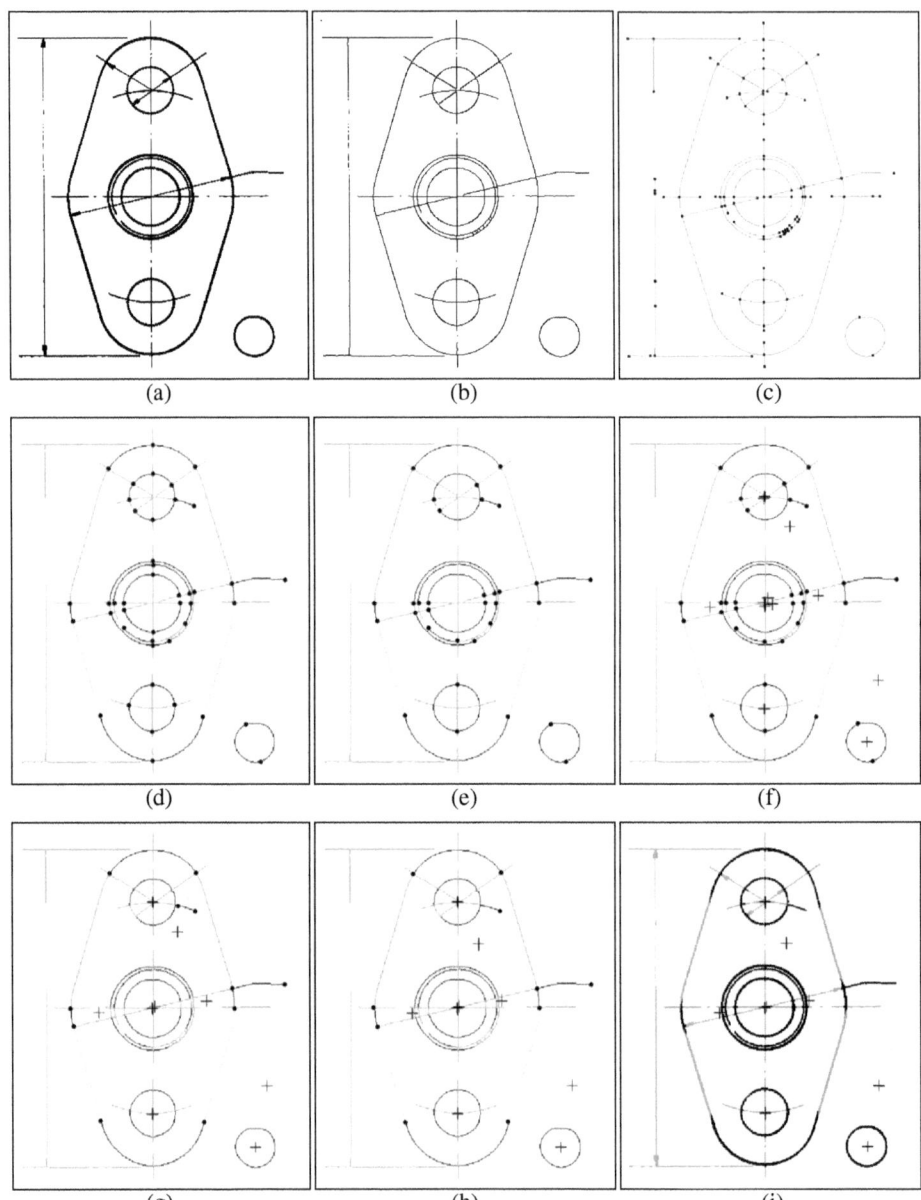

Fig. 4. Step-wise snapshots of our experiment on 2007-1.tif: (a) input; (b) after thinning; (c) intersection points and end points in $\mathcal{L}$; (d) detected circular arcs by *chord property*; (e) after combining adjacent arcs; (f) centers detected by *sagitta property*; (g) after merging circular arcs; (h) after applying RHT; (i) final result

taken as the number of points of the constituent arc. For example, some of the segments in Table 1 are combined into segments 1 and 2 (Fig. 3g), with the updated information being listed in Table 1. Next we apply RHT on each arc (Sec. 2.6). Resultant image is shown in Fig. 3(h) and the arcs are detailed out in Table 1. Finally, we consider the detected circular arcs in the original (i.e., input) image and for each pixel on a detected arc S, the object pixels in its 8-neighborhood are iteratively marked as pixels of the corresponding thick circular arc. Figure 3(i) shows the detected thick circular arcs of the input image.

4   Experimental Results

We have implemented the proposed algorithm in C on the openSUSE™ OS Release 11.0 HP xw4600 Workstation with Intel® Core™2 Duo, 3 GHz processor. We have performed tests on several database images. The results of the algorithm on some of these image files of thick digital curves are reported here. Fig. 4 shows the output of the experiment for each step on the image $2007-1$.tif, and Fig. 5 shows the results for 7.tif, $2007-2$.tif, and $2007-4$.tif. On comparing with the ground-truth images, it is evident that our algorithm has the desired efficiency and robustness (see results on images 6.tif in Fig. 3 and 7.tif in Fig. 5). In Table 2, the number of pixels on circular arcs in the ground-truth image and that in our output image are shown in third and fourth columns, respectively. The numbers of pixels corresponding to false acceptance and false rejection are listed in columns five and six, respectively. The execution time required in the proposed method is listed in the last column.

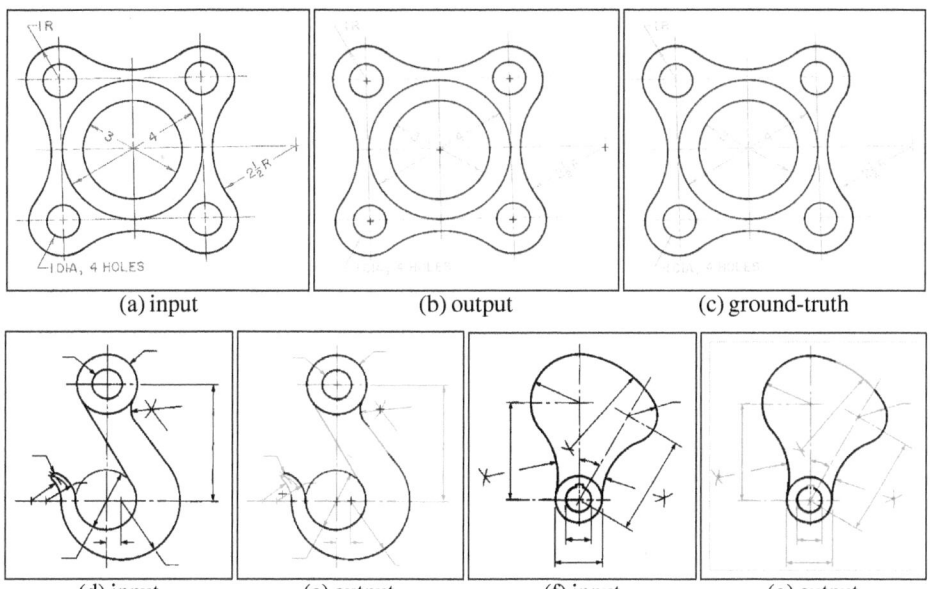

|(a) input | (b) output | (c) ground-truth|

|(d) input | (e) output | (f) input | (g) output|

Fig. 5. Results for a few more images: (a–c) 7.tif; (d, e) $2007-2$.tif; (f, g) $2007-4$.tif

5   Conclusion

In this paper, we have presented a new technique for identifying digital circles and circular arcs from a binary image using *chord property* and *sagitta property*. We have used a variant of the *chord property* of real circle to detect digital circles or circular arcs in an image. Based on the *chord property*, circular arcs are extracted and then using *sagitta property* centers and radii are computed. After merging the arcs of nearly equal radius and center, a complete circle or a larger circular arc is obtained. We have used restricted HT to improve the accuracy of centers and radii. Experimental results have been presented in support of the proposed method. Since for each circle or circular arc we have computed the center and radius before applying the HT, the size of Hough space is very small. Hence, the amount of accumulator memory and required computation time are reduced significantly. Estimation of the different parameter values for specific types of images in order to improve the results as well as to enhance the robustness and accuracy of the method may be studied as future research issues.

References

1. Bhowmick, P., Bhattacharya, B.B.: Fast polygonal approximation of digital curves using relaxed straightness properties. IEEE Trans. PAMI 29(9), 1590–1602 (2007)
2. Chen, T.C., Chung, K.L.: An efficient randomized algorithm for detecting circles. Computer Vision and Image Understanding 83(2), 172–191 (2001)
3. Chiu, S.H., Liaw, J.J.: An effective voting method for circle detection. Pattern Recognition Letters 26(2), 121–133 (2005)
4. Coeurjolly, D., et al.: An elementary algorithm for digital arc segmentation. Discrete Applied Mathematics 139, 31–50 (2004)
5. Davies, E.R.: A modified Hough scheme for general circle location. PR 7(1), 37–43 (1984)
6. Gonzalez, R.C., Woods, R.E.: Digital Image Processing. Addison-Wesley, Reading (1993)
7. Ho, C.T., Chen, L.H.: A fast ellipse/circle detector using geometric symmetry. Pattern Recognition 28(1), 117–124 (1995)
8. Illingworth, J., Kittler, J.: A survey of the Hough transform. CVGIP 44(1) (1988)
9. Kim, H.S., Kim, J.H.: A two-step circle detection algorithm from the intersecting chords. Pattern Recognition Letters 22, 787–798 (2001)
10. Kimme, C., Ballard, D., Sklansky, J.: Finding circles by an array of accumulators. ACM Commun. 18(2), 120–122 (1975)
11. Klette, R., Rosenfeld, A.: Digital Geometry: Geometric Methods for Digital Picture Analysis. Morgan Kaufmann, San Francisco (2004)
12. Klette, R., Rosenfeld, A.: Digital straightness: A review. Discrete Applied Mathematics 139(1-3), 197–230 (2004)
13. Leavers, V.: Survey: Which Hough transform? 58(2), 250–264 (September 1993)
14. Rosin, P.L.: Techniques for assessing polygonal approximation of curves. IEEE Trans. PAMI 19(6), 659–666 (1997)
15. Wall, K., Danielsson, P.-E.: A fast sequential method for polygonal approximation of digitized curves. CVGIP 28, 220–227 (1984)
16. Weisstein, E.W.: Sagitta. From MathWorld—A Wolfram web resource (1993), http://mathworld.wolfram.com/Sagitta.html
17. Xu, L., Oja, E.: Randomized Hough transform (RHT): Basic mechanisms, algorithms, and computational complexities. CVGIP 57(2), 131–154 (1993)
18. Yip, R., Tam, P., Leung, D.: Modification of Hough transform for circles and ellipses detection using a 2-dimensional array. Pattern Recognition 25(9), 1007–1022 (1992)

GOAL: Towards Understanding of Graphic Objects from Architectural to Line Drawings

Shyamosree Pal[1], Partha Bhowmick[1],
Arindam Biswas[2], and Bhargab B. Bhattacharya[3]

[1] Computer Science and Engineering Department
Indian Institute of Technology, Kharagpur, India
[2] Department of Information Technology
Bengal Engineering and Science University, Shibpur, India
[3] Advanced Computing and Microelectronics Unit
Indian Statistical Institute, Kolkata, India
shyamosree@cse.iitkgp.ernet.in, bhowmick@gmail.com,
abiswas@it.becs.ac.in, bhargab@isical.ac.in

Abstract. Understanding of graphic objects has become a problem of pertinence in today's context of digital documentation and document digitization, since graphic information in a document image may be present in several forms, such as engineering drawings, architectural plans, musical scores, tables, charts, extended objects, hand-drawn sketches, etc. There exist quite a few approaches for segmentation of graphics from text, and also a separate set of techniques for recognizing a graphics and its characteristic features. This paper introduces a novel geometric algorithm that performs the task of segmenting out all the graphic objects in a document image and subsequently also works as a high-level tool to classify various graphic types. Given a document image, it performs the text-graphics segmentation by analyzing the geometric features of the minimum-area isothetic polygonal covers of all the objects for varying grid spacing, g. As the shape and size of a polygonal cover depends on g, and each isothetic polygon is represented by an ordered sequence of its vertices, the spatial relationship of the polygons corresponding to a higher grid spacing with those corresponding to a lower spacing, is used for graphics segmentation and subsequent classification. Experimental results demonstrate its efficiency, elegance, and versatility.

1 Introduction

Graphics is a powerful form of expression as it can convey an overall idea at a glance. Expressions articulated from various graphic forms such as maps, architectural plans, engineering drawings, art designs, etc., are very clear, self-explanatory, and easily understandable in their way of representing their constituent elements and their spatial relation. This, in turn, aids in a comprehensive understanding of the underlying information directly without any ambiguity [16]. So, sometimes ideas and information are better communicated via graphical expressions rather than plain text. To do this, we need to have an easy exchange of graphical information

J.-M. Ogier, W. Liu, and J. Lladós (Eds.): GREC 2009, LNCS 6020, pp. 81–92, 2010.

between man and machine, and need to design an efficient mechanism for processing of graphical documents by the machine. Currently, a system inputs the graphical information from the user via the mouse/keyboard or some shape list provided in the toolbar. But a freehand drawing drawn using a mouse is not neat and the predefined shape list is liable to be too long and cumbersome [14].

While the text in a document image is a sequence of characters taken from a predefined alphabet, there is no limit to what can be labeled as graphics. Thus, graphics segmentation is practically a challenging problem, given that graphical information may be in the form of engineering drawings, architectural plans, musical scores, tables, charts, extended objects, hand-drawn sketches, etc. [16]. Hence, it requires not only the segmentation of text from graphics in a given document image but also the understanding of their different classes.

Most of the recent works that deal with the analysis and recognition of line drawings are mainly focused on symbol recognition, with a very little emphasis on the interpretation of various graphic images. These methods perform line or curve recognition in one or two stages without any higher level processing. Either they first perform a segmentation of the document image and then extract the primitive graphic entities, or they directly perform symbol recognition on the bitmap image in a single step [13]. In a more recent work [12], the authors have attempted to obtain a higher level interpretation of the graphical image. To do this, an intermediate stage is used in the graphical document analysis to provide a detailed description of all the shapes present in the given image. This information is then shared among the various specialist processes to interpret the graphical image. In another approach [12], the simpler graphical entities are interpreted first and then the complex entities are obtained in successive iterations for an intelligent way of understanding line drawings in graphical images.

A comprehensive understanding of document images aids in the hierarchical representation of their structure and content, which can be used for editing, browsing, indexing, and filing of document images [6,15]. In such document understanding systems, proper classification of the document into its various constituent parts or zones (text and graphics) is of great importance. The zone classification technique plays a key role in a document understanding system, which includes text extraction [17], OCR [9], math recognition [18], table understanding [7], logo detection [3,11], image and diagram extraction [4,10], etc. Recently, a zone classifier based on a decision tree and a hidden Markov model has been reported in [15]. An overview of existing works in document image classifcation in recent times with their performance evaluation is also presented in [15].

We present here a novel geometric algorithm for segmenting out all the graphic objects in a document image and subsequently recognizing various graphic types. Given a document image, our algorithm performs the text-graphics segmentation by analyzing the geometric features of the minimum-area isothetic polygonal covers of all the objects for varying grid spacing, g. As the shape and size of a polygonal cover depends on g, and each isothetic polygon is represented by an ordered sequence of its vertices, the spatial relationship of the polygons corresponding to a higher grid spacing with those corresponding to a lower grid spacing, is performed using an efficient geometric technique. The novelty of our algorithm lies in exploiting the features present in isothetic covers of elementary geometric primitives, such as lines and circles, which are not seen in existing works. For example, in [1], text-graphics classification was proposed using white spaces in the form of tiles. On the contrary, in our algorithm, each isothetic cover, being represented as a sequence of vertices and grid points whose internal angles are only 90^0 (denoted by '1'), 180^0 ('2'), and 270^0 ('3'), provides a way of analyzing the regularity properties, if any. Hence, a skewed straight piece is easily recognized by its isothetic cover as the vertex sequence follows a periodic pattern [8]. For example, . . . 222312223122231222. . . is digitally straight as the '2's are uniformly spaced and separated by '31'. Apart from a few other useful features, this idea is also used by us to recognize straight line segments, which are usually found in architectural plans and engineering drawings but not in hand-drawn sketches.

2   The Proposed Method

We consider an input document image, $\mathcal{I}$, which may contain both graphics and text components. Our method for understanding the type of graphics comprises of two stages:

Stage 1: Segmenting out all graphics object(s) in $\mathcal{I}$.
Stage 2: Analyzing the graphic object(s) identified in Stage 1 and performing its (their) classification.

2.1   Algorithm for an Isothetic Cover

In order to separate out graphics from text, we construct isothetic (i.e., axis-parallel) polygonal covers for the constituent objects in the input document image $\mathcal{I}$ using the algorithm IsoPoly (modified from TIPS [2] after applying the binarization [5]). Let Q_1, Q_2, Q_3, and Q_4 be the four square cells/quadrants incident at a grid point $p(i, j)$ (Fig. 1). To decide whether p is a vertex of some isothetic polygon, we need to check the combinatorial arrangement of object containment of the four quadrants incident at p. Depending on whether or not a quadrant has object containment, there exist $2^4 = 16$ different arrangements of these four quadrants. These sixteen arrangements, in turn, can be reduced to five

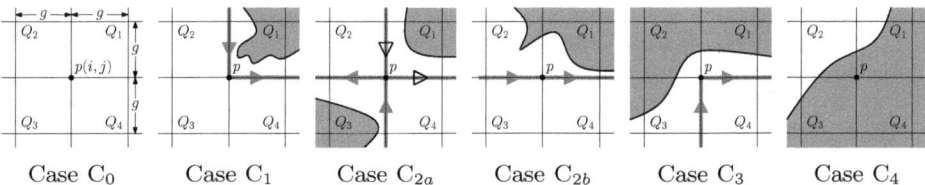

Case C_0 Case C_1 Case C_{2a} Case C_{2b} Case C_3 Case C_4

Fig. 1. Instances of five combinatorial cases to decide the vertex type of $p(i, j)$

cases, where, a case C_q $(q = 0, 1, \ldots, 4)$ includes all the arrangements for which exactly q square(s) has/have object containments, and the remaining (i.e., $4-q$) ones have not. If p befits Case C_1 or C_3, then p is always a vertex of an isothetic polygon. If p befits C_2, and the quadrants with object containment are diagonally opposite, then p occurs twice as a vertex (Subcase C_{2a}); otherwise, p is a non-vertex grid point lying on some edge of a (isothetic) polygon (Subcase C_{2b}). For C_0, p is just an ordinary grid point lying outside a polygon; whereas, for C_4, p is a grid point lying inside a polygon.

2.2   Segmenting out Graphics

The document image $\mathcal{I}$ is first provided as an input to the algorithm IsoPoly with an appropriately large grid spacing, $g = g_1$. The result is an output image, $\mathcal{I}_{out}^{(g_1)}$, with a set of isothetic polygonal covers, namely $S^{(g_1)}$, which tightly encloses all major components in $\mathcal{I}$. We then analyze the geometric nature of each isothetic polygon P_i of $S^{(g_1)}$ and classify P_i into one of the following two broad categories depending on its shape:

Type 1: Isothetic polygons having rectilinear shapes.
Type 2: Isothetic polygons having irregular shapes.

The inference from the above classification is that an irregularly shaped isothetic polygon (Type 2) is very unlikely to enclose some text material and hence can be marked as a graphics-containing polygonal cover. Based on this broad classification, nothing can be said in particular about the content of a Type 1 polygon, as it may contain either text or some graphic object(s) lying inside a bounding box or a rectilinear region. Hence, to examine further, we again feed the polygons of $S^{(g_1)}$ to IsoPoly with a lower grid spacing, g_2, and derive the set of polygons, $S_i^{(g_2, g_1)}$, corresponding to $\mathcal{I} \cap P_i$ for each polygon P_i of $S^{(g_1)}$. From $S_i^{(g_2, g_1)}$, we verify whether each Type 2 polygon of $S^{(g_1)}$, already identified to enclose graphic components, now has tighter isothetic polygon(s). On the contrary, if $P_i \in S^{(g_1)}$ is of Type 1 and $S_i^{(g_2, g_1)}$ has majority of its polygons with small perimeters and arranged in a rectilinear fashion with a uniform spacing between them, then P_i is identified as a text polygon; otherwise, P_i is a graphics polygon. The frequency plot of polygon perimeters for $g = g_2$ is used to decide whether a polygon in $S_i^{(g_2, g_1)}$ encloses a word.

Demonstration of the Segmentation. A demonstration is shown in Fig. 2. First, the set of isothetic polygonal covers, $S^{(12)} = \{P_1^{(12)}, P_2^{(12)}, \ldots, P_7^{(12)}\}$ (Fig. 2b: numbered from top to bottom, left to right), is constructed using the algorithm IsoPoly from the input document image (Fig. 2a) for the grid spacing $g_1 = 12$. We classify the polygons in $S^{(12)}$ into two types: Type 1 having a rectilinear shape, and Type 2 with an irregular shape. As a result, we find that only one polygon, namely $P_4^{(12)}$, is classified as a Type 2 polygon while the remaining six polygons are classified as Type 1 polygons. As explained in Sec. 2.2, it conforms to the idea that an irregularly shaped isothetic polygon (Type 2) is very unlikely to enclose some text material, which is usually laid in compact blocks containing words in a linear fashion. Since we cannot make any prediction about the contents of Type 1 at this stage, we again feed $S^{(12)}$ to IsoPoly with a lower grid spacing, $g_2 = 2$, and derive the set, $S_i^{(2,12)}$, corresponding to $\mathcal{I} \cap P_i$ for each $P_i \in S^{(12)}$. At this point, we find that the Type 2 polygon of $S^{(12)}$, i.e., $P_4^{(12)}$, already identified to enclose graphic components, now has tighter isothetic polygon(s), shown in dark gray in Fig. 2d. On the contrary, we find that for each of the three Type 1 polygons, namely $P_2^{(12)}$, $P_3^{(12)}$, and $P_5^{(12)}$, a large number of small rectilinear polygons appear inside the polygon. Figure 2c shows a high frequency of polygons having perimeters in $[1, 400]$ and a low frequency of polygons having perimeter greater than 400. So a polygon that has perimeter above 400 is labeled as a graphics component and that below 400 as a text component or word(s). Thus, with the aid of the frequency plot we are able to decide whether a polygon in $S_i^{(2,12)}$ encloses (a set of) words or a graphic element. For example,

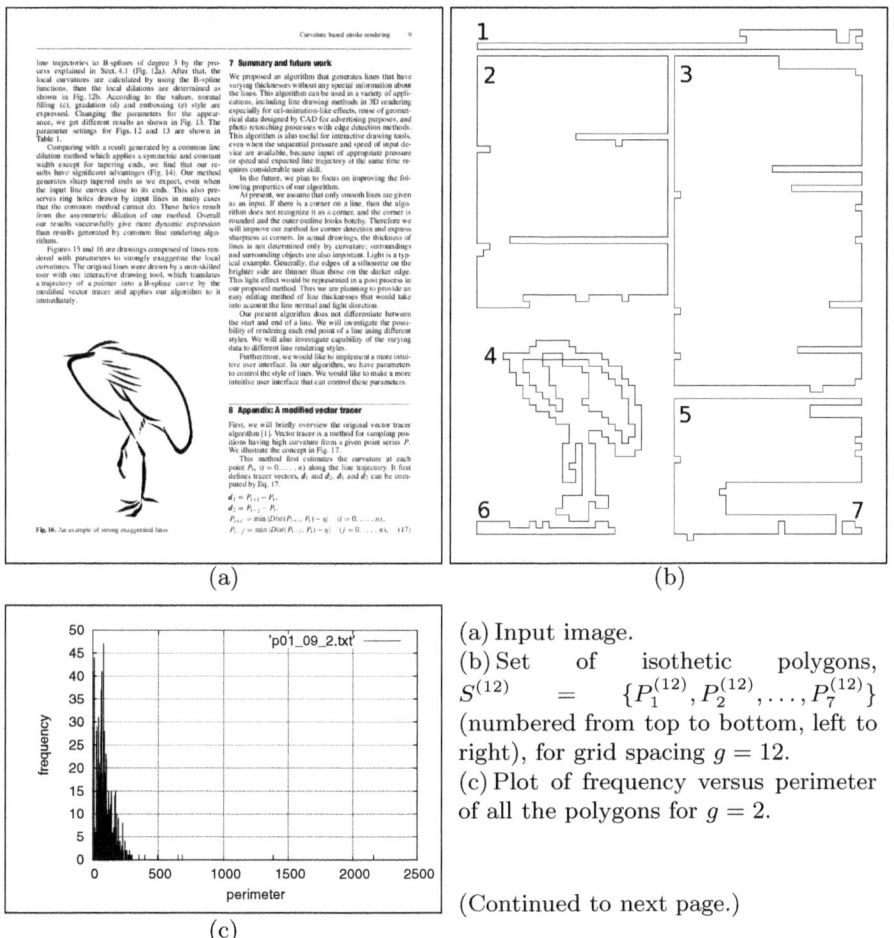

(a) Input image.

(b) Set of isothetic polygons, $S^{(12)} = \{P_1^{(12)}, P_2^{(12)}, \ldots, P_7^{(12)}\}$ (numbered from top to bottom, left to right), for grid spacing $g = 12$.

(c) Plot of frequency versus perimeter of all the polygons for $g = 2$.

(Continued to next page.)

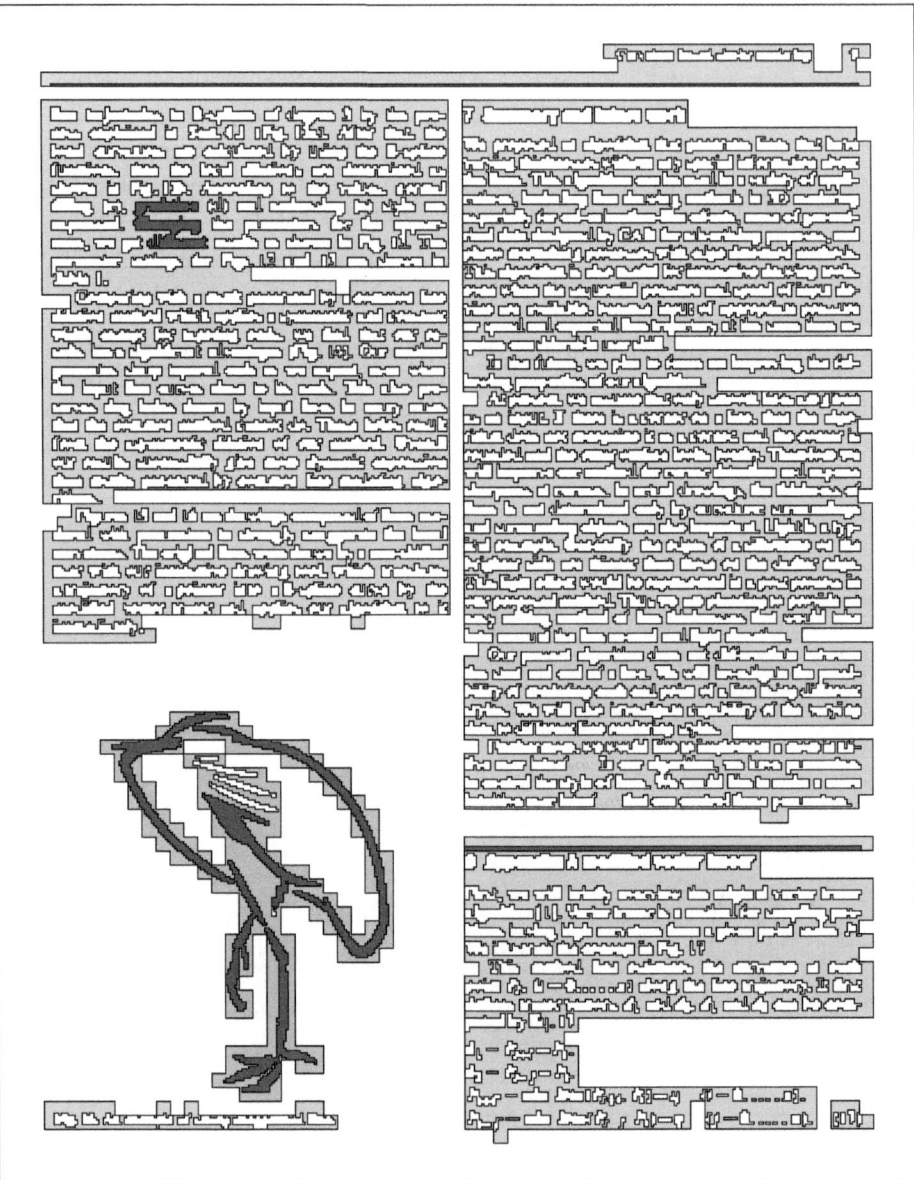

(d) Four small polygons (shown white) inside $P_4^{(12)}$ throw false alarms as text polygons, but are considered as graphic components as $P_4^{(12)}$ is a Type 2 polygon containing several large (graphic) components (shown in dark gray). Similarly, three dark gray polygons lying inside three Type 1 polygons ($P_1^{(12)}$, $P_2^{(12)}$, and $P_5^{(12)}$) are also recognized as false alarms because the predominant polygon types for $g = g_2(= 2)$ correspond to text. In fact, the two false alarms in $P_1^{(12)}$ and $P_5^{(12)}$ are two horizontal lines, and that in $P_2^{(12)}$ is a coalesced text.

Fig. 2. Segmentation of a graphic object in a document page

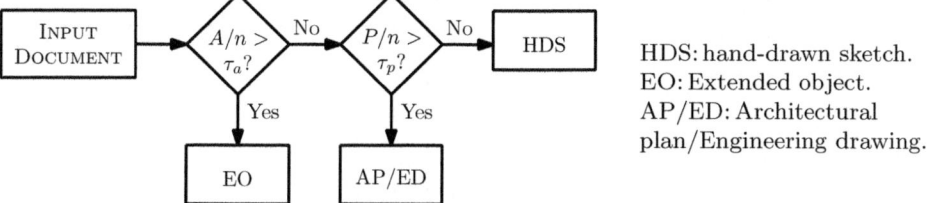

Fig. 3. The scheme of our algorithm ($\tau_a = 35$, $\tau_p = 7$ in our experiments)

the polygons with small perimeters in Fig. 2d (shown in white) are contained within the respective Type 1 polygons shown in light gray. These small polygons signify individual words or a few coalesced words, confirming that they are text.

2.3   Identifying Different Graphic Types

After doing the segmentation of graphics from the text for a given document image, our method is set to identify the type of each graphic component present in the document image. A brief schematic overview of our algorithm is given in Fig 3. We have used the features A/n and P/n to classify a graphics, where A and P denote the sum of areas and the sum of perimeters of all the polygons in $S_i^{(g_2,g_1)}$, and n denote their total number of vertices. The graphic types are determined by analyzing the shapes or/and arrangement of their enclosing isothetic polygons, as explained next.

Tabular Structures. Tables are graphic objects in which information is arranged in a regular manner along rows and columns. These rows and columns usually contain text or numerical data. Such a regularity in the arrangement of data in a table is actually missing in most of the other graphic objects. Hence, a tabular structure is distinguished from other graphic objects on the basis of its spatial arrangement of isothetic polygons, when it is initially identified as a Type 2 polygon. Although the contents (words or numbers) of a table are very much similar to those of a text box, the arrangement of the contents of the former is distinguishably different from that of the latter. Hence, when tables are contained within a bounding box and initially identified as a Type 1 polygon, then special care has to be taken to have it identified as a table and not as an ordinary text box. If $P_i \in S^{(g_1)}$ corresponds to a table, then $S_i^{(g_2,g_1)}$ contains mostly text polygons arranged along rows and columns, a feature which is absent in text where words are arranged linearly from left to right and top to bottom. Vertical or/and horizontal separators, if present, are recognized easily as long and thin rectangles. Figure 4 shows an input image of a table (Fig. 4a) followed by the intermediate output (Fig. 4b) segmenting it out from text and then the final output (Fig. 4c) showing the nature of the isothetic polygonal cover for a table.

Architectural Plans and Engineering Drawings. Architectural plans and engineering drawings mostly consist of straight line segments and circular arcs

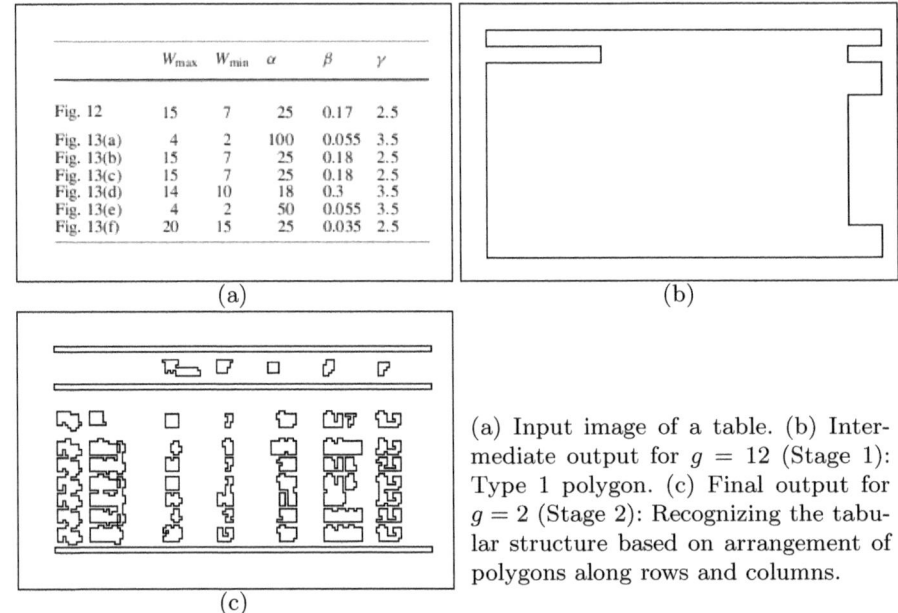

	W_{max}	W_{min}	α	β	γ
Fig. 12	15	7	25	0.17	2.5
Fig. 13(a)	4	2	100	0.055	3.5
Fig. 13(b)	15	7	25	0.18	2.5
Fig. 13(c)	15	7	25	0.18	2.5
Fig. 13(d)	14	10	18	0.3	3.5
Fig. 13(e)	4	2	50	0.055	3.5
Fig. 13(f)	20	15	25	0.035	2.5

(a)   (b)

(c)

(a) Input image of a table. (b) Intermediate output for $g = 12$ (Stage 1): Type 1 polygon. (c) Final output for $g = 2$ (Stage 2): Recognizing the tabular structure based on arrangement of polygons along rows and columns.

Fig. 4. Identifying a tabular structure

(and occasionally, smooth curves), usually bounding empty spaces. Hence, they can be distinguished from other graphic objects, e.g., a table, from its empty space content. Again, hand-drawn sketches may also have a lot of free or empty spaces but the lines contained in hand-drawn sketches are usually not regular (i.e., straight/circular) in nature (Sec. 2.3). Hence, plans and drawings can be very well distinguished from other graphic types using their characteristic features of low A/n and high P/n. Figure 5a shows an input document image $\mathcal{I}$ containing architectural plan and some text. Given the input image $\mathcal{I}$, we first segment out the graphic object from the text. The intermediate output in Fig. 5b shows the set $S^{(12)}$ of isothetic polygonal covers. So, if an isothetic polygon P_i encloses an architectural plan, then the outer and the inner polygons of $S_i^{(2,12)}$ have long straight edges in frequent succession enclosing empty spaces, which is evident in Fig. 5c. Similar results on engineering drawing are shown in Fig. 6.

Extended Objects. Extended objects are probably the easiest of the graphic types to be identified. As the name suggests, they are graphical objects that cover large areas of the canvas. They actually have the highest area-to-vertex (A/n) ratio of their isothetic polygonal covers compared to all other graphic objects, e.g., tables, architectural plans, hand-drawn sketches, etc. They may have isothetic polygonal covers of all kinds of possible shapes but these covers enclose very little or no empty space. As a result, when the algorithm IsoPoly is run on the extended object for a lower grid spacing g_2, there is no significant change in the number of isothetic polygonal covers appearing for the grid spacing

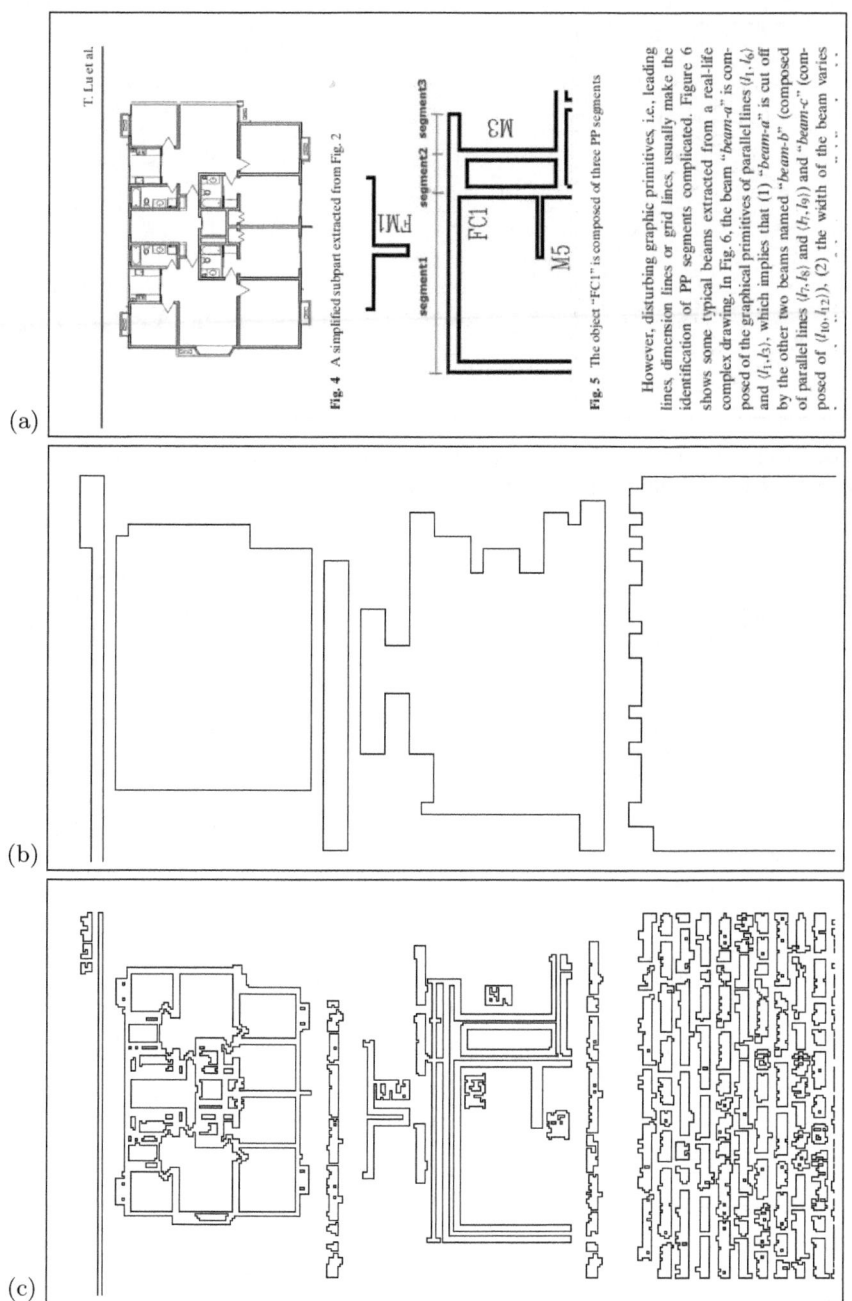

Fig. 5. Identifying architectural plans in a document image. (a) Input document having an architectural plan. (b) Intermediate output for $g = 12$ (Stage 1): Segmenting out graphics from text. (c) Final output for $g = 2$ (Stage 2): Recognizing architectural plans based on long straight edges in succession enclosing empty spaces. (Images rotated by 90^0 ↺.)

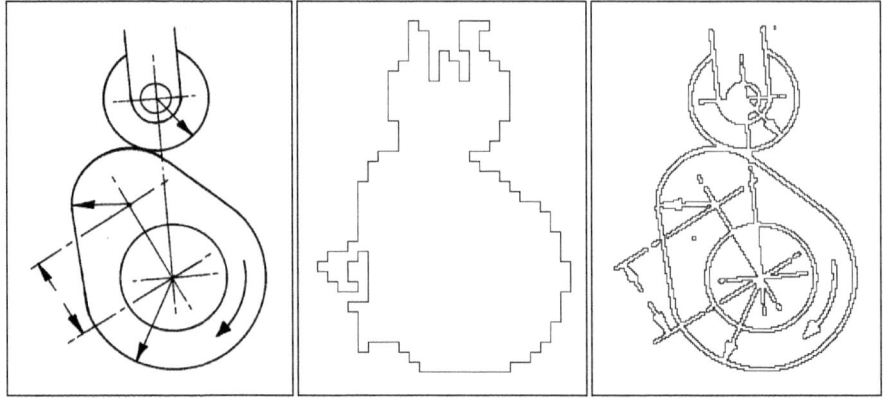

Fig. 6. Results for two engineering drawings. Left: Input. Middle: Intermediate output for $g = 12$. Right: Final output for $g = 2$ (Stage 2).

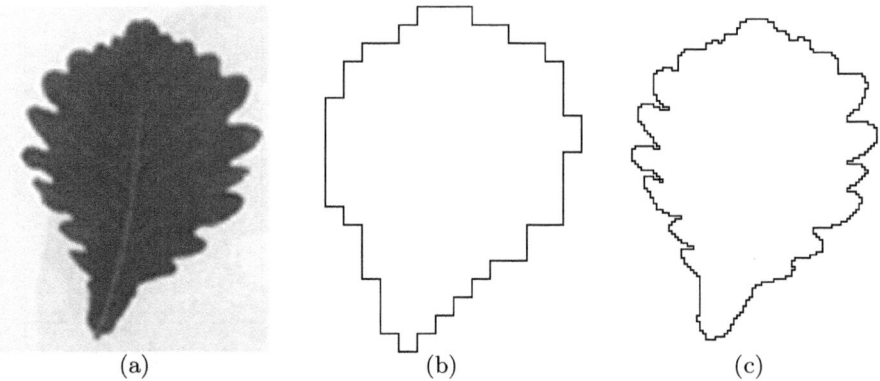

(a)   (b)   (c)

Fig. 7. Recognizing an extended graphic object (leaf)

g_2 compared to that for the higher grid spacing, g_1. This unique property easily differentiates it from all other graphic types. Figure 7 depicts the fact that extended objects are characterized by their high area-to-vertex ratio irrespective of the grid spacing, g.

Hand-drawn Sketches. The act of recognizing a hand-drawn sketch is probably the most complex task out of all graphic classification, since hand-drawn sketches usually possess lines/curves of varying thickness and length. These lines are of all kinds of shapes and sizes ranging from very long and thick ones to very short and thin ones. But they are all a bit wavering or irregular in nature and oriented in a nonuniform fashion. These irregular lines contained in a hand-drawn sketch act as its characteristic property using which we distinguish it from other graphic types. Due to the long and thin nature of a hand-drawn line, the resultant isothetic polygonal cover is also long and thin in shape, and hence

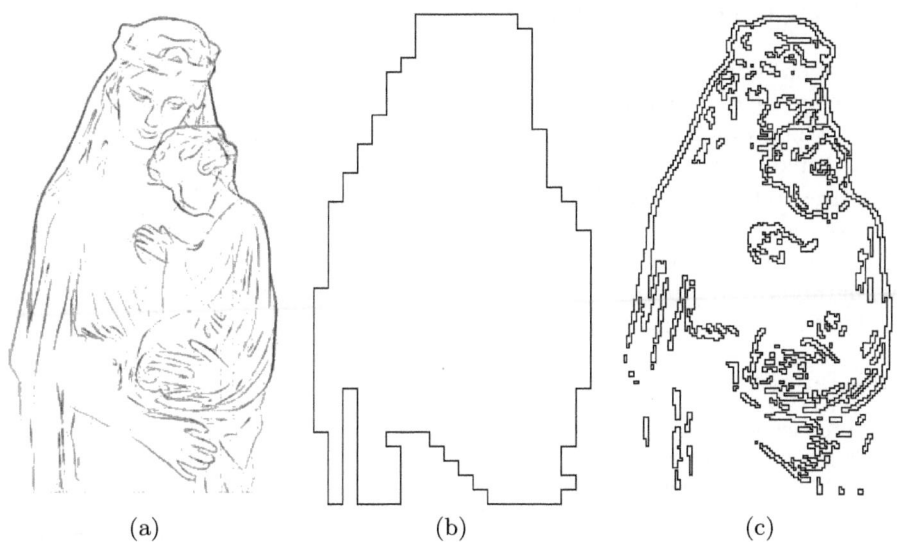

(a)   (b)   (c)

Fig. 8. Identifying a hand-drawn sketch. (a) Input image. (b) Intermediate output for $g = 12$. (c) Final output for $g = 2$.

characterized by its large number of vertices and small area (i.e., low A/n and P/n). Figure 8 shows a hand-drawn image and its respective output.

3   Experimental Results

We have implemented the algorithm in C on an Intel(R) Core(TM)2 Duo CPU E4500 2.20 GHz machine, the OS being Mandriva Linux Release 2008. We have tested the algorithm on University of Washington's document image database (UW1 and UW2) and http://www.iupr.org/arcseg2007. As explained in Sec. 2.3, various graphic types are successfully recognized from different document sets as evident in our experimental results, some of which are presented in this paper (Figs. 4, 5, 7, 2). Table 1 shows the performance measures (explained in [15]) of our algorithm on various graphic images taken from the above-mentioned databases.

Table 1. Performance of the proposed algorithm

	HDS	EO	T	AP/ED	CR(%)	MR(%)
HDS	22	0	0	1	95.45	4.55
EO	3	26	0	0	89.66	10.34
T	0	0	18	6	75.00	25.00
AP/ED	2	0	3	8	61.53	38.46
FR(%)	9.62	0	5.36	10.61	—	—

Acronyms [15]: HDS: hand-drawn sketch. EO: Extended object. T: Table. AP/ED: Architectural plan/Engineering drawing. CR: Correctness rate. MR: Miss rate. FR: False alarm rate.

4   Future Works and Conclusion

This paper shows how certain simple-yet-efficient geometric properties of isothetic covers can be used for segmentation of graphics from text in a document image. The geometric relation among these covers for appropriate grid spacings can be exploited to determine the type of graphic elements ranging from tabular structures to hand-drawn sketches. There are ample possibilities to explore various ways of characterizing the isothetic covers corresponding to different graphic types. We are presently working on designing efficient algorithms (with theoretical guarantees) to recognize different real-world graphic entities, reports of which will be presented in near future.

References

1. Antonacopoulos, A., Ritchings, R.T.: Representation and classification of complex-shaped printed regions using white tiles. In: Proc. ICDAR 1995, pp. 1132–1134 (1995)
2. Biswas, A., Bhowmick, P., Bhattacharya, B.B.: Construction of isothetic covers of a digital object: A combinatorial approach. JVCIR (in press, 2010)
3. Chen, J., Leung, M.K., Gao, Y.: Noisy logo recognition using line segment Hausdorff distance. Pattern Recognition 36(4), 943–955 (2003)
4. Futrelle, R.P., et al.: Extraction, layout analysis and classification of diagrams in PDF documents. In: ICDAR 2003, pp. 1007–1014 (2003)
5. Gonzalez, R.C., Woods, R.E.: Digital Image Processing. Addison-Wesley, California (1993)
6. Haralick, R.M.: Document image understanding: Geometric and logical layout. In: Proc. CVPR, pp. 385–390 (1994)
7. Hu, J., Kashi, R., Lopresti, D., Wilfong, G.: Evaluating the performance of table processing algorithms 4(3), 140–153 (2002)
8. Klette, R., Rosenfeld, A.: Digital Geometry: Geometric Methods for Digital Picture Analysis. Morgan Kaufmann, San Francisco (2004)
9. Kopec, G.E., Chou, P.A.: Document image decoding using Markov source models. IEEE TPAMI 16(6), 602–617 (1994)
10. Li, J., Najmi, A., Gray, R.M.: Image classification by a two-dimensional hidden Markov model. IEEE Trans. Signal Process 48(2), 517–533 (2000)
11. Pham, T.D.: Unconstrained logo detection in document images. Pattern Recognition 36(12), 3023–3025 (2003)
12. Ramel, J.-Y., Vincent, N.: Strategy for line drawing understanding. In: Lladós, J., Kwon, Y.-B. (eds.) GREC 2003. LNCS, vol. 3088, pp. 1–12. Springer, Heidelberg (2004)
13. Song, J., et al.: An object-oriented progresssive-simplification-based vectorization system for engineering drawings: Model, algorithm, and performance. IEEE TPAMI 24(8), 1048–1060 (2002)
14. Sun, Z., Wang, W., Zhang, L., Liu, J.: Sketch parameterization using curve approximation. In: Liu, W., Lladós, J. (eds.) GREC 2005. LNCS, vol. 3926, pp. 334–345. Springer, Heidelberg (2006)
15. Wang, Y., Phillips, I.T., Haralick, R.M.: Document zone content classification and its performance evaluation. Pattern Recognition 39, 57–73 (2006)
16. Wenyin, L.: On-line graphics recognition: State-of-the-art. In: Lladós, J., Kwon, Y.-B. (eds.) GREC 2003. LNCS, vol. 3088, pp. 291–304. Springer, Heidelberg (2004)
17. Xiao, Y., Yan, H.: Text region extraction in a document image based on the Delaunay tessellation. Pattern Recognition 36(3), 799–809 (2003)
18. Zanibbi, R., Blostein, D., Cordy, J.R.: Recognizing mathematical expressions using tree transformation. IEEE TPAMI 24(11), 1455–1467 (2002)

Extracting Road Vector Data from Raster Maps

Yao-Yi Chiang and Craig A. Knoblock

University of Southern California,
Department of Computer Science and Information Sciences Institute
4676 Admiralty Way, Marina del Rey, CA 90292, USA
{yaoyichi,knoblock}@isi.edu

Abstract. Raster maps are an important source of road information. Because of the overlapping map features (e.g., roads and text labels) and the varying image quality, extracting road vector data from raster maps usually requires significant user input to achieve accurate results. In this paper, we present an accurate road vectorization technique that minimizes user input by combining our previous work on extracting road pixels and road-intersection templates to extract accurate road vector data from raster maps. Our approach enables GIS applications to exploit the road information in raster maps for the areas where the road vector data are otherwise not easily accessible, such as the countries of the Middle East. We show that our approach requires minimal user input and achieves an average of 93.2% completeness and 95.6% correctness in an experiment using raster maps from various sources.

Keywords: GIS, raster maps, road vectorization, map processing.

1 Introduction

Humans have a long history of using maps. In particular, paper maps have been widely used since the early years for documenting geospatial information. Because of the availability of low cost and high-resolution scanners and the Internet, we can now obtain a huge number of scanned maps in raster format from various sources. Since maps commonly contain road networks, raster maps are an important source of road vector data for areas where road vector data are not readily available. Moreover, we can use the road vector data as features to register maps to other geospatial data, such as imagery, and create an integrated view of heterogeneous geospatial data sets [3].

Extracting road vector data from raster maps is a challenging task. First, the extraction of road pixels is difficult since raster maps very often contain noise from image compression and scanning processes and roads often overlap with other map features. Further, for converting the road pixels to vector format, the previous work commonly uses the thinning operator [11] or line grouping and parallel-line matching techniques [1] to identify the road centerlines. The thinning operator can produce distorted lines around intersections and hence the extracted road vector data are not accurate without manual adjustment [1]. The line grouping and parallel-line matching techniques require manual settings

J.-M. Ogier, W. Liu, and J. Lladós (Eds.): GREC 2009, LNCS 6020, pp. 93–105, 2010.

on various parameters to identify the accurate centerlines, such as the maximum difference between the slopes of two line segments to be merged [11] .

In this paper, we present a general technique that requires minimal user input for extracting accurate road vector data from raster maps with varying map complexity (e.g., overlapping features) and image quality. We exploit our previous work on extracting road pixels from raster maps [5] and utilize the thinning operator to determine the road centerlines. We then automatically correct the distortions near road intersections caused by the thinning operator using our previous techniques on extracting accurate road-intersection templates from raster maps [4; 6] to extract accurate road vector data. We tested our road vectorization technique on a variety of maps including scanned and digital maps from different sources and compared our results to a commercial map-digitizing product.

The remainder of this paper is organized as follows. Section 2 discusses related work on road extraction from maps. Section 3 presents our approach to extract the road pixels from raster maps. Section 4 describes our approach to generate the road vector data from the extracted road pixels. Section 5 reports on our experimental results, and Section 6 presents the conclusion and future work.

2   Related Work

Much research work has been performed in the field of extracting road information from raster maps, such as separating lines from text [2; 14], detecting road intersections [8], and extracting road vector data [1; 11] from raster maps. In the previous work on text/graphics separation from raster maps, Cao and Tan [2] and Li et al. [14] utilize preset grayscale thresholds to remove the background pixels from raster maps and then detect text labels from the remaining foreground layers of the maps. The road pixels are the by-product (i.e., only the road pixels are extracted) after the text pixels are identified. Since in their work, the main goal is to recognize the text labels, they do not process the raster maps further to extract the road vector data.

Some of the previous work assumes a simpler type of raster maps for their algorithms. Habib et al. [8] extract road intersections from raster maps that contain road lines only. Itonaga et al. [11] employ a stochastic relaxation approach to first extract the road areas from digitally-generated maps (i.e., not scanned maps) and then apply the thinning operator to extract the road vector data. The distorted lines around the road intersections are corrected based on the straightness of the roads, which is determined using user specified constraints, such as the road width. In comparison, our approach can process a variety of raster maps including scanned maps, and we avoid the distortion with no parameter settings. Bin et al. [1] work on scanned maps to extract the road vector data. Instead of using the thinning operator, as in [1], the medial lines of parallel road lines are first produced and then linked to generate the road vector data. In general, the vectorization results of utilizing the medial lines of parallel road lines can be very accurate for the lines around the intersections, but the extraction processes require more manually specified parameters than the thinning operator, such as the thresholds to group medial-line segments and to produce the road intersections.

In addition to the research work, a commercial product called R2V from Able Software is an automated raster-to-vector conversion software package specialized for digitizing raster maps. To vectorize roads in raster maps using R2V, the user needs to first manually provide samples of road pixels or select a set of color threshold to identify the road pixels. The manual work of providing samples of road pixels can be laborious, especially for scanned maps with numerous colors, and the color thresholding function does not work if one set of threshold cannot separate all of the road pixels from the other pixels. In comparison, our approach automatically identifies road colors from a few user labels for extracting the road pixels. After the road pixels are extracted, R2V can automatically trace the centerlines of the extracted road pixels and generate the road vector data. Our approach detects the road format and road width automatically and uses the detected road information to extract accurate road vector data. In our experiments, we tested R2V using our test maps and show that our automatic technique generates better results.

3   Extracting Road Pixels

Distinct colors commonly represent different layers (i.e., a set of pixels representing a particular geographic feature) in a raster map, such as roads, contour lines, and text labels. By identifying the colors that represent roads in a raster map, we can extract the road pixels from the map. However, raster maps usually contain numerous colors due to scanning and/or compression processes and the poor condition of the original documents (e.g., color variation from aging, shadows from folding lines, etc.). For example, Figure 1(a) shows a 200x200-pixels tile cropped from a scanned map. The tile has 20,822 distinct colors, which makes it difficult to select the road colors manually. To overcome this difficulty, many techniques have been developed to first group the colors of individual feature layers into clusters based on the assumption that the color variation within a feature layer is smaller than the variation between feature layers [5; 10; 12; 13]. Therefore, the feature layers can be extracted by selecting specific clusters. In this paper, we utilize our supervised map decomposition technique in [5] to extract the road pixels, which requires minimal user input and is capable of handling various types of raster maps, especially scanned maps.

The supervised map decomposition technique first employs two color quantization techniques to reduce the number of colors in the raster map. To preserve object edges while clustering the colors in a raster map, we first employ the Mean-shift algorithm [7], which merges two colors into one by considering their distance in the color space (we use a color distance of 25 in the red, blue, and green color space) as well as in the image space (we use a spatial distance of 3 pixels). The Mean-shift algorithm reduces the number of colors in Figure 1(a) by 72% as shown in Figure 1(b). To further merge similar colors in the raster maps for reducing the user input to select the road colors, we apply the K-means

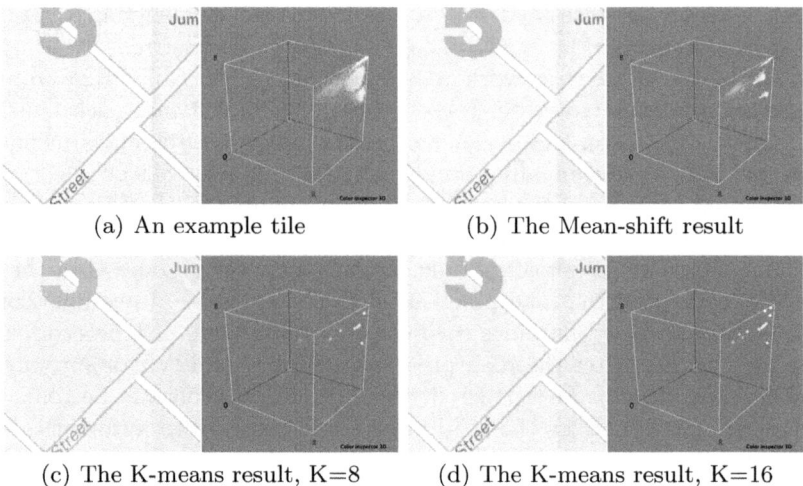

(a) An example tile　　　　　　(b) The Mean-shift result

(c) The K-means result, K=8　　(d) The K-means result, K=16

Fig. 1. An example map tile and the color quantization results with color cubes

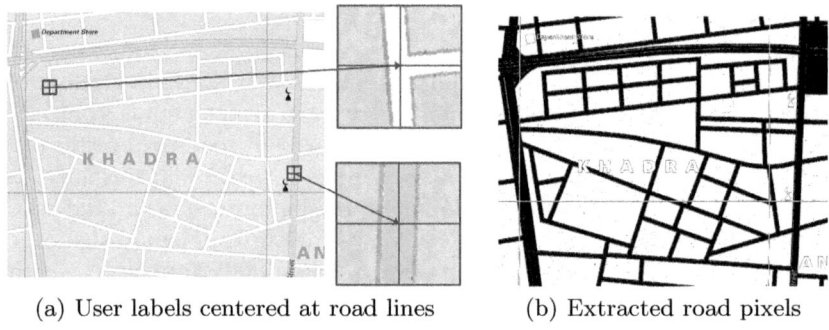

(a) User labels centered at road lines　　(b) Extracted road pixels

Fig. 2. Extracting road pixels using road color identified by analyzing user labels

algorithm with a user specified K to generate a quantized map image with at most K colors. The K-means algorithm can significantly reduce the number of colors in a raster map by maximizing the inter-cluster color variance; however, since the K-means algorithm considers only the color space, it is very likely that the resulting map has merged features with a small K. For example, Figure 1(c) shows the quantized map with K as 8 and the text labels have the same color as the road edges. Therefore, the user would need to select a larger K to separate different features, such as in the quantized map in Figure 1(d) with K as 16.

With the quantized map, the user provides labels of road areas such as the two user labels shown in Figure 2(a), and the map decomposition technique then exploits the fact that a user label is required to be centered at a road line or a road intersection to identify the road colors. Using this approach, the user only has to provide enough user labels to cover each road color in the raster map,

such as one for the white roads and one for the yellow roads in Figure 2(a).
Figure 2(b) shows the extracted road pixels by using the road colors identified
using these two user labels.

4   Vectorizing Road Pixels

Once we have the road pixels, we generate the road vector data by first determine the road centerlines and then vectorize the centerlines. Figure 3(a) and
Figure 3(b) show an example map tile from a scanned map and the road pixels
extracted from the map using the approach described in the previous section.
The extracted road pixels contain objects other than roads since they are drawn
using the same color as roads. In addition, some of the road lines in the extracted
road layer are broken since the missing pixels also belong to the text labels and
grid lines (i.e., overlapping features) and these pixels are not drawn using the
road colors. To separate the non-road features from the road pixels, we exploit
the distinctive geometric properties of road lines such as road lines are linear objects and are connected, to remove solid areas and small connected-components.
Next, we apply the closing operator to reconnect one-pixel wide gaps and fill
small holes. The closing operator first expands the foreground areas by one pixel
(i.e., one iteration of the dilation operation) and then expands the background
areas by one pixel (i.e., one iteration of the erosion operation). Figure 3(c) shows
the results after we apply the closing operator, where the red circles show that
some of the missing road pixels are filled if the missing parts are small, especially
in the places where the text labels overlap with roads.

In order to reconnect broken lines with larger gaps automatically, we expand
the areas of road pixels by utilizing the binary dilation operator as shown in
Figure 3(e). We determine the number of iterations of the dilation operator
(i.e., how far the foreground region should expend) using the road width and
road format (i.e., double-line and single-line roads) identified automatically by
the Parallel-Pattern Tracing algorithm [6]. In a road layer where road lines are
drawn as single lines (i.e., single-line format) as the example shown in Figure 3,
the detected road width is the thickness of the majority of the road lines in the
road layer as the dashed lines shown in Figure 3(d). If a road line is drawn using
two parallel lines (i.e., double-line format), the road width is the pixel distance
between corresponding road pixels on the parallel lines. During the thickening
process, we also merge parallel lines into thick single lines if the road layer is in
double-line format.

To generate the centerline representation of the thickened road lines, we apply
the binary erosion operator and the thinning operator as shown in Figure 3(f)
and Figure 3(g). We use the erosion operator to shrink the road areas before
we apply the thinning operator because the thinning operator distorts lines near
the intersections and the extent of the distortion depends on the thickness of
the lines before the thinning operator is applied. Although the binary erosion
operator helps to minimize the extent of the distortion caused by the thinning
operator, the road geometry near the intersections is still not accurate, especially

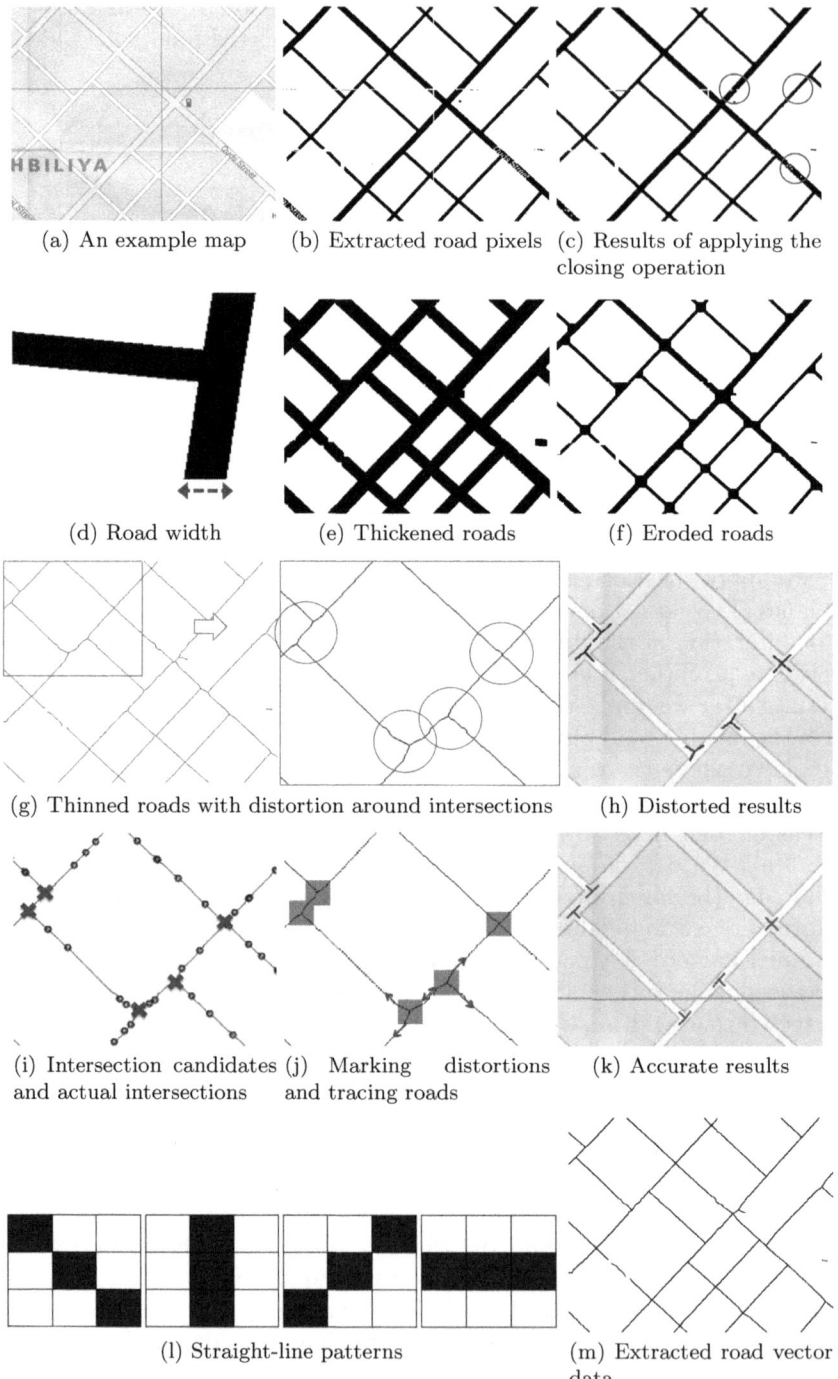

(a) An example map   (b) Extracted road pixels (c) Results of applying the closing operation

(d) Road width   (e) Thickened roads   (f) Eroded roads

(g) Thinned roads with distortion around intersections   (h) Distorted results

(i) Intersection candidates (j) Marking   distortions (k) Accurate results
and actual intersections   and tracing roads

(l) Straight-line patterns   (m) Extracted road vector data

Fig. 3. Extracting road vector data from an example map

```
CNList; // The connecting-node list (CN.x and CN.y are the pixel location)
road_vectors; // The line-segment list (a line segment contains two CN indexes)
start_id; end_id; // The IDs of the starting and ending CNs of the line segment we are currently tracing
```

```
void main() // Program starts here          Function void floodFill8(int x, int y)
  Foreach CN in the CNList {                   if (InsideImage(x,y) && NotVisited(x,y)
    start_id = CN.id;                             && NotBackground(x,y)) {
    SetVisited(CN.x, CN.y);                       if (IsCN(x,y)){ // We found a line
    floodFill8(CN.x, CN.y);                          end_id = GetCNID(x,y);
  }                                                  road_vectors.AddLine(start_id, end_id);
  // Correct the distortions                      } else {
  Foreach CN in the CNList {                         SetVisited(x,y);
    If (InsideGrayBox(CN.x, CN.y) {                   floodFill8(x + 1, y); floodFill8(x - 1, y - 1);
      // An intersection                             floodFill8(x, y + 1); floodFill8(x + 1, y - 1);
      CN.x = GetUpdatedIntersectionLocationX(CN.id);  floodFill8(x + 1, y + 1); floodFill8(x − 1, y);
      CN.y = GetUpdatedIntersectionLocationY(CN.id);  floodFill8(x - 1, y + 1); floodFill8(x, y - 1);
    }                                              }
  }                                            }
}
```

Fig. 4. Pseudo code for tracing line pixels

near T-shape intersections. Figure 3(g) shows the distorted examples of the road geometry around a T-shape intersection and Figure 3(h) shows the inaccurate results if the distorted lines are traced to generate the road vector data.

For correcting the distortion around the intersection points and generating accurate road vector data from the thinned-line image (Figure 3(g)), we first detect intersections of the thinned lines to mark potential distorted lines. We utilize the corner detector [15] to detect intersection candidates and then use the connectivity of the candidates to determine actual road intersections [6]. Figure 3(i) shows the intersection candidates in blue circles and the actual intersections with cross marks. Since the extent of the distortion around each intersection is determined by the thickness of the thickened lines (which is decided by the road width and the dilation operator), we can mark potential distorted thinned-lines near an intersection point using a gray box with the size as the thickness of the thickened lines as shown in Figure 3(j). We then trace the lines outside the gray boxes to generate accurate road orientations and update the positions of the road intersections based on the intersecting roads and their orientations. Figure 3(k) shows a portion of example extraction results. The road lines around the intersections are accurate despite the distortion of the thinned lines shown in Figure 3(g).

With the accurate positions of the road intersections and the knowledge of potential distorted areas, we start to trace the road pixels on the thinned-line image to generate the road vector data. The thinned-line image contains three types of pixels: the non-distorted road pixels, distorted road pixels, and background pixels, (as shown in Figure 3(j), they are the black pixels not covered by the gray boxes, black pixels in the gray boxes, and white pixels, respectively). We create a list of *connecting nodes* (CNs) of the road vector data. A CN is a point where two lines meet at different angles. We first add the detected road intersections into the CN list. Then, we identify the CNs among the non-distorted road pixels using a 3x3-pixels window to check if the pixel has any of the straight-line patterns shown in Figure 3(l). We add the pixel to the CN list if we *do not* detect a straight-line pattern since the road pixel is not on a straight line.

To determine the connectivity between the CNs, we trace the road pixels using an eight-connectivity flood-fill algorithm shown in Figure 4. The flood-fill algorithm starts from a CN, travels through the road pixels (both non-distorted and distorted ones), and stops at another CN. Finally, for the CNs that are road intersections, we use the previously updated road intersection positions as the CNs' positions. The CN list and their connectivity are the results of our extracted road vector data. Figure 3(m) shows the extracted road vector data. The road vector data around the road intersections are accurate since the distorted lines are not traced by the flood-fill algorithm and the intersection positions are updated using accurate road orientations.

5   Experiments

We evaluated our road vectorization approach using three raster maps produced from different sources. Two maps are scanned maps (350dpi) covering the city of Bagdad, Iraq published by Gecko Maps and International Travel Maps (ITM). We cropped and tested 10 map tiles (800x600 pixels each) from each of the scanned map. The paper maps have been folded, and the fold lines cause inevitable shadows and color differences between areas in the scanned maps, which enriches our test data since the cropped tiles from the same map have various color usage and image quality. In addition to the scanned maps, we tested a digitally generated map covering Afghanistan published by the United Nations (UN).[1] The digital map (3300x2550 pixels) shows the main and secondary roads, cities, political boundaries, airports, and railroads of the nation. We tested the digital map as a single tile in our experiments. For comparison, we also tested the automatic road vectorization function in R2V from Able Software.

We first applied our supervised map decomposition technique described in Section 2 to extract the road pixels from the test maps. We pre-processed the scanned map tiles using the Mean-shift and K-means algorithms with K as 8, 16, 24, and 32 to generate four quantized images for each map tile. The user started the user-labeling task from the quantized image containing eight colors. If the user cannot distinguish the road pixels from other map features (e.g., background) in the quantized image, the user will then select an image containing more colors (a higher K) for user labeling. We did not apply the color segmentation algorithms on the digital map before user labeling. This is because the digital map contains a smaller number of colors (i.e., 90 unique colors) and there is only one color representing both the major and secondary roads in the map. Table 1 shows the numbers of colors in the images used for user labeling and the numbers of user labels used for extracting the road pixels. The user-labeling task is the only process that requires user input in our experiments, and for all of the scanned map tiles, only two to four labels were needed.

We tested R2V on extracting the road pixels from the test maps. Since the scanned maps contain numerous colors, we need more than one set of color thresholds to extract the road pixels (R2V only allows one) or significant user effort to manually

---

[1] http://unama.unmissions.org/

Table 1. The number of colors in the image for user labeling of each tested map and the number of user labels for extracting the road pixels

Tile Number	ITM Map										Gecko Map										UN Map
	1	2	3	4	5	6	7	8	9	10	1	2	3	4	5	6	7	8	9	10	
Colors	8	16	8	16	16	16	16	16	16	8	8	8	16	16	16	8	16	16	16	8	90
User Labels	4	3	3	4	3	2	4	3	3	2	2	2	2	2	2	2	3	3	3	2	1

Table 2. Numeric results of the extracted road vector data from the scanned Gecko and ITM maps (four-pixel-wide buffer) using our approach

Tile	ITM	Gecko	ITM	Gecko	ITM	Gecko	ITM	Gecko	ITM	Gecko
	Completeness		Correctness		Quality		Redundancy		RMS Diff.	
1	98.7%	97.8%	96.7%	85.8%	95.5%	84.1%	0.07%	0%	2.34	1.69
2	99.3%	97.4%	93.6%	97.5%	92.9%	95%	0%	0%	1.23	3.51
3	98.1%	93.3%	75.8%	97.8%	74.7%	91.4%	0%	6.7%	1.52	2.46
4	91.7%	97.2%	96.0%	98.7%	88.3%	96%	0%	1.73%	2.57	1.61
5	92.0%	98.9%	94.7%	97.9%	87.5%	96.8%	0%	0%	2.79	1.32
6	92.7%	88.6%	99.0%	90.2%	91.9%	80.1%	0%	0.41%	2.50	3.2
7	97.5%	97.3%	99.2%	98%	96.7%	95.4%	3.34%	6.52%	1.81	1.65
8	95.1%	93.4%	97.1%	94%	92.5%	88.2%	0%	0%	2.02	2.56
9	93.7%	99.0%	94.6%	83.3%	88.9%	82.6%	0%	0%	2.21	1.54
10	97.1%	98.7%	85.9%	94%	83.7%	92.9%	0.7%	1.7%	2.20	1.47
Avg.	95.6%	96.2%	93.3%	93.7%	89.3%	90.3%	0.6%	1.7%	2.12	2.1

specify sample pixels for each of the road colors. Therefore, we did not successfully extract the road pixels from the scanned maps using R2V. For the digital map, we used one set of color threshold to extract the road pixels using R2V.

For the extracted road vector data, we report the accuracy of the extraction results using the road extraction metrics proposed in [9], which include the completeness, correctness, quality, redundancy, and the root-mean-square (RMS) difference. We manually drew the centerline of every road line in the maps as the ground truth. The completeness and correctness represent how complete/correct the extracted road vector data are (the optimum is 100%). The quality is a combination metric of completeness and correctness (the optimum is 100%). The redundancy shows the difference in percentage between the correctly extracted lines and the matched ground truth (the optimum is 0). The RMS difference is the average distance between the extracted lines and the ground truth, which represents the geometrical accuracy of the extracted road vector data. To generate these metrics, the authors in [9] suggest using a buffer width as half of the road width in the test data. In our test maps, the roads are five and eight pixels wide in the digital map and are seven to ten pixels wide in the scanned maps. We used a buffer width of four pixels.

Table 2 and Table 3 show the numeric results. The average completeness are from 87.9% to 95.6%, the average correctness are from 93.7% to 99.9%, and the

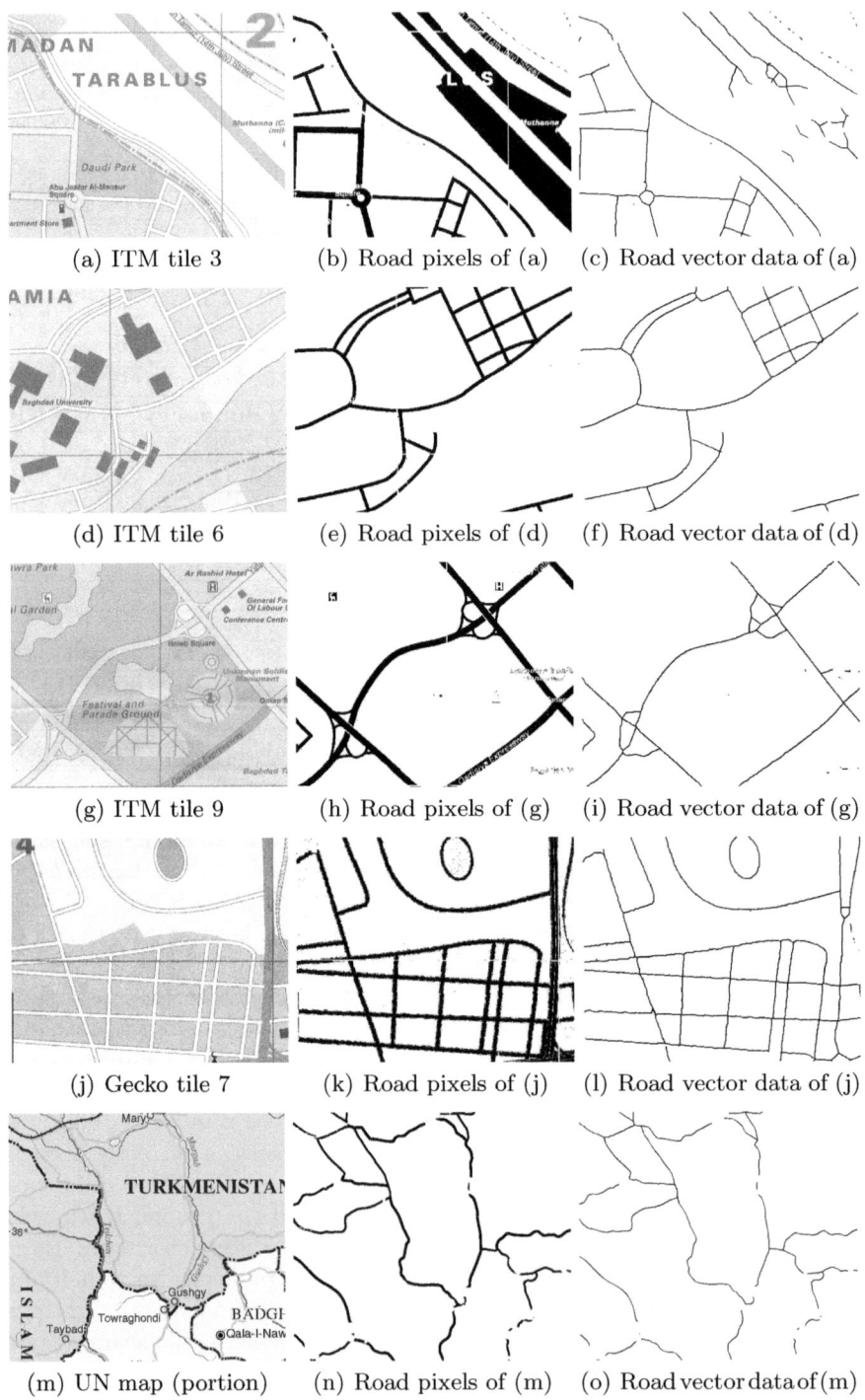

(a) ITM tile 3   (b) Road pixels of (a)   (c) Road vector data of (a)

(d) ITM tile 6   (e) Road pixels of (d)   (f) Road vector data of (d)

(g) ITM tile 9   (h) Road pixels of (g)   (i) Road vector data of (g)

(j) Gecko tile 7   (k) Road pixels of (j)   (l) Road vector data of (j)

(m) UN map (portion)   (n) Road pixels of (m)   (o) Road vector data of (m)

Fig. 5. Examples of the road vectorization results

Table 3. Numeric results of the extracted road vector data from the UN digital map (four-pixel-wide buffer) using our approach and R2V

Tested Technique	Completeness	Correctness	Quality	Redundancy	RMS Diff.
This Paper	87.9%	99.9%	87.8%	0%	3.75
Able R2V	76.1%	96.7%	74.2%	18.92%	3.91

average redundancy are from 0% to 1.7% for the scanned and digital map using our approach. Figure 5 shows some example results, where the geometry of the extracted road vector data are very close to the road centerlines in the maps. Some broken lines are not connected (causing lower completeness numbers, such as for the digital map) since the gaps are larger than the iterations of the dilation operations (we automatically detected the road format as single-line roads and used three iterations of the dilation operator to fix the gaps smaller than six pixels). The broken lines could be reconnected with post-processing on the road vector data since the gaps are now smaller than they were in the extracted road layers resulting from the dilation operations. The tiles 3 and 10 of the ITM map and tiles 1 and 9 of the Gecko map have lower correctness since parts of non-road features are also extracted using the identified road colors and those parts contribute to false-positive road vector data. Figure 5(a) to Figure 5(c) show the ITM tile 3 where the runways are represented using the same color as the white roads and hence are extracted as road pixels. This type of false-positives could be further removed by including a user validation step after the road pixels were extracted. Some tiles have higher redundancy numbers such as the Gecko tiles 3 and 7, which is because some of the straight road lines in these tiles were extracted as shorter line segments with a small orientation variation and their buffers overlap with each other. The average RMS differences are under three pixels for scanned maps and under four for the digital map, which shows that the thinning operator and our approach to correct the distortion results in good quality road geometry. Table 3 shows our approach achieved better results than R2V.[2] The lower completeness of R2V is because R2V did not automatically connect broken road pixels. The lower correctness and high redundancy of R2V is because R2V generated small line segments instead of long and smooth lines and did not generate accurate road lines near the intersections.

For the computation time, we built our test system using Microsoft Visual Studio 2008 running on a Windows XP Professional Virtual Machine installed on 2.4 GHz Intel Core 2 machine with one GB of memory. The average processing time for vectorizing the road pixels for a scanned map tile (800x600 pixels) is 13 seconds. The dominant factors of the computation time are the image size, the number of road pixels in the raster map, and the number of road intersections in the road layer.

---

[2] We used the "Auto Vectorize" function in R2V without manual post-processing.

6   Conclusion and Future Work

We present a general technique that extracts accurate road vector data from heterogeneous raster maps with minimal user input. We utilize our previous work [5] to handle raster maps with varying image quality and exploit the accurate road-intersection templates [4; 6] to prevent distorted extraction results. We show that our technique extracts accurate road vector data from three raster maps with varying color usage and image quality. In the future, we plan to test our approach on more maps from various sources and test to include post-processing on the road vector data to improve the results.

Acknowledgments

This research is based upon work supported in part by the University of Southern California under the Viterbi School Doctoral Fellowship, and in part by the United States Air Force under contract number FA9550-08-C-0010. The U.S. Government is authorized to reproduce and distribute reports for Governmental purposes notwithstanding any copyright annotation thereon. The views and conclusions contained herein are those of the authors and should not be interpreted as necessarily representing the official policies or endorsements, either expressed or implied, of any of the above organizations or any person connected with them.

References

[1] Bin, D., Cheong, W.K.: A system for automatic extraction of road network from maps. In: Proceedings of the IEEE International Joint Symposia on Intelligence and Systems, pp. 359–366 (1998)

[2] Cao, R., Tan, C.L.: Text/graphics separation in maps. In: Blostein, D., Kwon, Y.-B. (eds.) GREC 2001. LNCS, vol. 2390, pp. 167–177. Springer, Heidelberg (2002)

[3] Chen, C.-C., Knoblock, C.A., Shahabi, C.: Automatically and accurately conflating raster maps with orthoimagery. GeoInformatica 12(3), 377–410 (2008)

[4] Chiang, Y.-Y., Knoblock, C.A.: Automatic extraction of road intersection position, connectivity, and orientations from raster maps. In: Proceedings of the 16th ACM GIS, pp. 1–10 (2008)

[5] Chiang, Y.-Y., Knoblock, C.A.: A method for automatically extracting road layers from raster maps. In: Proceedings of the Tenth ICDAR (2009)

[6] Chiang, Y.-Y., Knoblock, C.A., Shahabi, C., Chen, C.-C.: Automatic and accurate extraction of road intersections from raster maps. GeoInformatica 13(2), 121–157 (2008)

[7] Comaniciu, D., Meer, P.: Mean shift: a robust approach toward feature space analysis. IEEE Transactions on PAMI 24(5), 603–619 (2002)

[8] Habib, A., Uebbing, R., Asmamaw, A.: Automatic extraction of road intersections from raster maps. Project Report, Center for Mapping, The Ohio State University (1999)

[9] Heipke, C., Mayer, H., Wiedemann, C., Jamet, O.: Evaluation of automatic road extraction. In: International Archives of Photogrammetry and Remote Sensing, pp. 47–56 (1997)

[10] Henderson, T.C., Linton, T., Potupchik, S., Ostanin, A.: Automatic segmentation of semantic classes in raster map images. In: The Eighth GREC Workshop (2009)

[11] Itonaga, W., Matsuda, I., Yoneyama, N., Ito, S.: Automatic extraction of road networks from map images. Electronics and Communications in Japan (Part II: Electronics) 86(4), 62–72 (2003)

[12] Lacroix, V.: Automatic palette identification of colored graphics. In: The Eighth GREC Workshop (2009)

[13] Leyk, S., Boesch, R.: Colors of the past: color image segmentation in historical topographic maps based on homogeneity. GeoInformatica 14(1), 1–21 (2010)

[14] Li, L., Nagy, G., Samal, A., Seth, S.C., Xu, Y.: Integrated text and line-art extraction from a topographic map. IJDAR 2(4), 177–185 (2000)

[15] Shi, J., Tomasi, C.: Good features to track. In: Proceedings of the IEEE Conference on Computer Vision and Pattern Recognition, pp. 593–600 (1994)

Human Perception in Segmentation of Sketches

Pedro Company, Peter A.C. Varley, Ana Piquer, Margarita Vergara,
and Jaime Sánchez-Rubio

Department of Mechanical Engineering and Construction, Universitat Jaume I,
12071 Castellon, Spain
{pcompany,varley,Ana.Piquer,vergara,al013546}@uji.es

Abstract. In this paper, we study the segmentation of sketched engineering drawings into a set of straight and curved segments. Our immediate objective is to produce a benchmarking method for segmentation algorithms. The criterion is to minimise the differences between what the algorithm detects and what human beings perceive. We have created a set of sketched drawings and have asked people to segment them. By analysis of the produced segmentations, we have obtained the number and locations of the segmentation points which people perceive. Evidence collected during our experiments supports useful hypotheses, for example that not all kinds of segmentation points are equally difficult to perceive. The resulting methodology can be repeated with other drawings to obtain a set of sketches and segmentation data which could be used as a benchmark for segmentation algorithms, to evaluate their capability to emulate human perception of sketches.

Keywords: Sketch recognition, Low level ink processing and pen stroke segmentation, Engineering Graphics, Segmentation Ability.

1 Presentation

Our interest is computer-based recognition of sketched engineering drawings, such as would allow automated conversion of engineering sketches into CAD representations. Segmentation of the drawing is a critical stage, and one which has received much attention over the years. Some important aspects of segmentation still remain unsolved, perhaps because (as [1] shows), segmentation is not, in fact, a single problem, but a set of similar problems. In this paper, we consider one such unsolved aspect: the benchmarking of computer-based segmentation of sketches.

Recognition of the object portrayed in engineering drawings is a topical subfield of graphics recognition, which deals more generally with how computers can interpret semi-structured drawings which contain both freeform elements and symbols defined by convention. In the case of creating 2D or 3D CAD models of engineering objects from single or multiple drawings, it is the freeform elements, lines and curves, which portray the surfaces of the object, and it is these which we wish to identify. At this stage of processing, the conventional symbols (like dimensions and hatching) are *clutter*, and should be removed (and perhaps stored for later use). Our objective is thus to segment engineering drawings into *lines*, *curves* and *clutter*.

J.-M. Ogier, W. Liu, and J. Lladós (Eds.): GREC 2009, LNCS 6020, pp. 106–117, 2010.
© Springer-Verlag Berlin Heidelberg 2010

When evaluating new segmentation approaches, one common strategy is simply comparing the number of segmentation points obtained by the new approach with the number of segmentation points which the "theoretical" shape possesses (by "theoretical" we mean the ideal primitives obtained from a line drawing by applying a well-defined set of topological and geometrical constraints). This strategy assumes that the new approach should detect those properties which the theoretical shape should possess, regardless of whether or not the actual drawing used as input really does possess them.

In reality, we cannot assume that a sketched line drawing on paper will always contain exactly the same number and type of segments as the "perfect" line drawing which existed only in the mind's eye of the drawing's creator. The total number of segments may vary, both because of imperfections in the sketch itself and because of differences between geometrical and perceptual interpretation of sketches (such as the well-known perceptual illusions described by Hoffmann [2] or Palmer [3]). The types of perceived segments may also vary: for example, a sketched arc of large radius may be perceived as a straight line.

Another common strategic deficiency is not paying attention to the locations of the segmentation points. As a result, a new approach may be considered good simply because it finds the ideal number of segmentation points, even though their actual locations are far from ideal (see, for example, [4]).

If we are to evaluate a sketch recognition algorithm realistically, we should compare the differences between what the algorithm detects and what human beings perceive when parsing the same sketch. The comparison should consider not merely "how many?" but also "how close?". We must also bear in mind that perhaps not all segmentation points are equally difficult to find. In such case, recognising many "easy" segmentation points should not be considered as a measure of success.

To this end, we have performed experiments aimed at discovering: which segmentation points people perceive; where the segmentation points are located; and what geometrical flexibility in the locations of segmentation points can be tolerated.

The paper is organised as follows. We first explain our motivation and hypothesis. Then we describe the design of the experiment so that the procedure we have developed may be used by other researchers to obtain segmentations of different sketches. In the subsequent section, we analyse our results and how they validate, modify or refute our hypothesis. The paper finishes with lessons learned and main conclusions.

2 Motivation

Most sketch-based modelling approaches need line drawings as input for the model reconstruction stage. Freehand sketches must be converted into "tidy" line drawings [5]. The two main problems of this process are segmentation and overtracing. Overtracing is the use of multiple strokes to represent a single line. Readers interested in this topic can find a recent contribution by Ku et al. [6]. Segmentation is the process of dividing a complex stroke into its geometrical primitives. Segmentation of sketches is an open problem in the process of sketch recognition. One recent contribution can be found in [4].

We can note, in passing, that even segmentation of line drawings remains an unsolved problem. For example, arc segmentation is a classical process related to vectorisation and line drawing interpretation. Starting in 2001, the GREC workshops (organized by IAPR) have included contests focused on arc segmentation. Those contests test the abilities of participating algorithms to detect arcs from raster images.

Segmentation of freehand sketches presents further difficulties due to the inherent imperfections of such sketches. E.g., it is often difficult to determine whether small variations from perfect geometry in the sketch are intentional, and should be detected during segmentation, or are simply the accidental consequence of hasty drawing.

Most of the approaches described in the literature ([5], [7], [8], [9]) attempt to solve this problem by requiring the user to provide additional information. However, humans are able to segment sketches without requiring such extra information. It is reasonable to foresee, and prepare for, the day when advances in cognitive science result in automated approaches which come close to matching human performance. When they do, we shall require benchmarking criteria to evaluate such approaches.

2.1 Hypothesis

Our initial hypothesis was that four different aspects affect the segmentation process done by humans:

- Input quality. We hypothesise that sketches can be roughly graded as *good*, *average* or *bad*. Given *good* sketches, everybody will find the same segmentations (with, perhaps, meaningless differences). Given *bad* drawings, humans will not reach a consensus on how to interpret them. Thus neither *good* drawings *nor* bad drawings are appropriate for benchmarking. Only in *average* drawings will there be some obvious segmentation points upon which everyone will agree, but other segmentation points upon which opinions diverge.
- Other lines. We hypothesise that some auxiliary lines (e.g. axes and dimensions) will help people to find the best segmentation, while others (e.g. grids) will disturb them. Perhaps, some lines will be neutral (e.g. hatching).
- Clutter. We hypothesise that clutter (including auxiliary lines) will disturb people much less than it currently disturbs computer segmentation algorithms.
- 2D versus 3D. We hypothesise that drawings of two dimensional shapes are easier to segment, as segmentation is not mixed with other problems. People perceive the image as "flat" and try to find its segmentation points without first trying to create a mind's eye 3D model of the object portrayed in the image. However, the segmentations they produce after perceiving 3D shapes are more constrained, as they may never contradict their mind's eye model.

Another aspect of this problem is whether we should use *natural* or *wireframe* drawings. From the strict point of view of segmentation, this should make no difference, but if we assume that perception of 3D and perception of segmentation affect one another, then we should test the two modes separately. Ideally, we should produce test drawings in both styles.

However, in fixing the limits of our current research, we decided that this initial investigation will consider neither wireframe drawings nor drawings containing representations of "scenes" (assemblies of several parts designed to function together). We limit our study to natural drawings depicting single parts.

3   Design of the Experiment

Since our experiments are aimed at finding how humans segment sketched drawings, the core of our experiment is of necessity to produce a set of drawings and to ask people to segment them.

In order to investigate our hypothesis given above, we distinguish three types of drawings:

- Single orthographic views. These are not used as input in any existing sketch-based modeling application, but they nevertheless constitute a segmentation problem. They have the advantage of simplicity, and are useful for detecting very bad segmentation strategies and/ or approaches.
- Multiple orthographic views. This is the input format used in some existing Sketch-Based Interfaces and Modelling (SBIM) systems. For example, we can hypothesise that segmentation strategies which combine the views and analyse the resulting 3D shape will be more successful than those which simply scrutinise the separate views.
- Axonometric or perspective views. This is the input format used in most existing SBIM systems and includes several segmentations point types which can not be found in single orthographic views.

Consequently, three different experiments are required. Each experiment consists of three main stages: (a) production of sketches, (b) segmentation and (c) measurement.

3.1   Production of Sketches

As discussed above, we require sketches which meet the following criteria:

- the sketch must not be too simple: if segmentation is easy, any reasonable approach will process it correctly, and the benchmark is meaningless,
- the sketch must not be too complex: if the majority of humans cannot agree on an interpretation, there is no "human performance" to be duplicated,
- the sketch must not be perfect: we are interested in the human ability to interpret freehand sketches, not in the application of simple geometrical rules,
- the sketch must not be too imperfect: we must be able to reach a consensus as to whether an imperfection is deliberate or accidental,
- the sketches must, as a set, contain examples of all of the common cases where curves meet planar faces (see, for example chapter 7 of Cooper's book [10])
- the sketches must be representative of real engineering drawings: to avoid the problem of "gaming the system", where an approach obtains high benchmark scores but does not perform well with a larger set of real drawings.

The production process was divided into two steps: (a) choosing the suitable drawings; and (b) obtaining versions of different quality.

To choose suitable drawings, we first reviewed figures from the literature and created our own large initial set of figures, using our experience as teachers of engineering graphics and researchers in the field of SBIM to select figures which met our criteria. By circulating them to all of the members of the research team for comment we obtained a reduced but diverse set. After some iterations of this step, we finally reduced the test set to five CAD drawings (see figures 1, 2 and 3).

To obtain versions of different quality, we asked other people from the research team to draw sketches reproducing the CAD drawings obtained in the previous step. All the sketches were drawn in standard sheets marked with a 15 x 15 cm square frame, in order to encourage the sketchers to draw sketches with similar sizes and proportions. The same frame was later useful as a reference system to measure the location of segmentation points. In order to evaluate the effects of input quality, the members of the research team evaluated the quality of the sketches and scored them from bad to good (fig. 1- 2). From the resulting set of drawings, we selected those we needed for the three experiments. Each volunteer segmenter was given only one of the three chain plate sketches and only one of the three pipe flange sketches.

For comparison purposes, the segmenters were also asked to segment line drawings of both the chain plate and the pipe flange. The line drawings were given to the segmenters only after they had finished segmenting the sketches, to avoid those images influencing their perception of the sketches.

In order to evaluate the influence of other lines, we compared the differences in perception of a drawing containing only edges, and the same drawing containing auxiliary lines (axis, hatching, dimensions, etc). For this test, we chose an *average* quality sketch of multiple orthographic views, and deleted auxiliary lines to obtain a "clean" version (fig 3, top). Half of the segmenters were asked to segment the original sketch, while the other half were asked to segment the "clean" version.

In order to evaluate understanding of axonometric views, *average* quality versions of the two selected drawings were given to the segmenters (fig. 3, bottom).

3.2 Segmentation

During the segmentation of the final set of sketches, each segmenter was asked to segment a small subset of the full set of sketches, in order to avoid wearying the subject. The figures assigned to each particular subject were chosen randomly, to avoid subjective grouping of similar or dissimilar figures.

We asked the segmenters to segment the sketches by marking the exact position of each segmentation points and indicating the type of each resulting segment. These instructions were refined in each experiment. More information about the instructions given to answer the test can be found in [11].

3.3 Measurements

For the first experiment, segmenters were chosen from different profiles: from 11 to 69 years old, males and females, and a variety of technical drawing knowledge acquired in different formal education levels, ranging from primary school to university professors.

The information contained in the tests was collated and recorded in: identification of the subject (sex and age), level of technical drawing knowledge, number of segmentation points marked, and (x,y) coordinate pairs of each segmentation point.

The process we followed to obtain the coordinates was: a) scan the image as a bitmap; b) import the image into a CAD application and align its origin and the horizontal axis with those of the coordinates of the CAD application; c) mark the locations of the segmentation points and save their coordinates.

Before storing the coordinates, we first had to decide which segmentation points they belonged to. To do this, we first analysed all the answers and produced templates containing the different segmentation points, using frequency and position as our two criteria. More details on measurements can be found in [11].

4 Analysis

Qualitative results for the "chain plate" experiment are shown in figure 1, where every segmentation point of the chain plate marked by any of the segmenters has been superimposed. Analysing the results of the experiment 1, we can conclude that our first hypothesis is valid, as quality of sketches has clearly influenced the perception of segmentation points.

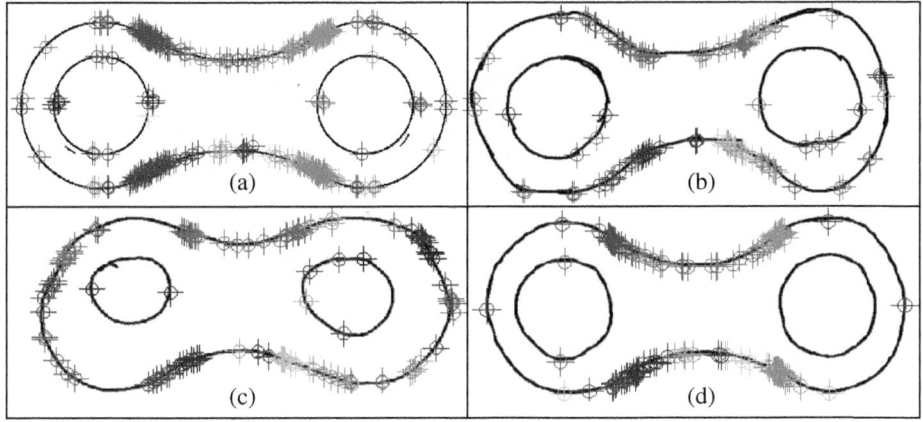

Fig. 1. Superimpositon of every segmentation point marked by any segmenter in the four classified drawings: (a) line drawing, (b) poor, (c) average and (d)good quality sketch

However, our results suggest that the grading (*poor, average* or *good*) which we gave to the three selected sketches does not always fit with the dispersion in the segmentation points found by the segmenters. The chain plate sketch considered as *average* (upper right in figure 1) was marked with more erroneous segmentation points that the sketch graded as *poor* (lower left). We conclude that some of the grading criteria we followed were wrong. For example, we assumed that overtracing makes a sketch more difficult to perceive, but this seems not to be so. On the other hand, greater the topology and geometry distortions appear to be distracting for the segmenters. More studies are needed to determine a method of grading the quality of sketches aimed particularly at the segmentation process.

We can note in passing that the exact line drawing (Figure 1 a) also led to in some erroneous segmentation points. This result appears to contradict our hypothesis that in *good* drawings everyone should perceive the same segmentation points. However, most of these erroneous points come from a misunderstanding: some segmenters said that they had thought that full circles were not arcs, and should thus be segmented, and decided to break the circles into two halves or four quadrants. It is interesting to note that, having segmented full circles, some of them propagated their segmentation points to the surrounding concentric external arcs, perhaps because these too encompassed more than 180°.

The same results can be confirmed through qualitative analysis of "pipe flange" segmentation as shown in figure 2.

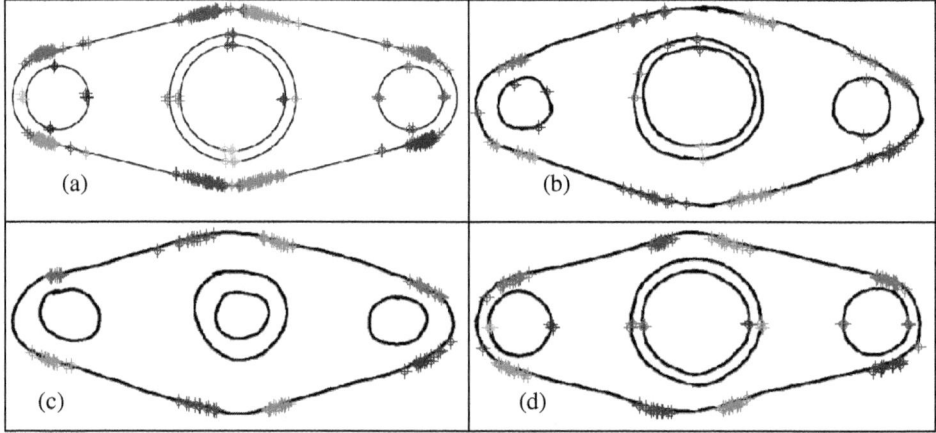

Fig. 2. Superimpositon of every segmentation point for the pipe flange drawing in the four classified drawings of it: (a) line drawing, (b) poor, (c) average and (d) good quality sketch

As can be seen in figure 3 (on the top), the second experiment clearly validates one aspect of our second and third hypotheses: no significant differences can be found between segmentations of the rocker arm with and without auxiliary lines. Only a few segmenters marked some intersections between edges and dimensions (e.g. the upper arrow of the diameter 45 dimension in the left side). However, one main question which remains unanswered is whether or not the prior perception of the 3D shape depicted in the drawing is important when uncoupling edges from the remaining lines. Another factor which could have contributed to the result is previous knowledge of the meaning of those symbols—all of the segmenters for this experiment had some exposure to technical drawing. Finally, more experiments should be required to determine the exact impact of noise in the segmentation process.

The results of our third experiment, fork and hinge segmentations shown in figure 3 (bottom), cannot be used to validate our fourth hypothesis. It certainly seems that segmentation points are as dispersed (and possibly *more* dispersed) in the flat

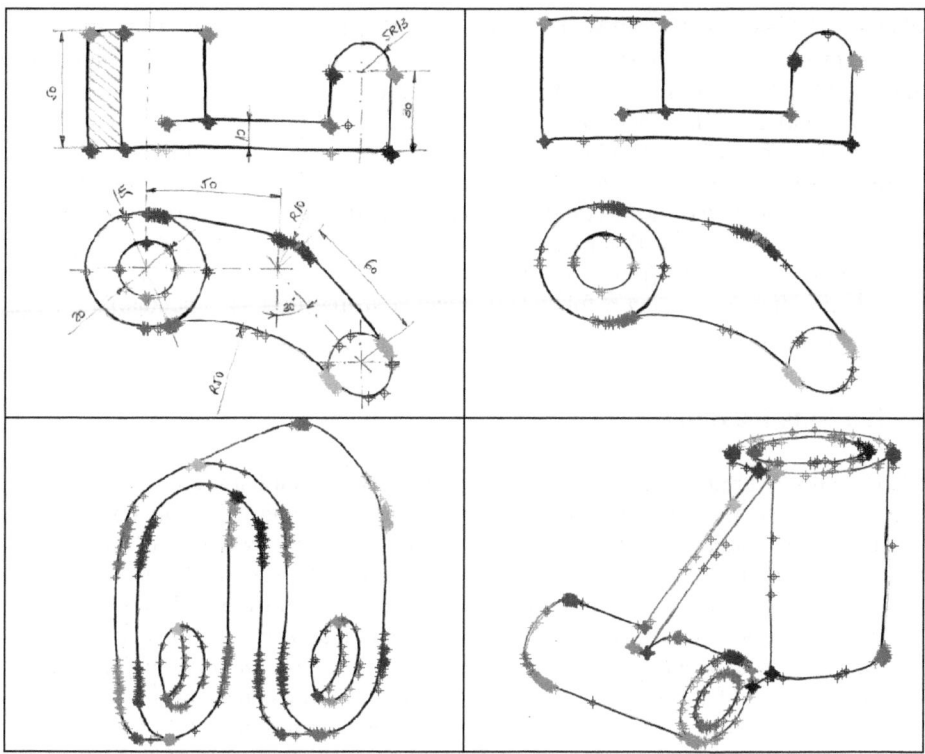

Fig. 3. Superimpositon of every segmentation point for the second (up) and the third (low) experiment

drawings of the first experiment as in the 3D shapes depicted in the second and third experiments. This would be what we would expect.

However, there are methodological problems with the third experiment which could "pollute" our results. Firstly, the instructions explicitly asked the segmenters to perceive a three dimensional shape for the second and third experiment. Secondly, while many of the segmenters in the first experiment had no technological background, all of the segmenters in the second and third experiments had had previous exposure to technical drawings. We believe that training and practice are more important for interpreting multi-view drawings than for axonometric drawings.

Thus, although we believe that, left to themselves, people would first perceive the 3D shape and then produce a segmentation influenced by this perception, we cannot as yet claim any experimental evidence to validate this belief.

We can, however, confirm one result which was partially observed in the previous experiments. Segmentation points located on those junctions where two or more straight segments meet are perceived by almost all segmenters, and the location of those points is very precise (dispersion is very low). Segmentation points located on junctions where more than two lines meet are also readily perceived, irrespective of whether the lines are straight segments or arcs. We regard this result as conclusive—no further studies are required about segmentation points located on junctions where three or more lines meet, or two straight lines meet.

Segmentation points located at tangential junctions of two arcs or one straight segment and one arc seem to be more difficult to perceive, as a significant part of the segmenters failed to mark them. The dispersion in the location of those points is high.

We have calculated the average positions of all the segmentation points (figure 4 left), and the average position of those points that are perceived by most than 50% of the interviewed subjects (figure 4 right). These images are indicative of the results we will obtain after processing a full set of sketches to be used as benchmarks for segmentation algorithms.

5 Lessons Learned

We have discovered small distortions in size and orientation between the paper sheets that we gave to the interviewed subjects and the electronic copies that we used to process the data after scanning the paper sheets. Although they have had no influence in the current qualitative analysis of results, this problem should be resolved before proceeding to a fine measurement of average location and tolerable deviations for those segmentation points where significant dispersion appears.

Although we detected some misunderstanding of the task due to ambiguities in the explanatory text of the first experiment and tried to correct them in the explanatory texts of the subsequent experiments, some misunderstandings nevertheless occurred.

We have to detect the origin of the misunderstandings and produce a clearer set of instructions in order to ensure that future segmenters understand clearly the task they are supposed to do.

For example, some segmenters included an excessive number of segmentation points. Although their answers are not statistically significant, one of their repeated comments is valuable. When they were asked why they had done so, their replies were as follows: I *perceive* what you intend to represent in the drawing, but, as you have asked me to segment what I can *see*, I have had to mark what I know that are actual imperfections due to mistakes in the sketching process, or even due to the printing process (i.e. serrations).

The comment raises the important distinction between what can be seen but should be ignored, and what is really important because it corresponds to the perceived purpose or message of the image. Obviously, humans are able to perceive the latter, and it is this complex ability which should be emulated by computer applications.

The procedure we followed for measuring the coordinates of the segmentation points is tedious and should be automated. Even more importantly, measuring Cartesian coordinates with reference to an external origin is not an ideal strategy. Firstly, they are statistically awkward to process. Secondly, (x,y) coordinates are particularly bad choice as they are paper-relative, not drawing-relative. If the results are scanned obliquely or at an offset, the (x,y) coordinates of the drawing itself change. What is needed is a single-parameter parameterisation of locations where a hand-drawn annotation intersects a pre-existing hand-drawn sketch.

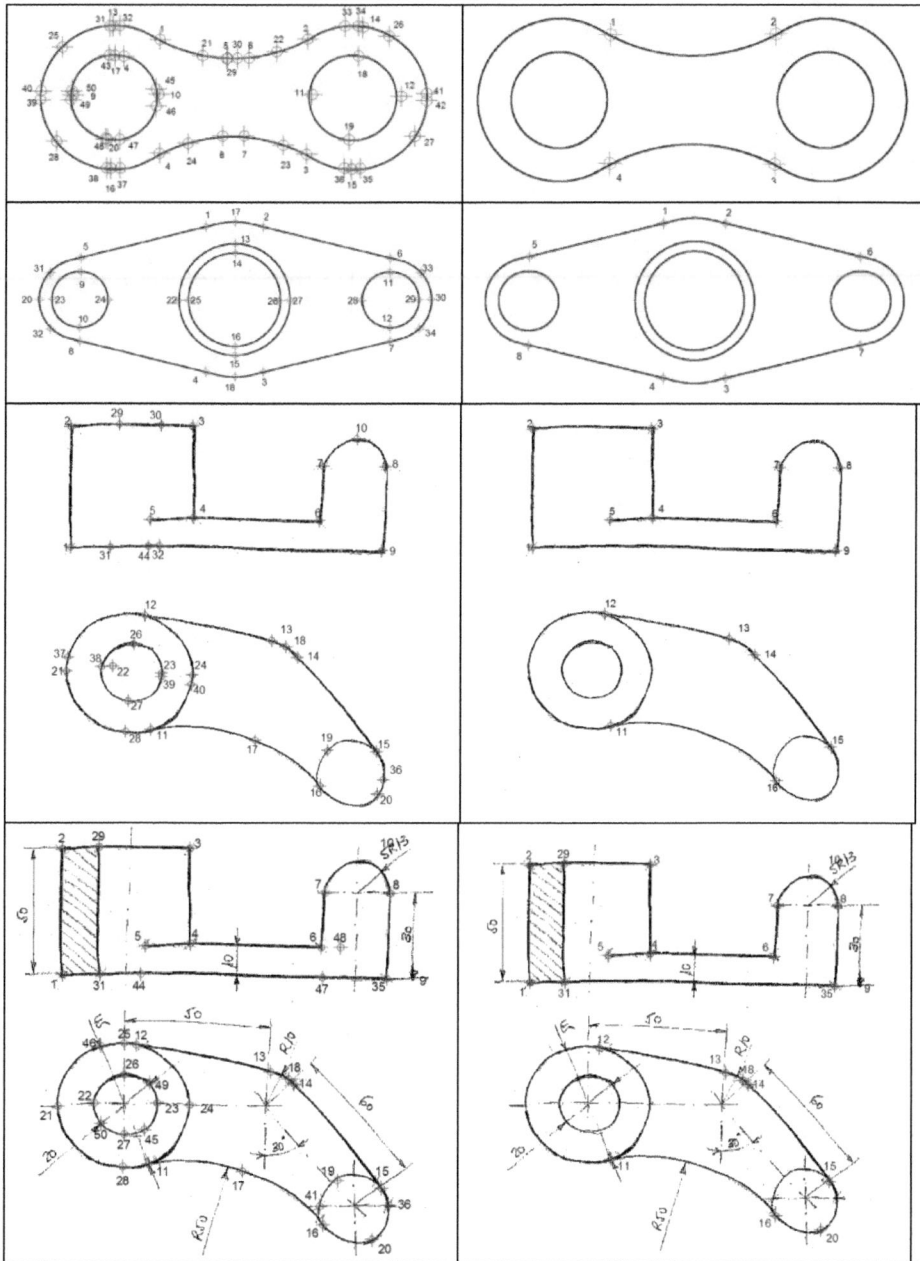

Fig. 4. Average locations of all segmentation points (left) and those perceived by more than 50% of the segmenters (right)

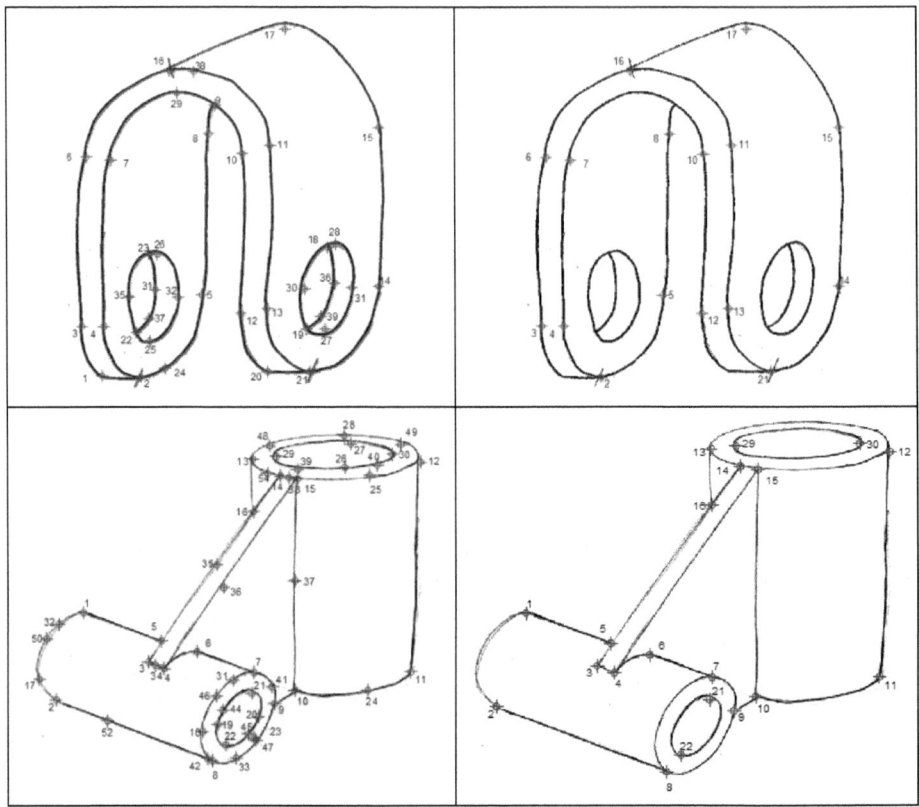

Fig. 4(cont). Average locations of all segmentation points (left) and those perceived by more than 50% of the segmenters (right)

The main requirement of the single-parameter parameterisation is that it must be object-relative: the coordinates must be relative to fixed features of the object. In future, we intend to fit the sketches to parametric curves (for example, by applying some variant of the approach described in [4]), and use natural coordinates for coordinates and statistical values.

6 Conclusion

We have defined and evaluated a procedure for obtaining a set of benchmark sketches that will be useful for evaluating the quality of segmentation approaches, with regard to their capability to emulate human perception of sketches.

The procedure includes criteria for selecting the drawings, and an approach for extracting and analysing the information.

In spite of its apparent simplicity, even examples like the fork have proved to be challenging, as they contain many segmentation points whose locations are difficult to fix. This indicates that the procedure followed when choosing the benchmarking drawings was appropriate. The evidence collected during our experiments supports the hypothesis that not all kinds of segmentation points are equally difficult to perceive. As a consequence, we should ensure that the final set of benchmarking sketches will contain a balanced set of different kinds of segmentation points, as it is important to consider levels of difficulty of segmentation points rather than merely their number.

The procedure should be refined to avoid the inconveniences described in Section 5 and a large set of drawings should be processed to obtain a benchmarking set with segmentation points statistically validated as being those which people perceive.

Acknowledgments. The Spanish Ministry of Science and Education and the European Union partially supported this work: DPI2007-66755-C02-01 (CUESKETCH: multi-agents based recognition of ideation sketches). The support of the Ramon y Cajal Scholarship Programme is also acknowledged with gratitude. We are also grateful to Jorge Domingo who helped us in processing the tests information.

References

1. Tombre, K.: Analysis of Engineering Drawings: State of the Art and Challenges. In: Chhabra, A.K., Tombre, K. (eds.) GREC 1997. LNCS, vol. 1389, pp. 257–264. Springer, Heidelberg (1998)
2. Hoffmann, D.: Visual Intelligence. How we create what we see. Norton Publishing, New York (1998)
3. Palmer, S.E.: Vision science. Photons to phenomenology. The MIT Press, Cambridge (1999)
4. Pu, J., Gur, D.: Automated Freehand Sketch Segmentation Using Radial Basis Functions. Computer-Aided Design 41(12), 857–864 (2009)
5. Jenkins, D.L., Martin, R.R.: Applying constraints to enforce users' intentions in free-hand 2-D sketches. Intelligent System Engineering 1(1), 31–49 (1992)
6. Ku, D.C., Qin, S.F., Wright, D.K.: Interpretation of Overtracing Freehand Sketching for Geometric Shapes. In: WSCG 2006 (2006)
7. Stahovich, T.F.: Segmentation of pen-strokes using pen speed. In: AAAI Fall Symposium - Technical Report FS-04-06, pp. 152–158 (2006)
8. Gennari, L., Kara, L.B., Stahovich, T.F., Shimada, K.: Combining geometry and domain knowledge to interpret hand-drawn diagrams. Computers and Graphics 29(4), 547–562 (2005)
9. Pu, J., Ramani, K.: Implicit geometric constraint detection in freehand sketches using relative shape histogram. In: Sketch-Based Interfaces and Modeling 2007 - ACM SIGGRAPH/Eurographics Symposium Proceedings, pp. 107–114 (2007)
10. Cooper, M.: Line Drawing Interpretation. Springer, Heidelberg (2008)
11. Company, P., Varley, P.A.C., Piquer, A.: Benchmarking for Computer-based Segmentation of Sketches. Technical Report Ref. 06/2010 Regeo. Geometric Reconstruction Group, http://www.regeo.uji.es

SSP: Sketching Slide Presentations, a Syntactic Approach

Joan Mas, Gemma Sanchez, and Josep Lladós

Computer Science Dept., Centre de Visio per Computador, Edifici O Campus UAB ,
Bellaterra, Spain
{jmas,gemma,josep}@cvc.uab.es

Abstract. The design of a slide presentation is a creative process. In this process first, humans visualize in their minds what they want to explain. Then, they have to be able to represent this knowledge in an understandable way. There exists a lot of commercial software that allows to create our own slide presentations but the creativity of the user is rather limited. In this article we present an application that allows the user to create and visualize a slide presentation from a sketch. A slide may be seen as a graphical document or a diagram where its elements are placed in a particular spatial arrangement. To describe and recognize slides a syntactic approach is proposed. This approach is based on an Adjacency Grammar and a parsing methodology to cope with this kind of grammars. The experimental evaluation shows the performance of our methodology from a qualitative and a quantitative point of view. Six different slides containing different number of symbols, from 4 to 7, have been given to the users and they have drawn them without restrictions in the order of the elements. The quantitative results give an idea on how suitable is our methodology to describe and recognize the different elements in a slide.

1 Introduction

The design of a slide presentation is a creative process where a user tries to express a certain knowledge in an understandable way. Commercial software exists allowing the design of slide presentations but the creativity of the user is rather limited. Normally, this software forces the user to use some predefined widgets and constrains the spatial arrangements. Most of the times the process of designing becomes an arduous task. On the other hand, most of the times when attending to a slide presentation the audience becomes bored. This fact is due to two factors, the slide presentation and the speaker. Concerning the speaker they normally read what is written on the slide and use a monotone speech without emphasizing the relevant parts. Concerning the presentation, slides use to contain excessive text or a lot of formulas where the speaker explains the minimum parameter of them. Nowadays a new trend of slide presentations has become. It is known as *zen presentations* [12]. This trend tries to design slides with the important concepts to show associating images to them. Summarizing the creativity in the design becomes a helpful point in a good way of doing a presentation.

In terms of an automatic interpretation task, two fields are involved in the design of slides, namely Graphics Recognition (GR) and Human Computer Interaction (HCI).

J.-M. Ogier, W. Liu, and J. Lladós (Eds.): GREC 2009, LNCS 6020, pp. 118–129, 2010.

From the point of view of Graphics Recognition, a slide can be seen as a graphical document where each of its widgets may be seen as a graphical symbol with an specific meaning. There exists wide variety of techniques to recognize the different widgets in a slide as the presented in [6,11]. Moreover, these widgets follow spatial arrangements. These facts make us to plan to use a syntactic approach to describe and recognize a slide. Then, the syntactic approach defines a vocabulary of symbols referring to the widgets in a slide and the parser, given a slide, will recognize the different elements of the slide. Taking up again the concept of creativity and viewing a slide as a graphical document, we may consider that the use of sketches to design a slide is a powerful tool. In the last decades the advent of new devices has increased the focus of graphic recognition in recognizing hand drawn symbols. Different techniques have been applied to work with this kind of symbols like syntactic approaches as the work of Mace and Anquetil [7]. They use a derivation of Constraint Multiset Grammars to interpret electrical circuits in an on-line input form. Statistical approaches have also been used, as Zernike Moments in the work of Hse and Newston [4] or 2D dynamic programming in the work of Feng et al. [3].

From the point of view of Human-Computer Interaction, due to the appearance of new devices such as Tablet PC's, PDA's and pen & paper protocols new paradigms of interaction have appeared as the proposed in [9,1]. In these paradigms the user interacts with the computer by means of a pen input. A pen input is composed by a set of *strokes*. A stroke is a set of points captured between two consecutive pen down and pen up users actions. A gesture is a stroke with an associated meaning involving a user action. Then, in these paradigms the user draws in a display or a board and waits for a computer response, depending on whether this response adapts to his/her intention or not, the user continues or edit it to achieve the user intention. The literature is prolific in sketch recognition applications from these point of view. Davis in [2] summarizes of the works done by the MIT during the last decade, DAI in [1] presents a new paradigm of Human Computer Interaction based on Pen Based User Interfaces (PUI).

The aim of our work is to develop an application to create a slide presentation raising the creativity of the user in the process. To do so the application recognizes the elements in the slide and produces the corresponding latex source. The recognition of the graphical part is done by means of a syntactic approach consisting in an Adjacency Grammar and a Directional Parser. The grammar allows to describe the different elements without constraining an order in the appearance of them. The parser analyzes the input while the user is drawing it. Analyzing 2D structures, at each new primitive the parser has to re-scan all the previous input to search for candidates to produce a valid reduction. To reduce the searching space a spatial division based on a grid has been defined in the parser. The results show in a quantitative way the performance of our application to create a slide presentation.

The rest of the article is organized as follows: next section describes the technical core of the application composed by the grammar and the parser. Section 3 is devoted to describe the application and to discuss its performance evaluation. Section 4 shows in a theoretically way how the user may edit his/her slides. Finally, section 5 presents the conclusions and possible future lines of the work.

2 Technical Core

This section is devoted to present the technical details behind the application presented in this paper. As we may conceive a slide as a diagram where the different elements composing it are described by means of a graphical alphabet of widgets and which follow spatial arrangements, we decided to use a syntactic approach to describe and recognize the different widgets of a slide. This syntactic approach is based on an adjacency grammar and a parsing algorithm.

2.1 Adjacency Grammars

Adjacency Grammars have been first introduced by Jorge in [5]. This kind of grammars belongs to the attributed multiset family, i.e, Adjacency Grammars express their productions as a multiset of terminal and non-terminal symbols. The fact they use a multiset instead of a list takes profit in the use of multiple instances of the same type of symbol and the order of appearance is not restricted. Among all the representative grammars in this family, Adjacency Grammars have been chosen because they are context-free in opposite to Constraint Multiset Grammars [8].

Formally our Adjacency Grammar is defined as a 6-tuple $G = \{V_T, V_N, S, R, A, P\}$ where:

- V_T is the finite set of terminal symbols.
- V_N is the finite set of non-terminal symbols.
- S is the initial symbol of the grammar.
- R is a set of predefined constraints.
- A is a finite set of attributes defined in the symbols of the grammar.
- P is the set of productions of the grammar of the form:

$$\alpha \rightarrow \{\beta_1, \ldots, \beta_j\} \; if \; r_1(\Gamma_1, c_1), \ldots, r_k(\Gamma_k, c_k) \tag{1}$$

with $\alpha \in V_N$, $\beta_1, \ldots, \beta_j \in V_T \bigcup V_N$ and for all $i = 1 \ldots k$, $r_i \in R$, $\Gamma_i \subseteq \{\beta_1, \ldots, \beta_j\}$ and c_i represents the distortion measure associated to each constraint r_i. The sum of all these measures gives an idea of the distortion from the ideal shape.

This kind of grammars has been adapted to cope with the slide presentation problem. The terminal alphabet of the grammar is depicted to cope with graphical languages and is formed by *segments* and *arcs*, i.e. $V_T = \{segment, arc\}$. Depending on whether a primitive is a segment or an arc its attributes will be the tuples (x_o, y_o) and (x_f, y_f) representing the endpoints of the segment or the triplet (x_c, y_c, r) describing the x and y coordinates of the center of the arc and its radius respectively. The different possible elements in the composition of a slide are described by the productions of the grammar, i.e. $V_N = \{text, equation, image, item, \ldots\}$. Figure 1 presents different productions of our grammatical approach, e.g., the production *Image* is composed by a terminal symbol arc and a non-terminal symbol triangle which has a vertex inside of the arc. Similarly, the production *Item* is defined as a non-terminal square and a segment that has a vertex inside the square.

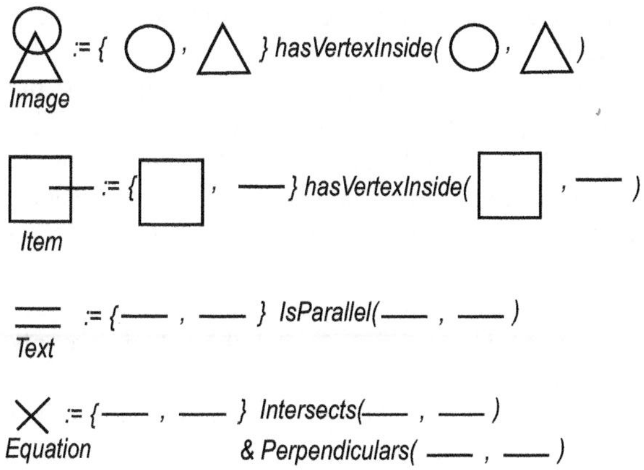

Fig. 1. Grammatical productions describing the visual alphabet of the SSP application

2.2 Parsing a Slide

A syntactic approach, as mentioned above, consists of a grammar and a parser or syntactic analyzer. The parser is the process that given an instance and a grammar, decides if the input belongs to the language generated by the grammar. As a result, the parser constructs a parse tree where the root is the start symbol of the grammar and the leafs correspond to the basic primitives.

A parser or syntactic analyzer is categorized by two features: the way the input is analyzed and how the parse tree is constructed. The former defines a parser as a directional or non-directional parser, while the latter describes it as a bottom-up or top-down. The syntactic analyzer presented in this work is a directional incremental bottom-up parser.

The parser takes the input symbol by symbol in an incremental way while the user is drawing it. At each step, it applies a reduction using the primitive being analyzed and the previous scanned symbols. Thus, the parse tree is constructed from the leafs (basic primitives) to the root (Start symbol of the grammar). At each level, we found all the possible non-terminals that the parser could reduce.

When analyzing an input primitive, the parser must decide which of the possible productions it will apply. There are two ways to choose these productions, Depth First Search and Breadth First Search. The former entails more complexity since the parser selects a production and tries to apply it until no valid reduction is possible. Then, the parser has to go back to a previous step and re-trace the parse tree. The second, opens all the possible productions simultaneously, eliminating those that belong to invalid ones when analyzing a basic primitive. The parser presented in this works proceeds in a Breadth First Search.

The analysis of two dimensional inputs, at the contrary of linear ones, implies that the parser could not assume that the primitive being analyzed forms a symbol with the previous analyzed ones. Thus, when analyzing a primitive, the parser must search among the previous analyzed symbols for candidates to produce a valid reduction. This search entails more time consuming as the input increases it size, and the possibility of reduce a production with primitives that are so far away to be considered as a part of the same symbol. To solve this problem we have proposed an indexing structure based on a spatial tessellation.

Let $C = \{C_1, \ldots, C_n\}$ be the set of cells obtained by dividing the space into n parts and let O be a symbol representing a graphical element in a diagram. We define as $CO(O) \mid CO(O) \subseteq C$ the set of cells containing the symbol and $NC(O) \mid NC(O) \subset C$ represents the set of neighbouring cells for each of the cells in CO. We define the influence zone Ψ of a symbol O as the union of the sets CO and NC, $\Psi(O) = CO(O) \bigcup NC(O)$.

Given a non-terminal symbol of the grammar α defined by a production $\alpha \rightarrow \{\beta_1, \ldots, \beta_j\}$, we denote as $F_\alpha \subseteq \{\beta_1, \ldots, \beta_j\}$ the set of symbols forming α which have been already reduced or analyzed, and as $M_\alpha \subset \{\beta_1, \ldots, \beta_j\}$ or $M_\alpha = \emptyset$, the set of symbols missing in α to be completely reduced. For a given symbol α, $F_\alpha \bigcup M_\alpha = \{\beta_1, \ldots, \beta_j\}$ and $F_\alpha \bigcap M_\alpha = \emptyset$. In the following we refer as *under reduction* symbols to non-terminal symbols where $M_\alpha \neq \emptyset$, and as *reduced* symbols to the non-terminal symbols with $M_\alpha = \emptyset$.

Algorithm 1 describes the parsing method. The parser works as follows: Given an input symbol $x \in V_T$, the parser allocates it into the grid and searches for candidates to produce a valid reduction in its influence zone. Depending on whether the candidate is an under reduction production or a reduced production the parser proceeds as follows:

- For each under reduction symbol α, the parser has to check:
  - If the symbol x is one of the remaining symbols of the under reduction one $x \in M_\alpha$ and satisfies the constraints in the production, then $\{x\} \bigcap F_\alpha \subseteq \{\beta_1, \ldots, \beta_j\}$. $M_\alpha = M_\alpha \setminus \{x\}$ and $F_\alpha = F_\alpha \bigcup \{x\}$.
  - The symbol x invalidates an under reduction symbol: $x \in M_\alpha$ but not satisfies the constraints in the production, or $x \notin M_\alpha$ and invalidates possible successive relations between the symbol α and other possible terminals.
- For each new symbol that x can generate:
  - The symbol x and its neighbours, NB, produce a valid partial reduction of a production creating an under reduction symbol α'. I.e., $\{x\} \bigcup NB \subset F_{\alpha'}$.
  - The symbol x and its neighbours, NB, produce a valid reduction of a production creating a reduced symbol α'. I.e., $\{x\} \bigcup NB = F_{\alpha'}$.

3   Experimental Framework

This section presents an overview of the application developed to create slide presentations and the experimental part in terms of graphics recognition and the distortion measure.

3.1   Application Overview

In this work we present an application to create slide presentations. We have developed a sketch based application where the user interacts with the computer drawing the different elements that are present in a slide. The aim of the application, based on a

---

Algorithm 1. Incremental parsing algorithm.

Require: $C = \{C_1, \ldots, C_n\}$ be a space tessellation.
   x a terminal symbol in the input.
   O' a non-terminal symbol of the grammar.
   $SI = \{Y_1, \ldots, Y_m\}$ be a set of previous parsed symbols.
   Parse(x)
   for all $Y_j \in SI | j = 1, \ldots, m$ & $CO(Y_j) \bigcap \Psi(x) \neq \emptyset$ **do**
      if $Finished(Y_j)$ **then**
         Find productions P' with x and Y_j
         for all $p \in P'$ **do**
            $O' = Validate(p, x, Y_j)$
            insert O' into $NC(O')$
         end for
      else
         $O' = Finalize(Y_j, x)$
      end if
   end for

---

visual vocabulary, is to produce the latex source and finally, the slide presentation that it represents. The application follows the architecture of Fig. 2. A user draws a stroke and it is pre-processed by the application. Once the stroke is pre-processed, the application sends it to the parser that tries to recognize it in terms of the grammar productions and returns to the application the possible object. Once the user has completely drawn the slide it may generate whole presentation. The application sends the corresponding source latex code to the latex which, with the help of the prosper class, generates the slides. Figure 3 presents a snapshot of our application. As we can see, the application consists of a drawing area and a set of buttons that allow to the user to insert a new slide in the presentation, go forward and backward in the created slides, preview the current slide and run the whole presentation.

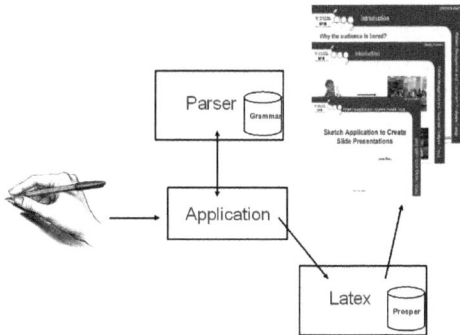

Fig. 2. Schema of the SSP application

Fig. 3. Snapshot of the SSP application

3.2   Experimental Evaluation

The experimental part of this paper is driven to show the performance of our methodology in a quantitative way. To achieve that six different slides containing different number of widgets, from 4 to 7, has been presented to 7 different people to draw it. The components of the slide can be drawn in an unrestricted order. Figure 5 presents in a graphical way the six slides given to the users in this experiment. To evaluate this experiment in a quantitative way let us define the concept of *ideal parse tree*. The ideal parse tree can be defined as the parse tree from an instance of the slides in Fig. 5 without containing any kind of distortion, i.e., the ideal parse tree of Fig. 5(a) is shown in Fig. 4. As we may observe at the bottom part of the tree we found the different primitives compounding the slide and at each level of the tree we find the different elements: two text, one equation and one image. To reduce the size of the tree width we have renamed the primitive segment to *Seg*. The number in the primitives, Seg, $Arcs$, describes the temporal order of them.

Figure 6 presents the percentiles of the distortion of the different elements compounding the slides. The values in the y-axis represents the distortion measure computed during the parsing. This measure value comes from 0 to 1, being 0 an element with high distortion and 1 an element near the ideal shape. The box represents the 50%

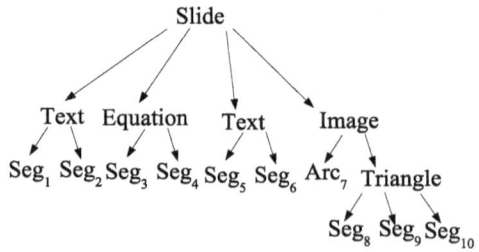

Fig. 4. Ideal Parse Tree Corresponding to the Slide of Fig. 5(a)

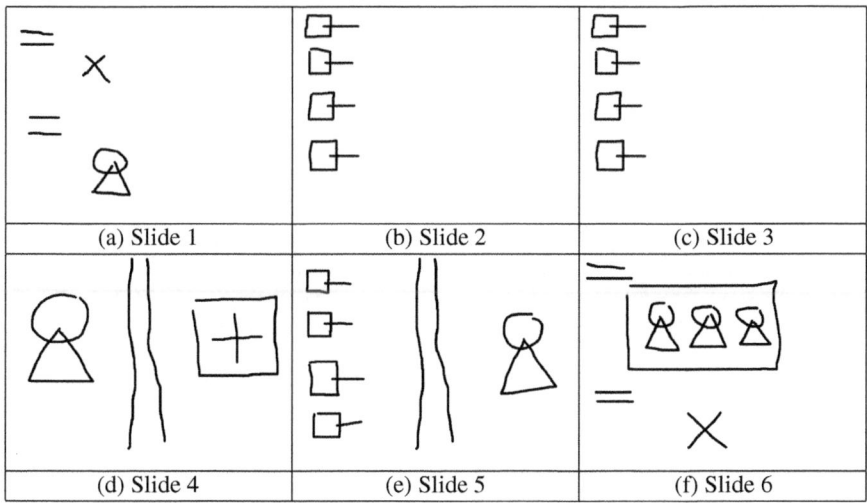

Fig. 5. Slide samples in the first experiment

of the values being the extremes of the box the 1st and 3rd percentile. The line inside the box represents the median value. Figure 6(c) shows the percentiles for the image element. This element is composed by an arc and a triangle. The triangle is a non-terminal composed by three segments. As it is sketched there are three outliers in this element but with a deviation of 0.2 at the most. The element that presents major distortion in its instances is the $item$ element, see Fig. 6(b), the distortion measure comes from 0.53 to 0.93. The IQR rate is of 0.15 between the values 0.62 and 0.77 we found the 50% of the instances. The distribution is not symmetric, but it is negative asymmetric, i.e., most of the values in the distribution are below the median. The elements of Text, Fig. 6(a), and Minipage, Fig. 6(e) are the elements with the small distortion in their instances. To conclude, as the symbol contains more primitives it increases the distortion, see Fig. 6(b) which the symbol is represented by a $QUAD$ and a $line$. The $line$ contains a vertex inside of the $QUAD$. $Text$ and $Minpage$ symbol contains a low distortion rate since they are described by two parallel $lines$ in different orientation for each symbol.

Figure 7 presents the percentiles of the recognition rate for each of the slides in the experiment. To estimate a recognition rate we propose a measure based in the number of elements in a instance parse tree that appears in the ideal parse tree. The values in the y-axis represent the recognition rate for each slide. In the x-axis the numbers coincide with the slides of Fig. 5. As we may observe the slides 1 and 3, Fig. 5(a) and (c) respectively, are the slides containing the best recognition rates while slide 6, Fig. 5(f) presents the worst results. This is due to the parser does not recognizes the symbol $imatges$ in most of the instances of this slide. This symbol is composed by a square containing two or more images. The grammatical production describing the symbol is as follows:

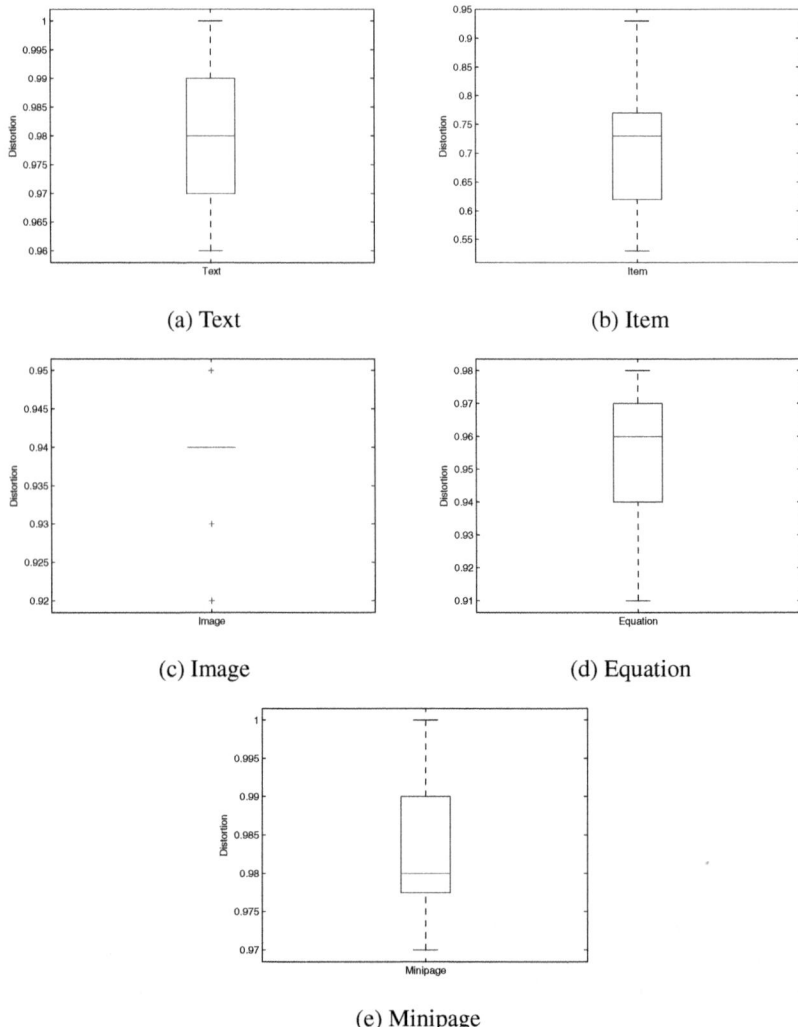

(a) Text   (b) Item

(c) Image   (d) Equation

(e) Minipage

Fig. 6. Percentiles of the distortion of the elements in the experiments: (a) Text , (b) Item, (c) Image, (d) Equation and (e) Minipage

```
Imatges --> {Square, Image1, Image2} IsInside(Square,Image1) &
IsInside(Square, Image2)

Imatges --> {Imatges, Image1} IsInside(Imatges,Image1)
```

In most of the instances the square is not recognized. The worst recognition rate belongs to slide 2 with a value of 0.25%, it is considered as an outlier in the distribution of Fig. 7(b). The median of the recognition result is over the 0.7% showing a good performance of our method to the application domain where it is used. Diverse errors have been occurred during the experiment. We may classify them into three types:

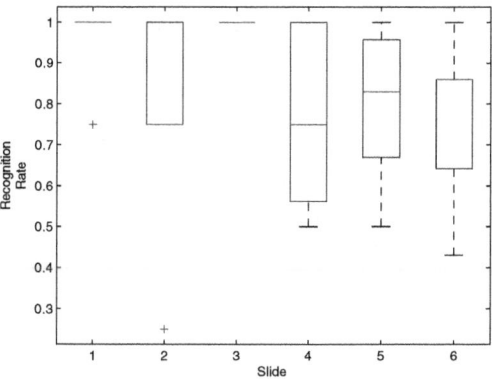

Fig. 7. Percentiles representing the recognition rate for each of the slides in Fig. 5

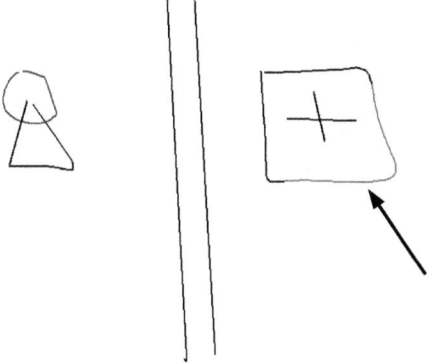

Fig. 8. A Slide containing an error of Type 1

- Type 1: Miss-classification of basic primitives.
- Type 2: Additional primitives.
- Type 3: High distortion in the constraints between the primitives of a symbol.

Figure 8 presents an instance of the Slide 4 of Fig. 5. This instance contains an error of type 1. The two segments marked with arrow are recognized as an arc. The parser has recognized two *OverlayedSlide* symbols rather than recognizing a *QUAD* which is the corresponding one.

4 User Interaction: Adding the Capability of Editing the Slides

This section describes how the user is able to edit a slide. First of all, this part of the work is in an embryonic state. The main idea is to present a set of gestures that allow to the user edit their slides such as, delete, copy, move, etc. In this case, rotation is not necessary due to the symbols will be placed always in the same orientation. The user has to change to edition mode to be able to edit the slides.

The main drawback of the user interaction becomes the deletion of part of the slide. In this case, we have to take into account the parse tree and the grid in this state. Previously to the deletion we have to save the state of the parser and the grid in order to recover in a future time if it is necessary. Then, the parse tree and the grid should be modified deleting those elements forming the symbol or symbols the user wants to delete. This action may leave the slide in an inconsistent state. In that case, the slide generator should inform to the user that the slide contains erroneous symbols.

5   Conclusions

In this work we have presented an application to create a slide application from the user sketches. To describe and recognize the different elements of a slide a syntactic approach has been presented. The syntactic approach is based on Adjacency Grammars and a directional parser improved with a spatial tessellation. The grid helps the parser to reduce the search space when a new primitive is being analyzed. The results show the performance of the proposed approach from a quantitative point of view. Focusing in the values of recognition rate and distortion rate we may conclude that our method solves the problem with a high accuracy level. Nevertheless, the addition of the user into the loop of recognition will increase the performance of the parser. Adding some gestures, the user is able to modify the bad recognized symbols. The future work can be devoted to improve the application to allow to the user to create its own style of slides.

Acknowledgements

This work has been partially supported by the Spanish projects TIN2008-04998, TIN2009-14633-C03-03 and CONSOLIDER - INGENIO 2010 (CSD2007-00018).

References

1. Dai, G., Wang, H.: Physical Objects Icons Buttons Gestures (PIBG): A New Iteraction Paradigm with Pen. In: Shen, W.-m., Lin, Z., Barthès, J.-P.A., Li, T.-Q. (eds.) CSCWD 2004. LNCS, vol. 3168, pp. 11–20. Springer, Heidelberg (2005)
2. Davis, R.: Magic paper: sketch understanding research. Computer 40(9), 34–41 (2007)
3. Feng, G., Viard-Gaudin, C., Sun, Z.: On-line Hand-drawn Electric Circuit Diagram Recognition using 2D Dynamic Programming. Pattern Recognition, doi:10.1016/j.patcog.2009.01.031
4. Hse, H., Newton, A.R.: Sketched Symbol Recognition using Zernike Moments. In: Proc. of the 17th International Conference on Pattern Recognition, ICPR 2004, Cambridge, vol. 1, pp. 367–370 (2004)
5. Jorge, J.A.P., Glinert, E.P.: Online Parsing of Visual Languages Using Adjacency Grammars. In: Proc. of the 11th Intl. IEEE Symposium on Visual Languages, Darmstadt, pp. 250–277 (1995)

6. Llados, J., Valveny, E., Sanchez, G., Marti, E.: Symbol Recognition: Current Advances and Perspectives. In: Blostein, D., Kwon, Y.-B. (eds.) GREC 2001. LNCS, vol. 2390, pp. 104–127. Springer, Heidelberg (2002)
7. Mace, S., Anquetil, E.: Eager Interpretation of On-line Hand-drawn Structured Documents. Pattern Recognition, doi:10.1016/j.patcog.2008.10.018
8. Marriot, K.: Constraint multiset grammars. In: IEEE Symposium on Visual Languages, St. Louis, pp. 118–125 (1994)
9. Narayanan, N., Hübscher, R.: Visual Language Theory: Towards a Human Computer Interaction Perspective. In: Visual Language Theory, pp. 85–127 (1998)
10. Weitzman, L., Wittenburg, K.: Automatic presentation of multimedia documents using relational grammars. In: Proc. of the 2nd ACM Intl. Conf. on Multimedia, San Francisco, pp. 443–451 (1994)
11. Liu, W.: On-Line graphics recognition: state-of-the-art. In: Lladós, J., Kwon, Y.-B. (eds.) GREC 2003. LNCS, vol. 3088, pp. 291–304. Springer, Heidelberg (2004)
12. http://www.presentationzen.com

QuickDiagram: A System for Online Sketching and Understanding of Diagrams

Liu Wenyin[1,2], Xiangfei Kong[2], Yiming Wang[1], Chester Wan[2], Cheuk-Yin Ho[2], Tong Lu[1], and Zhengxing Sun[1]

[1] State Key Lab. for Novel Software Technology, Nanjing University,
Nanjing, 210093, China
[2] Department of Computer Science, City University of Hong Kong,
Hong Kong SAR, PR China
csliuwy@cityu.edu.hk, lutong@nju.edu.cn

Abstract. In this paper, a system named *QuickDiagram* is proposed for quick diagram input and understanding. With a user sketching a (complete or partial) component/symbol or a wire (connecting two components) of the diagram, the system can recognize and beautify it immediately. After the entire diagram is finished, certain understandings can be obtained. Especially, the following two methods are used to interpret the recognized diagram: 1) Nodal Analysis on resistive circuits, and 2) generation of PSpice codes from the recognized diagrams. Experiments on a few sketched circuit diagrams show that the results are robust and accurate for both recognition and understanding.

1 Introduction

Currently, most interactive graphical editing systems face difficulties on how to further improve the efficiency and reduce human burdens. Such a system has to ask a user to input graphical objects by a series of mouse/keyboard operations, making design ideas interruptive. Hence, as a customary and natural expression method for human beings, sketching with a pen is more preferable and is especially helpful and suitable for creative design. However, the freehand sketches are not elegant in appearance and not compact for representation or storage. Moreover, the sketchy objects are difficult for further automatic processing, including automatic understanding.

Real-time recognition of on-line sketches provides an effective approach to converting the freehand sketches to the regular ones, which can also be stored in a neat way. Moreover, the diagram input process can be greatly facilitated in a productive way since the recognized results can be used as instant feedbacks for real-time correction or positioning. In addition, if the recognized diagram is understood and analyzed immediately, the semantic errors and/or deficiency found from the diagram can be used to correct and improve the design.

In this paper, we present a new on-line system, *QuickDiagram*, to recognize the instant sketching input of electronic circuits and make certain analyses of the recognized circuits. When a user freely designs a circuit diagram by sketching it,

J.-M. Ogier, W. Liu, and J. Lladós (Eds.): GREC 2009, LNCS 6020, pp. 130–141, 2010.
© Springer-Verlag Berlin Heidelberg 2010

QuickDiagram can recognize all diagram components and connections immediately, and then perform real-time circuit analysis simultaneously. In this way, the efficiency of an interactive circuit design system can be greatly improved.

Before we explain our work in more detail, we first present related definitions. An electronic circuit diagram is comprised of the following three types of elements:

1) Device (or component): a meaningful electronic symbol, such as a voltage source, a resistor, a diode, an inductance or a transistor. An electronic device/component has at least one connection point (terminal), which is used to connect to another device.

2) Connection: the wire (usually in the form of a line segment or a polyline) used to connect two neighboring devices at their connection points. Each connection is between two connection points.

3) Node: a congregation point of connections, which can be at a connection point or at a merged point connecting with multiple connection points.

QuickDiagram consists of two main modules: sketch recognition and diagram understanding. The sketch recognition module first distinguishes connection strokes (with at least one end point touching a connection point) from symbol strokes. Connection strokes are segmented and beautified into regularized lines or polylines. Symbol strokes are segmented into regular segments (lines or arcs) and then grouped together and matched with the candidate model symbols (devices/components) in a predefined database. After the devices are recognized and connected, the understanding module analyzes the entire recognized circuit diagram and obtains its semantics with the guidance of domain knowledge. In this module, we use the following two approaches to automatically understand the recognized circuit diagrams: 1) for resistive circuit diagrams, we adopt Nodal Analysis [24] to find all the relevant equations about the voltage of the recognized devices and the current passing through the devices, then we collect all the equations and solve them by Gaussian elimination to obtain the unknowns; and 2) for the certain types of circuit diagrams, we generate PSpice codes, which can be read by PSpice for simulation.

The rest of this paper is organized as follows. Section 2 presents related work. Section 3 briefly presents the diagram recognition processes of our system. Section 4 presents the two understanding methods to interpret circuit diagrams. Experimental results are described in Section 5. Finally, Section 6 presents concluding remarks and future research.

2 Related Work

Research works on sketch understanding may date back to the birth of pen-based interaction, starting with Sutherland's "SketchPad" [30]. Due to limitations of the technology at that time, its functions and performance are far from what we call now "understanding" but only a method for human-computer interaction. In 1990', diagram-understanding has reached a new level in both researches and applications, with the evidence of electronic whiteboards mainly used for sketching during informal workgroup meetings, such as "Tivoli" [26] and "Flatland" [9]. In the last decade, quite a few sketch understanding systems have been developed and reported by Davis' group, who also presented a very good survey of the state-of-the-art research of sketching recognition and understanding [7]. In this section, we only briefly mention the main problems and some of their solutions: stroke processing, symbol recognition, and understanding based on domain knowledge.

(1) A user may want to draw only one primitive geometric shape (e.g., line or arc) or at most a connected sequence of shapes in one stroke. However, the stroke usually consists of many intermediate and redundant, and sometimes, even noisy points. These non-critical should be removed during stroke pre-processing and the entire stroke can be segmented and fit into a few primitive shapes, which can be either line segments, polylines, or arcs [28][5]. In addition to high curvature, low speed is also a good feature to locate segmentation points [7]. While most systems do primitive shape recognition after the user finishes a stroke, the stroke can be continuously morphed to the predicted primitive shape in some systems [4].

(2) Many symbol recognition methods have been reported in literature. In certain offline recognition systems statistic features are extracted [37] while some others use structural methods, which convert symbols to specific structures like trees [21] or graphs [34]. A comparison between these features and those of the model symbols in the database are computed, using certain algorithms such as the Hausdorff distance and Tanimoto coefficient [17]. Certain hybrid approaches may use statistics of structural features and thus lower down the computation [33][12]. Almost all offline methods can be applied to online recognition, but the later can be utilizes more information. For example, after one primitive shape is just recognized, it is combined with previously inputted components together (based on their spatial relationship) as one component to form a query for searching for similar objects in the symbol database [21]. All these approaches could be categorized as appearance-based methods. Other approaches include the "definition-based" method [27][14], which recognizes a symbol described in terms of its component shapes and their spatial relationship (geometric constraints) using a well-defined language, and the "drawn sequence-based" method, which recognizes shapes based on the orders of the strokes [29]. In addition to symbols previously known and described, recent studies can learn descriptions of new symbols for future recognition from their examples and recognition feedback [32][15].

(3) Researches on sketching understanding mainly focus on certain domain(s) where graphical lexicon, syntax, and semantics can be well-modelled, such as electronic engineering (circuits) [3], mechanic engineering [2], architectural engineering [20], and chemistry (understanding of chemical compounds [25]). ASSIST (which stands for A Shrewd Sketch Interpretation and Simulation Tool) [2] can understand mechanical engineering sketches and then interpret their mechanisms using simulation. ESQUISE [16] can capture and understand architectural sketches and then evaluate the energy needs of the project. Sketch-based 3D construction or understanding uses similar techniques at the 2D level of recognition and understanding. However, more techniques specific to the 3D level are required. Especially, Zeleznik et al. invent an interface to input 3D sketchy shapes by recognizing the predefined patterns of some 2D shapes that represent certain sketchy solid shapes [36], and Matsuda et al. present a freehand sketch system for 3D geometric modelling [22]. Other well-known application domains of sketch recognition and understanding include flowcharts [13], mathematic expressions [31], music notations [11], and UML diagrams [19].

In the context of sketching recognition and understanding for Spice, Narayanas-wamy developed an interface for SPICE which uses hard-coded recognizers [23]. However, it assumes a fixed drawing order in sketching and requires the user to pause between symbols. Gennari also developed such an interface for electric circuit analysis program called AC-SPAEC [12]. De silva et al. employed this interface as the core part in the development of their system – Kirchhoff's pen, a pen based circuit analysis tutor [8].

Applications for creative and conceptual design are among the most useful applications of sketching recognition and understanding. Although some works at this level have been reported in certain limited domains [20][2], currently, sketching understanding is still a challenging problem requiring domain knowledge and integration of multidisciplinary technologies, e.g., both graphics recognition and imaginal thinking research. More research works should be done such that reasoning and prediction of the user's intentions can be made from the sketches he/she draws in order to support and facilitate his/her conceptual design, e.g., in creative design tasks.

3 Instant Sketchy Diagram Recognition

After a sketchy stroke is drawn, it is first classified as either a connection stroke or a symbol stroke. If an end point of the stroke touches a connection point of any recognized symbol/device, it is a connection stroke. Otherwise, it is a symbol stroke. The connection strokes are segmented and beautified into regularized lines or polylines. The symbol strokes are segmented into regular segments (lines or arcs) by the stroke segmentation module. We developed a so-called Quick Penalty Based Dynamic Programming (QPBDP) [18] method for stroke segmentation. QPBDP is an extension of the dynamic programming framework with a customizable penalty function. It measures the correctness of splitting a stroke at a particular point. With the help of the penalty function, the proposed dynamic programming framework can finish the stroke segmentation process without any prior knowledge, like the number and/or the type of the segments contained in the sketchy stroke. Moreover, its response time is short enough for online applications, even for long strokes.

After each symbol stroke is input and segmented, we try to recognize the current symbol/device even before it is finished. We have developed quite a few algorithms for quickly filtering [33] and recognizing [37] complete or incomplete symbols [34][21]. Although the vectorial signature/descriptor [33] we proposed is mainly for rapid discrimination and filtration of symbols, it still preserves high recognition accuracy. This is due to the utilization of the complete set of the primitive-pair relationships in a symbol as the signature for the symbol. It can therefore act as a pre-processing step to reduce the computational load if a full recognition process is applied rather. After quick filtration, the remaining candidates can undergo a full recognition process with even higher recognition accuracy, e.g., the statistical method based on kernel densities of the symbols [37]. A syntactical/structural method [34][21] can also be used, which is usually based on matching of the tree/graph representations of the relationships among the symbol components. This kind of methods can recognize online sketching symbols of varied stroke orders and stroke numbers, or even before they are completely sketched. Due to page limit, we refer the readers to these papers for more details of symbol/device recognition and focus more on diagram understanding methods in this paper.

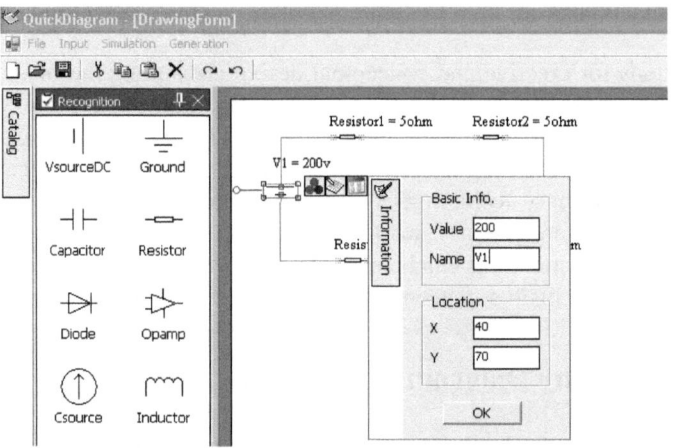

Fig. 1. An User Interface of QuickDiagram

4   Circuit Diagram Understanding

Currently, QuickDiagram can only understand certain types of circuit diagrams, in which each device is modelled in its regular form with its connection points also marked. With more devices/components modelled, QuickDiagram can be extended to understand other types of circuit diagrams and diagrams in other domains. After recognition, each device is assigned with a default name automatically numbered and displayed beside the device in the diagram. The default value of its parameters, e.g., R1=5 ohm, is also assigned and displayed next to its name. Its name and value can also be changed in its property dialog box, as show in the user interface of Figure 1. The connection points of the devices are also numbered such that each connection point in the entire diagram has a unique name. After all the individual devices and their connections are recognized, the entire diagram is displayed in a neat form prior to further analysis and understanding. In this paper, we focus on understanding of the recognized circuit diagrams and present two approaches: Nodal Analysis and generation of PSpice codes, based on semantics of the recognized devices, connections, and nodes of certain types of circuit diagrams. The key step of such understanding is the recognition of individual devices, the connections among them, and the connection nodes. After that, the semantics of each individual device, each connection, and each node can be applied and many kinds of understanding can then be obtained.

4.1 Semantics Modeling

The semantics and syntax of each device are the basis for circuit diagram understanding, just like the lexical model is the basis of the recognition of each device/symbol. Take a resistor named R_i as example, its semantics is actually an equation following Ohm's Law ($V_{i1} - V_{i2} = I_i * R_i$), where V_{i1} is the voltage of its first connection point, V_{i2} is the voltage of its second connection point, I_i is the current passing from the first connection point to the second connection, and R_i is its parameter value, resistance. For the example of a voltage source named V_j, its semantics is also an equation: $V_{j1} - V_{j2} = V_j$, where V_{j1} is the voltage of its first connection point, V_{j2} is the voltage of its second connection point, and V_j is its parameter value, the voltage source value.

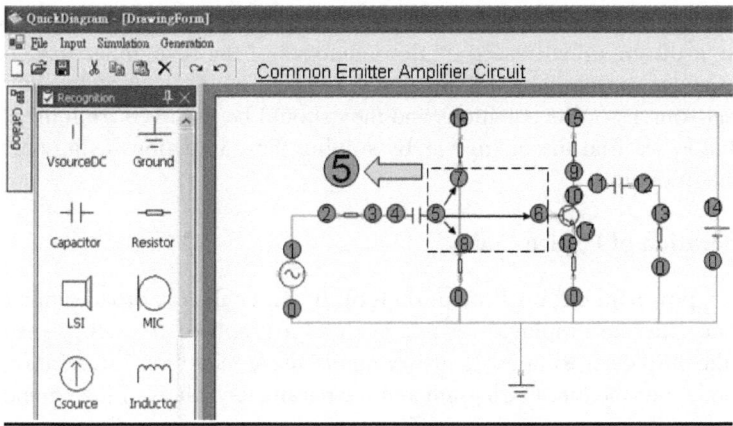

Fig. 2. How nodes are merged in a common emitter transistor amplifier circuit

The semantics of each connection is also an equation: $V_{k1} - V_{k2} = 0$, where V_{k1} and V_{k2} are the voltages of the two connection points at the two ends of the connection. The connection points of each connection can be merged into one node and multiple connection points connected together can be merged into one node as well. In fact, each connection point can be considered as a kind of node before merging. For each merge, the nodes before and after merging are kept in one merge record revealing the equivalence or aliasing relationships among them, as we can from Figure 2. For example, merged node 5 is the result of merging node 5, 6, 7 and 8 on the same connection and can then represent them in certain equations.

The semantics of each node following the Kirchoff's Current Law (KCL: The current flowing into the node is equal to the current flowing out of the node) is then applied to obtain an equation: $\sum I_n = 0$, where I_n is the current flowing from the node along each edge (branch).

Recognition of these individual devices, their connections and the merged nodes are the bases for full understanding of a circuit diagram, as we will show in both of the following methods.

4.2 Nodal Analysis for Resistive Circuits

Nodal Analysis is applied in pure resistive circuits to find the unknowns (i.e., the voltage of the devices and the current passing through the devices). In this paper, we choose the method of "Nodal Analysis" as an illustration of diagram understanding of QuickDiagram because it is a very basic method in circuits and the users can benefit a lot when they are teaching students or learning about KCL, Kirchhoff's Voltage Law, and Ohm's Law. In such a circuit diagram, we first have to declare a reference node, which can be any node in the circuit. The reference node is considered to have a voltage value zero and can be marked (or connected) with a ground symbol/device. In fact, the semantics of a ground device is exactly such. The voltage of any node is then the voltage difference between the node and the reference one. Next, we build up the equations about the voltage of the devices and the current passing through them based on the semantics of the recognized devices, connections, and nodes we mentioned in Section 4.1. There might be duplicate equations due to different sources of the equations and they should be removed from the set of equations. Finally, we find the unknowns by solving these equations using the method of Gaussian elimination.

4.3 Generation of PSpice Codes

PSpice is a powerful circuit simulation tool. It can analyze a broad range of circuits. It can also generate graphical results, such as a graph of frequency response. The code of the PSpice is simple. It just contains the connection information (via each merged nodes) of the circuit diagram and the parameter values of the components.

In the current prototype of QuickDiagram, we can generate PSpice codes for circuits consisting of the following 10 common types of devices: resistor, inductor, capacitor, DC voltage source, AC voltage source, current source, op-amp, diode, JFET and BJT. In the PSpice code, each line describes one device about its connection information and parameter information. Every device's code has its own format [6]. Generally speaking, the format is "<name> <node1> <node2> <node3> <value>", where the node should be a merged node shared with other devices. E.g., for resister, it could be "R1 1 3 600", which means R1 is connected to node 1 and 3 and its resistance value is 600. QuickDiagram combines all the lines of the code and save them into a PSpice (.cir file), which can be read by PSpice for circuit simulation. In addition, it can also display the text format in the output form of QuickDiagram so that the user can verify if the code is satisfying.

5 Experiments

We have done experiments on quite various types of circuits using the two understanding methods, which can generate accurate results. Due to space limit in this paper, we only show one example for each method. Readers can visit

http://www.cs.cityu.edu.hk/~liuwy/QuickDiagram/QD_Understanding_Experiments.htm   to
see other examples. The circuits QuickDiagram can recognize in our experiments
include series resistive circuits, parallel resistive circuits, series-parallel combination circuits, low pass filters, high pass filters, elliptic filters, common emitter
amplifier transistor circuits, NPN transistor circuits [1], and Op-amp saw tooth
oscillator [10]. The Nodal Analysis method we implemented in QuickDiagram can
analyze circuits constructed by resistors and voltage sources connected in both
series and parallel modes. The PSpice code generation approach can generate
PSpice code correctly and gives correct analysis results for the recognized circuits.

Figure 3 shows a resistive circuit containing 8 resistors connected in a series-
parallel combination circuit. Applying Nodal Analysis on it, QuickDiagram
can analyze and generate all the necessary equations for calculation of the
unknowns (the voltages of each components and the current passing through them in
the circuit).

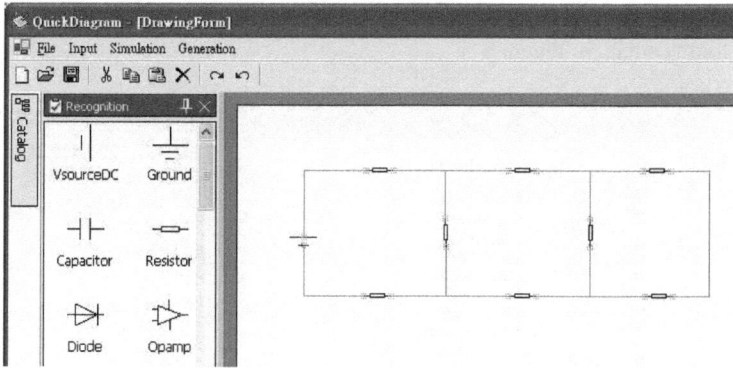

Fig. 3. A resistive circuit containing 8 resistors

The parameter values of corresponding components can be input using the UI
shown in Figure 1. We summarize all the parameter values of these devices in a form
in Figure 4(a). The generated equations and solved unknowns are demonstrated in
Figure 4(b) and Figure 4(c), respectively. All the values are correct as checked by
hand and by PSpice.

The common emitter transistor amplifier circuit is a very common amplifier circuit which usually appears in the IC design. Such circuit as shown in Figure 2 is used
to test QuickDiagram's function to generate the PSpice code, which can be input to
PSpice to analyze the frequency response of the circuit. The unit of the resistor is set
to be ohm and the unit of both DC and AC voltage source are set to be Volt by default. The parameter values of the devices are summarized in Figure 5(a). Figure 5(b)
displays the generated PSpice code which represents the whole diagram connections
and the parameter value of each device.

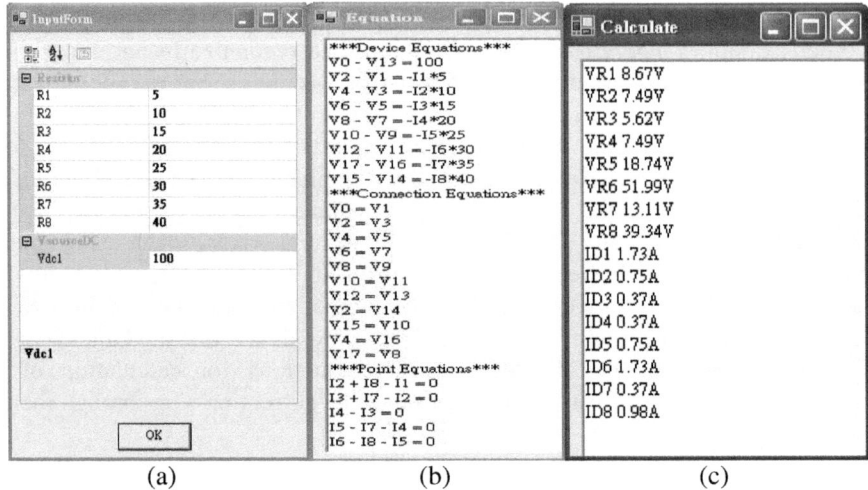

Fig. 4. The values of the devices (a), the generated equations (b) and the values of unknowns calculated for the circuit in Figure 3

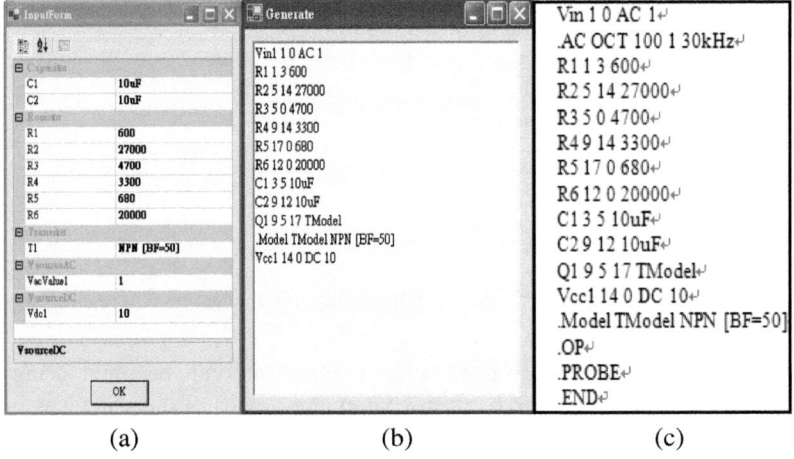

Fig. 5. The parameter values of the devices (a), the generated code (b), and modified code (c) for the circuit shown in Figure 2

Before being read by PSpice for further analysis, the generated code needs to be customized to meet our specific task. In this example, we set the model of the transistor to be NPN with the current gain equal to 50. The frequency analysis range is set to 1HZ to 30 kHz with 100 frequency points on a logarithmic scale. The ".Probe" command is to tell the system to plot the graph. The ".OP" command is to discover the bias point and the various currents of the amplifier transistor. ".End" is to indicate the end of the circuit file. The modified code is shown in Figure 5(c). The modified code is then read by the PSpice system, which can generate the graphs of the frequency response of the circuit, as shown in Figure 6.

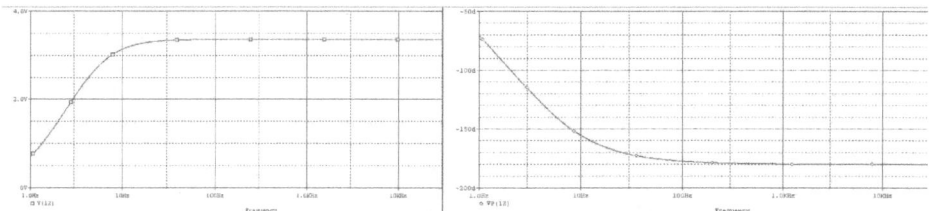

Fig. 6. Frequency response of the circuit of Figure 2: (a) gain, (b) phase

6   Concluding Remarks

In this paper, we present an integrated system QuickDiagram for online sketching and understanding of electronic circuit diagrams. The system is composed of a recognition module and an understanding module. Currently, the symbol database has more than 600 symbols, some of which are for electronic circuit devices. We also have a separate system to model more types of symbols to further extend the symbol database. While we refer readers to some of our papers for the symbol recognition methods, we focus on the understanding methods in this paper. We developed two methods: Nodal Analysis and generation of PSpice codes to interpret the recognized diagrams of certain types of circuits. Especially, the Nodal Analysis can be applied on "resistive circuits", while generation of PSpice codes can be applied to quite a wide range of circuits. Experimental results show that the results are robust and accurate for both recognition and understanding.

We hope QuickDiagram can understand more types of diagrams in the future when more domain knowledge and semantics are modelled. We also wish to continue to improve the QuickDiagram system as an open source project, which is hosted at http://code.google.com/p/quickdiagram/. We hope all researchers and developers interested in this project can help improve it in any aspect, including the stroke pre-processing algorithms, graphic recognition algorithms, diagram understanding algorithms for more domains, and the general UI. Suggestions for new functions, or even bug reports or bug fixing suggestions will also be appreciated.

Acknowledgement

The work described in this paper was fully supported by a grant from City University of Hong Kong (Project No. 7100247), a grant from the Research Grants Council of the Hong Kong Special Administrative Region, China [Project No. CityU 1147/04E], the National Grand Fundamental Research 973 Program of China under Grant No. 2010CB327903, and the Natural Science Foundation of China under Grant No 60603086 and 60723003.

References

1. Al-Hashimi, B.: The art of simulation using PSPICE. CRC Press, Boca Raton (2000)
2. Alvarado, C., Davis, R.: Resolving ambiguities to create a natural computer-based Sketching Environment. In: Proceedings of IJCAI 2001, pp. 1365–1371 (2001)
3. Alvarado, C., Davis, R.: SketchREAD: a multi-domain sketch recognition engine. In: Proceedings of UIST 2004, pp. 23–32 (2004)
4. Arvo, J., Novins, K.: Fluid sketches: continuous recognition and morphing of simple hand-drawn shapes. In: Proceedings of ACM Symposium on UIST, San Diego, California (2000)
5. Calhoun, C., Stahovich, T.F., Kurtoglu, T., Kara, L.M.: Recognizing multi-stroke symbols. In: Proceedings of AAAI Spring Symposium Series–Sketch Understanding (2002)
6. Candence: PSpice Student Edition with the manual, http://www.electronics-lab.com/downloads/schematic/013/
7. Davis, R.: Magic paper: sketch understanding research. Computer 40(9), 34–41 (2007)
8. de Silva, R., Bischel, D.T., Lee, W., Peterson, E.J., Calfee, R.C., Stahovich, T.F.: Kirchhoff's pen: a pen-based circuit analysis tutor. In: Proceedings of Eurographics Workshop on Sketch-Based Interfaces and Modeling, Annecy, France, June 11–13 (2007)
9. Mynatt, E.D., Igarashi, T., Edwards, W.K., LaMarca, A.: Flatland: new dimensions in office whiteboards. In: Proceedings of SIGCHI Conf. Human Factors in Computing Systems: The CHI is the Limit, pp. 346–353 (1999)
10. Fenical, L.H.: PSpice: a tutorial, pp. 202–204. Regents/Prentice Hall (1995)
11. Forsberg, A., Dieterich, M., Zeleznik, R.: The music notepad. In: Proceedings of ACM Symposium on UIST, San Francisco, CA (1998)
12. Gennari, L., Kara, L.B., Stahovich, T.F., Shimada, K.: Combining geometry and domain knowledge to interpret hand-drawn diagrams. Computers & Graphics 29(4), 547–562 (2005)
13. Gross, M.D.: Stretch-A-Sketch: a dynamic diagrammer. In: Proceedings of IEEE Sym. on Visual Languages, pp. 232–238 (1994)
14. Hammond, T., Davis, R.: Ladder: A language to describe drawing, display, and editing in sketch recognition. In: Proceedings of IJCAI, pp. 461–467. AAAI Press, Menlo Park (2003)
15. Hammond, T., Davis, R.: Interactive learning of structural shape descriptions from automatically generated near-Miss examples. In: Proceedings of Intelligent User Interfaces, pp. 37–40. ACM Press, New York (2006)
16. Juchmes, R., Leclercq, P., Azar, S.: A freehand-sketch environment for architectural design supported by a multi-agent system. Computers and Graphics 29(6), 905–915 (2005)
17. Kara, L.B., Stahovich, T.F.: An image-based, trainable symbol recognizer for hand-drawn sketches. Computers & Graphics 29(4), 501–517 (2005)
18. Liu, Y., Yu, Y., Liu, W.: Online segmentation of freehand stroke by dynamic programming. In: Proceedings of ICDAR 2005, pp. 197–201 (2005)
19. Lank, E., Thorley, J., Chen, S., Blostein, D.: On-line recognition of UML diagrams. In: Proceedings of 6th ICDAR, pp. 356–360 (2001)
20. Leclercq, P.: Invisible sketch interface in architectural engineering. In: Lladós, J., Kwon, Y.-B. (eds.) GREC 2003. LNCS, vol. 3088, pp. 353–363. Springer, Heidelberg (2004)
21. Liu, Y., Liu, W., Jiang, C.: A structural approach to recognizing incomplete graphic objects. In: Proceedings of ICPR, vol. 1, pp. 371–375 (2004)
22. Matsuda, K., et al.: Freehand sketch system for 3D geometric modelling. In: Proceedings of International Conference on Shape Modeling and Applications, pp. 55–62 (1997)

23. Narayanaswamy, S.: Pen and speech recognition in the user interface for mobile multimedia terminals. Ph.D. thesis, University of California, Berkeley (1996)
24. Nodal analysis of electric circuits, http://mathonweb.com/help/backgd5.htm
25. Ouyang, T.Y., Davis, R.: Recognition of hand drawn chemical diagrams. In: Proceedings of the Twenty-Second AAAI Conference on Artificial Intelligence, Vancouver, British Columbia, Canada, July 22-26, pp. 846–851 (2007)
26. Pedersen, E.R.: Tivoli: an electronic whiteboard for informal workgroup meetings. In: Proceedings of SIGCHI 1993, pp. 391–398 (1993)
27. Pinto-Albuquerque, M., Fonseca, M.J., Jorge, J.A.: Visual languages for sketching documents. In: Proceedings of IEEE Int. Sym. on Visual Languages 2000, pp. 225–232 (2000)
28. Sezgin, T.M., Stahovich, T., Davis, R.: Sketch based interfaces: early processing for sketch understanding. In: Workshop on Perceptive User Interfaces, Orlando, FL (2001)
29. Sezgin, T.M.: Sketch interpretation using multiscale stochastic models of temporal patterns, PhD thesis, Dept. of Electrical Eng., Massachusetts Institute of Technology (2006)
30. Sutherland, I.E.: SketchPad: a man-machine graphic communication system. PhD Thesis, Massachusetts Institute of Technology (1963)
31. Tapia, E., Rojas, R.: Recognition of on-line handwritten mathematical expressions using a minimum spanning tree construction and symbol dominance. In: Lladós, J., Kwon, Y.-B. (eds.) GREC 2003. LNCS, vol. 3088, pp. 329–340. Springer, Heidelberg (2004)
32. Veselova, O., Davis, R.: Perceptually based learning of shape descriptions. In: Proceedings of Intelligent User Interfaces, pp. 37–40. ACM Press, New York (2006)
33. Wan, Z., Liu, W.: A new vectorial signature for quick symbol spotting, filtering and recognition. In: Proceedings of ICDAR 2007, pp. 516–520 (2007)
34. Xu, X., Sun, Z., Peng, B., Jin, X., Liu, W.: An online composite graphics recognition approach based on matching of spatial relation graphs. International Journal on Document Analysis and Recognition 7(1), 44–55 (2004)
35. Yang, S.: Symbol recognition via statistical integration of pixel-level constraint histograms: a new descriptor. IEEE Trans. on PAMI 27, 278–281 (2005)
36. Zeleznik, R.C., Herndon, K.P., Hughes, J.F.: SKETCH: an interface for sketching 3D scenes. In: Proceedings of SIGGRAPH, New Orleans, pp. 163–170 (1996)
37. Zhang, W., Liu, W., Zhang, K.: Symbol recognition with kernel density matching. IEEE Trans. on PAMI 28(12), 2020–2024 (2006)

Segmenting and Indexing Old Documents Using a Letter Extraction

Mickael Coustaty, Sloven Dubois, Jean-Marc Ogier, and Michel Menard

L3i Laboratory
Avenue Michel Crepeau, 17042 La Rochelle, France
{mcoustat,sduboi01,jmogier,mmenard}@univ-lr.fr

Abstract. This paper presents a new method to extract areas of interest in drop caps and particularly the most important shape: Letter itself. This method relies on a combination of a Aujol and Chambolle algorithm and a Segmentation using a Zipf Law and can be enhanced as a three-step process: 1)Decomposition in layers 2)Segmentation using a Zipf Law 3)Selection of the connected components.

1 Introduction

With the improvement of printing technology since the 15th century, there is a huge amount of printed documents published and distributed. Since that time, books have been falling into decay and degrading. This means not only books themselves are disappearing, but also the knowledge of our ancestors. Therefore, there are a lot of attempts to keep, organize and restore ancient printed documents. With the improving digital technology, one of the preservation methods of these old documents is the digitization. However, digitized documents will be less beneficial without the ability to retrieve and extract the information from them which could be done by using techniques of document analysis and recognition. This paper presents a new method to improve old document images description using segmentation and characterization of letter inside.

1.1 NaviDoMass

NAVigation Into DOcuments MASSes is a french collaborative projec, financed by the National French Research Agency, with the challenge to index ancient documents. With the collaboration of seven laboratories in France, the global objective of this project is to build a framework to derive benefit from historical documents. It aims to preserve and provide public accessibility to this national heritage and is established on four principles: anywhere (global access), anyone (public and multilingual), anytime and any media (accessible through various channels such as world wide web, smartphone, etc.). The focus of NAVIDOMASS is on five studies: (1) user requirement, participative design and ground truthing, (2) document layout analysis and structure based indexing, (3) information spotting, (4) structuring the feature space [HSO+08, JT08] and (5) interactive extraction and relevance feedback. As a part of NAVIDOMASS project, this

J.-M. Ogier, W. Liu, and J. Lladós (Eds.): GREC 2009, LNCS 6020, pp. 142–149, 2010.

paper focuses on the graphics part : graphics indexing and CBIR. However, the main interest of this study is based on specific graphics called drop caps, and on the extraction of shapes in drop caps and particularly on the most important shape : the letter itself. This work is inspired by [PV06] and [ULDO05] which used a Zipf law and a Wold decomposition to extract elements of drop caps.

1.2   Drop Caps in Details

The images of documents of the inheritance are heterogeneous and damaged by time. Drop caps (decorative capital letters also named drop caps or drop cap) belong to the images to index. These images are made up of two principal elements: the letter and the background. (See Fig. 1). An important step in the recognition process of the drop caps consists in segmenting the letter and the elements of the background to characterize them using a signature. This signature will allow a simple and fast comparison for our indexing process of great masses of data. This paper presents in details the various stages of our method: 1) Simplification of the images using layers 2) Extraction of shapes from one of these layers 3) Selection of these shapes.

Fig. 1. Drop Caps Examples

2   Simplification of Images Using Layers

Decomposing an image into meaningful components appears as one of major aims in recent development in image processing. The first goal was image restoration and denoising; but following the ideas of Yves Meyer [Mey01], in total variation minimization framework of L. Rudin, S. Osher and E. Fatemi [LSE92], image decomposition into geometrical and oscillatory (i.e texture) components appears an useful and very interesting way in computer vision and image analysis. There is a very large literature and also recent advances on image decomposition models, image regularization and texture extraction and modeling. So, we only cite, among many others, most recent works which appear most relevant and useful paper. In this way, reader can refer to the work of Stark et al. [SED05], Aujol et al. [AAFC05], [AGCO06], Aujol and Chambolle [AC05], Aujol and Ha Kang [AK06], Vese and Osher [OSV03], [VO04], [VO06] and more recently Bresson and Chan [BC07] and Duval et al. [DAV08] to cover the most recent and relevant advances.

2.1   The Developed Method

Images of drop caps are very complex and very rich images in terms of information and requires to be simplified. These images are mainly made up of lines, unsuitable for usual texture methods. We thus use an approach developed by Dubois and Lugiez [DLPM08] to separate original image in several layers of information, easier to process. This decomposition relies on minimization of a functional calculus F:

$$\inf_{(u,v,w)\in X^3/f=u+v+w} F(u,v,w) =$$

$$\underbrace{J(u)}_{\substack{\text{Regularization} \\ \text{TV}}} + \underbrace{J^*\left(\frac{v}{\mu}\right)}_{\substack{\text{Texture} \\ \text{extraction}}} + \underbrace{B^*\left(\frac{w}{\delta}\right)}_{\substack{\text{Noise extraction by:} \\ \text{shrinkage}}} + \underbrace{\frac{1}{2\lambda}\|f-u-v-w\|_X^2}_{\text{Residual part}} \tag{1}$$

where each element of the functional represents a layer of information and corresponds to a type of information in the image. B^* can be seen as a wavelet soft threshold, J^* a computation of a gradient and J a linear computation between the original image and the two precedent elements. For deeper explanation about notations and each element, one can refer to [DLPM08].

2.2   Layers in Details

We are aiming to catch the pure geometrical component in an image independently of texture and noise to extract shapes. So, we are studying here how to decompose images into three components:

- The Regularized Layer corresponds to the area of image which has low fluctuation of gray level. This layer permits to highlight geometry which corresponds to shapes in the image. In the following of this paper, we will name this layer the "*Shape Layer*".
- Oscillating Layer which corresponds to the oscillating element of the image. In our case, this layer highlights texture from drop caps and in the following of this paper, we will name this layer the "*Texture Layer*".
- Highly Oscillating Layer which corresponds to noise in image. in fact, this layer retrieves all that do not belong to the two first layers. So, we can find in this layer noise, text of background and problem of ageing. Our goal is to recognize old document images while being robust toward noise variations. That is why we will not use this layer in the next of this work.

An example of decomposition applied to the first image of Fig. 1 is given in Fig. 2.

3   Extraction of Shapes

Regularized layer obtained by decomposition contain all shapes. In order to extract them and to select the most interesting, we used a *Zipf Law*. Zipf Law was empirically defined by *George Kingsley Zipf* and relies on the frequency and on the rank of

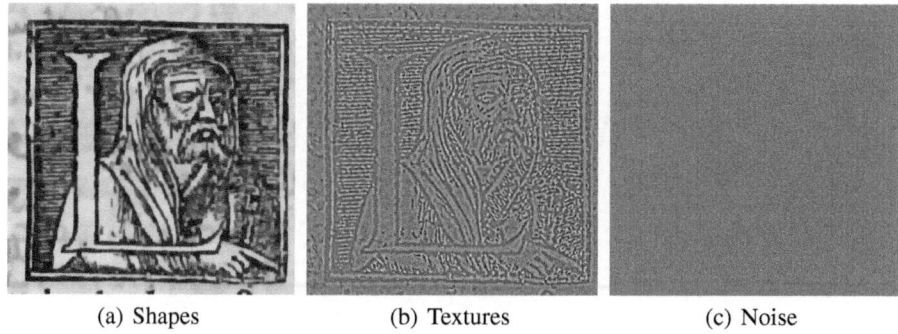

(a) Shapes   (b) Textures   (c) Noise

Fig. 2. An example of drop cap decomposition using Aujol and Chambolle algorithm

appearance of words in a text. This law has been transposed on images by [PV06] by taking subimages as patterns and by calculating frequency and rank of these patterns. This method is a three steps process:

- Simplification of image applying a 3-means on gray level histogram to reduce number of patterns
- Seek for patterns of size three by three to obtain their frequency and their rank
- Classification of patterns in three classes according to the evolution law of the frequency compared to their rank.

3.1 Simplification of Images

Images provided by historians are composed of 256 gray levels. A huge amount of three by three patterns are possible (theoretically 256^9 different patterns). Indeed, if all patterns are represented only once, the model that is deduced from the pattern frequencies would not be reliable, the statistics would lose their significance. Then it is necessary to restrict the number of perceived patterns to give sense to the model.

To decrease the number of graylevels in the image, we have made use of k-means clustering algorithm [McQ67]. As images we are dealing with in this study are composed of three elements (background, foreground and motive), we decided to keep only three gray levels. Moreover, this reduction is made without loosing too much information.

3.2 Patterns Research

Once the number of gray levels have been reduced, a simple count of each pattern permit to know their frequency and their rank. This step is essential to build the Zipf curve which represent the evolution law of the frequency compared to their rank. From this curve, three straight lines are computed to estimate three main parameters of Zipf laws that interfere. The splitting points are defined as the furthest points from the straight line linking the two extreme points of the curve. The first line, which correspond to the most frequent patterns, represent shapes of image (uniform areas). We have extracted pixels from each layer and we display them in an image. Figure. 3 show an example

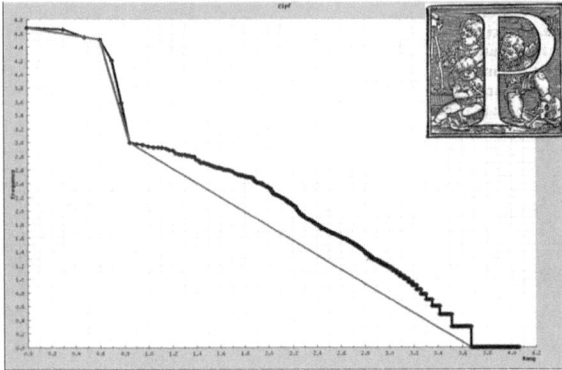

Fig. 3. Example of a Drop Cap and its Zipf plot where are indicated the different straight zones extracted

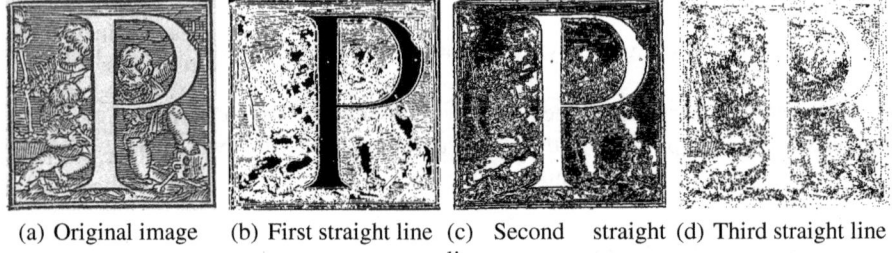

(a) Original image   (b) First straight line (c) Second   straight (d) Third straight line
                                                        line

Fig. 4. Example of a Drop Cap and images corresponding to the straight lines of Zipf'curve

of a Zipf curve with its drop cap while Figure. 4 show an example of binarized images obtained with each straight line.

3.3   Shapes Extration

Once shapes have been extracted, one can seek connected components of binarized image. When we observe all the connected components in Figure 5(b), we can see that the most important shapes have particular characteristics (based on size, location, center of mass and eccentricity). A selection of connected components in accordance with these parameters permit to obtain region of interest of drop caps. An example of extracted connected components can be seen in Figure 5(c). Finally, with an accurate selection on these parameters, the most important connected component for historian can be extracted: the letter. This one can be obtained by selecting the bigger connected component which center of mass is centered and which don't touch borders of image. An example of result can be seen in Figure 5.

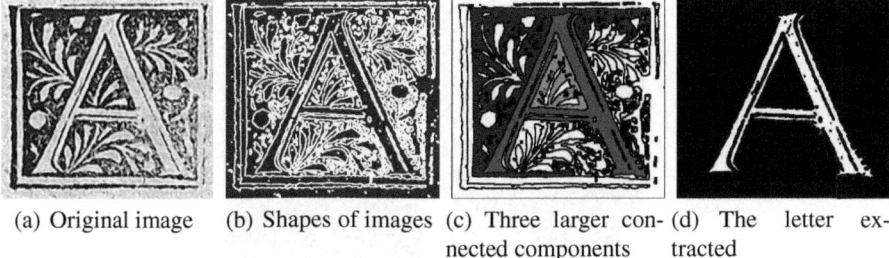

(a) Original image (b) Shapes of images (c) Three larger con- (d) The letter ex-
                                            nected components   tracted

Fig. 5. An example of treatment on a drop cap

4 Experimentations and Validation

The evaluation of such a system is a fundamental point because it guarantees its usability by the users, and because it permits to have an objective regard on the system. In the context of such a project, the implementation of an objective evaluation device is quite difficult, because of the variability of the user requirements: historian researchers, netsurfers, are likely to retrieve many different information which can be very different the ones from the others.

In the context of NAVIDOMASS project, and more specifically for this objective of drop caps indexing, we have decided to evaluate the quality of our system by considering the purpose of "Letter Based Retrieval". This choice is motivated by the fact that many historians want to be able to retrieve drop caps in regard with this criterion. As a consequence, the evaluation of our system relies on the application of an OCR system at the issue of the letter segmentation. Considering these aspects, the classification rate is the main performance evaluation criterion of our system.

Table 1. Recognition rate of drop caps using two kinds of OCR

	FineReader	Tesseract
Classification Rate	72,8%	67,9%

For the evaluation, we have used commercial OCR systems, as well as open source system. In order to implement the evaluation, we have used FineReader on the one hand, and Tesseract on the other hand. We have experimented the approach on an image database containing 4500 images. 1500 of these images were considered for the training set, while 3000 were considered for the tests. The results are summarized in the Table 1. As one can see the obtained results are still unsatifying, but very encouraging. We are working on the improvement of the processing chain, as one can see in the conclusion and perspective part. However, there is not such existing system dealing with this problem, and historians researchers are satisfied to use our system for the classification of their graphic images. The cases for which our system fails correspond to very difficult images, as one can see an example in Fig. 6.

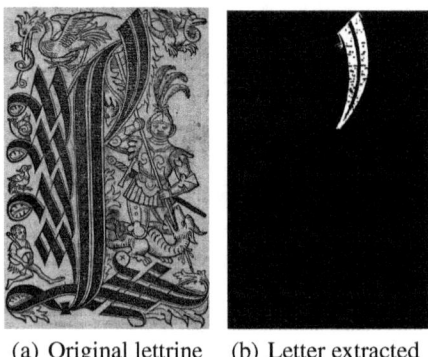

(a) Original lettrine   (b) Letter extracted

Fig. 6. An example of very difficult letter extraction

5   Conclusions

This paper presents a new method to extract informations in drop caps. It relies on a combination of two decomposition. The first one simplifies image to only extract shapes of original image while the second one, a Zipf Law' decomposition, realize a background-foreground segmentation. From this segmentation, a selection of shapes segmented permit to extract some interesting shapes and particularly the letter itself. The first experimentations are encouraging and we are actually working on improvements of this global process.

References

[AAFC05]   Aujol, J.F., Aubert, G., Blanc Feraud, L., Chambolle, A.: Image decomposition into a bounded variation component and an oscillating component. Journal of Mathematical Imaging and Vision 22(1), 71–88 (2005)

[AC05]   Aujol, J.-F., Chambolle, A.: Dual norms and image decomposition models. International Journal of Computer Vision 63(1), 85–104 (2005)

[AGCO06]   Aujol, J.-F., Gilboa, G., Chan, T., Osher, S.: Structure-texture image decomposition - modeling, algorithms, and parameter selection. International Journal of Computer Vision 67(1), 111–136 (2006)

[AK06]   Aujol, J.-F., Kang, S.H.: Color image decomposition and restoration. J. Visual Communication and Image Representation 17(4), 916–928 (2006)

[BC07]   Bresson, X., Chan, T.: Fast minimization of the vectorial total variation norm and applications to color image processing. SIAM Journal on Imaging Sciences (SIIMS) (submitted, 2007)

[DAV08]   Duval, V., Aujol, J.-F., Vese, L.: A projected gradient algorithm for color image decomposition. Technical report, CMLA Preprint 2008-21 (2008)

[DLPM08]   Dubois, S., Lugiez, M., Péteri, R., Ménard, M.: Adding a noise component to a color decomposition model for improving color texture extraction. In: CGIV 2008 and MCS 2008 Final Program and Proceedings, pp. 394–398 (2008)

[HSO⁺08]   Chouaib, H., Tabbone, S., Ramos, O., Cloppet, F., Vincent, N.: Feature selection combining genetic algorithm and adaboost classifiers. In: ICPR 2008, Florida (2008)

[JT08]   Jouili, S., Tabbone, S.: Applications des graphes en traitement d'images. In: ROG-ICS 2008, Mahdia Tunisia, pp. 434–442. University of Ottawa, University of Sfax, Canada, Tunisia (2008)

[LSE92]   Rudin, L., Osher, S., Fatemi, E.: Nonlinear total variation based noise removal. Physica D 60, 259–269 (1992)

[McQ67]   McQueen, J.B.: Some methods for classification and analysis of multivariate observations. In: University of California Press (ed.) Proceedings of 5th Berkeley Symposium on Mathematical Statistics and Probability, Berkeley, vol. 1, pp. 281–297 (1967)

[Mey01]   Meyer, Y.: Oscillating patterns in image processing and nonlinear evolution equations. The fifteenth dean jacqueline B. Lewis Memorial Lectures (2001)

[OSV03]   Osher, S.J., Sole, A., Vese, L.A.: Image decomposition, image restoration, and texture modeling using total variation minimization and the H-1 norm. In: International Conference on Image Processing, pp. 689–692 (2003)

[PV06]   Pareti, R., Vincent, N.: Ancient initial letters indexing. In: ICPR 2006: Proceedings of the 18th International Conference on Pattern Recognition, Washington, DC, USA, pp. 756–759. IEEE Computer Society, Los Alamitos (2006)

[SED05]   Starck, J.L., Elad, M., Donoho, D.L.: Image decomposition via the combination of sparse representations and a variational approach. IEEE Trans. Image Processing 14(10), 1570–1582 (2005)

[ULDO05]   Uttama, S., Loonis, P., Delalandre, M., Ogier, J.-M.: Segmentation and retrieval of ancient graphic documents. In: Liu, W., Lladós, J. (eds.) GREC 2005. LNCS, vol. 3926, pp. 88–98. Springer, Heidelberg (2006)

[VO04]   Vese, L.A., Osher, S.J.: Image denoising and decomposition with total variation minimization and oscillatory functions. Journal of Mathematical Imaging and Vision 20(1-2), 7–18 (2004)

[VO06]   Vese, L.A., Osher, S.: Color texture modeling and color image decomposition in a variational-PDE approach. In: SYNASC, pp. 103–110. IEEE Computer Society, Los Alamitos (2006)

A New Minimum Trees-Based Approach for Shape Matching with Improved Time Computing: Application to Graphical Symbols Recognition

Patrick Franco[1], Jean-Marc Ogier[1], Pierre Loonis[2], and Rémy Mullot[1]

[1] Laboratoire Informatique, Image, Interaction (L3I)
UPRES EA 2118
Université de La Rochelle
17042 La Rochelle Cedex 1, France
{patrick.franco,jean-marc.ogier,...}@univ-lr.fr
[2] Laboratoire Electronique, Informatique et Image (LE2I)
UMR CNRS 5158
Université de Bourgogne
58027 Nevers cedex, France
pierre.loonis_isat@u-bourgogne.fr

Abstract. Recently we have developed a model for shape description and matching. Based on minimum spanning trees construction and specifics stages like the mixture, it seems to have many desirable properties. Recognition invariance in front shift, rotated and noisy shape was checked through median scale tests related to GREC symbol reference database. Even if extracting the topology of a shape by mapping the shortest path connecting all the pixels seems to be powerful, the construction of graph induces an expensive algorithmic cost. In this article we discuss on the ways to reduce time computing. An alternative solution based on image compression concepts is provided and evaluated. The model no longer operates in the image space but in a compact space, namely the Discrete Cosine space. The use of block discrete cosine transform is discussed and justified. The experimental results led on the GREC2003 database show that the proposed method is characterized by a good discrimination power, a real robustness to noise with an acceptable time computing.

Keywords: Document analysis, Graphics Recognition, Region Based Shape Descriptor, Feature extraction, Minimum Spanning Tree, Discrete Cosine Transform, Image Compression.

1 Motivations

From the last ten years an intensive campaign of document digitalization as been lead all around the world. A large part of documents handled are graphics document (Technical Drawings, Maps etc). i.e. documents dominated by graphic components (signs or shapes) which can be classified as symbols. Because they bring complementary information than purely textual, these entities are often used as information support. So the symbol recognition process plays a central role in the framework of automatic document

J.-M. Ogier, W. Liu, and J. Lladós (Eds.): GREC 2009, LNCS 6020, pp. 150–162, 2010.

interpretation and decision making. In other side, various models for symbol character-ization have been developed, especially in the field of technical documents. Usually a partition is made between structural and statistical approaches. A complete state of the art can be find in [16], some original methods can be find in [30,9] and the perspectives for the next ten years are mentioned in [32]. In front of the proliferation of the number of models , one could claim that it is not necessary to develop new ones. Many factors contribute to the persistance of the characterization problem :

- The task of representation is intimately constrained by the recognition process and it can be difficult to decorrelate from any solution of this second classification stage. That's why neural network solutions were widely used [4,22].
- Links between image segmentation and object recognition play an important role, as was emphasized by K. Tombre and B. Lamiroy in [31]. We need to develop a model tolerant to segmentation errors or any approach allowing to simultaneously combine a segmentation/recognition stage.
- Practical aspects should not be forgotten. The impact of time learning models es-pecially for future web applications - such as "an on line user select on the fly an arbitrary symbol in a technical document and queries for similar symbols" - could not be neglected. Noise resistance it's also a key component of an efficient shape descriptor. When levels of degradation induce the loss of the connectedness of the shape many reference models [23,14] are not powerful any more. Our approach is globally inscribed in the framework of these items.

Recently, we have developed a class of algorithms [5,6] based on the capture of the topological properties of a shape via Minimum Spanning Tree Lengths. This approach was evaluated in previous works through the problem of graphical symbols recognition. Medium scale tests on reference symbols database (GREC 2003) have been lead and the position with a standard approach (Zernike moments) was also investigated. A brief review of the key stages of the technique, completed by an example of application are provided in Section 2. The results seem to be conclusive but the approach still has an expensive algorithmic cost. In this article, the main ways to reduce it are discussed and an alternative is provided (Section 3). The proposed solution is guided by image compression concepts. The target object is expressed in the Discrete Cosine space. The use of blocked Discrete Cosine Transform (DCT) is discussed and justified (cf Section 3.3). This track leads to a substantial time computing reduction whilst conserving a significant level of inter-object discrimination (cf. Section 4).

2   Using Minimum Spanning Trees for Shape Description: Previous Algorithm

Any object is described in the image system coordinates as a multi-dimensional points set with a specific topology. By mapping all the points (completely connected tree), with non oriented edges (not directed tree), without cycle (tree and not graph) under the constrainst to build a minimal path (tree with minimal length), we define a measure of the object's topology.

MSTs are widely exploited by many algorithms for varied operations dealing with image processing. For example, the MST-based image segmentation task can be found in [37]. In the field of pattern recognition some works investigate the use of MST to line image matching [20] or fingerprint matching [21]. By definition, the construction of MST is a natural way to extract topology properties from a multi-dimensional data distribution. This idea is well traducted in data clustering [36], where clusters are localized in the MST by thresholding the edge lengths (connected component labeling). Moreover, the MST length is linked to an entropy measure called Rényi's entropy [27]. The Recent work of A. Hero and O. Michel [11] shows that a family of trees[1] which satisfies the Redmond's quasi-additivity property [26], built on one d -dimensional points distribution, is a robust estimator of Rényi's entropy of this distribution. These results can shed a new light on previous works using MST for pattern recognition [33] or clustering [38].

We have shown that the MST length can define a metric capable of discriminating between a prototype and an unknown object. A mixture of both objects (union points sets) is realized before computing the variation of the lengths from to the prototype length. This stage removes any risk of ambiguity in the recognition process. In the real world problem the objects can be multi-oriented. So we can iteratively apply geometrical transformations on the input object, in order to really estimate their similarity. When the prototype completely matches the unknown object, the topology defined by their union point sets is identical to the prototype one. In this case, the two MST have the same length and the error measure is minimum. This rule is synthesized by the Algorithm 1. O_1 being the prototype and O_2 being a given object and γ is an order length of the tree ($\gamma \in]0, d[$).

2.1   An Example of Application

The approach described above is implemented within the framework of graphical symbol recognition. The samples test are extracted from the reference database [1] used in the Symbol Recognition Contest of GREC'2003 (Barcelona, Spain). Two kind of symbols have been considered, Architectural prototypes (cf Figure 1) and Electrical ones. In order to show the characteristics of our method, the original prototypes have been translated, rotated and then degraded by 30% "salt and pepper" noise (cf. Figure 2). Median filtering (arbitrary mask 3×3) and a segmentation task enabled the objects to be classified to be isolated (cf. Figure 3). This artificial process results in the objects to be classified substantially differing from the prototypes. For example, even full regions of the "sink" class no longer appear in "object 1" which is supposed to belong to this class (cf. Figure 3). Numerous shape connexity breaks can also be observed, perturbing the pixels neighbourhood relationships[2] that is fundamental for the construction of the spanning tree. We can see that our algorithm is quite robust to this kind of deterioration and this is thanks to the mixture. We explain this relative resistance in the following way. The MST defines the dominant skeleton of the symbol by mapping the shortest path of

---

[1] The Steiner tree, the minimal spanning tree, and the trees related to Traveling Salesman Problem.

[2] That's why the "salt and pepper" noise model was chosen.

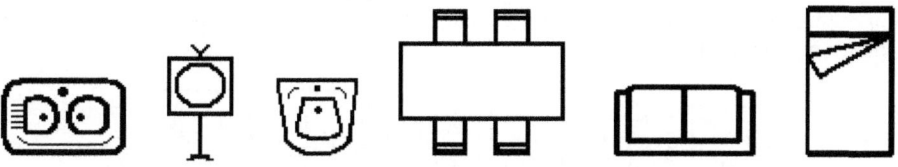

Fig. 1. Original symbols: 6 classes. From left to right respectively : Sink, Television, Washbasin, Table, Sofa and Bed.

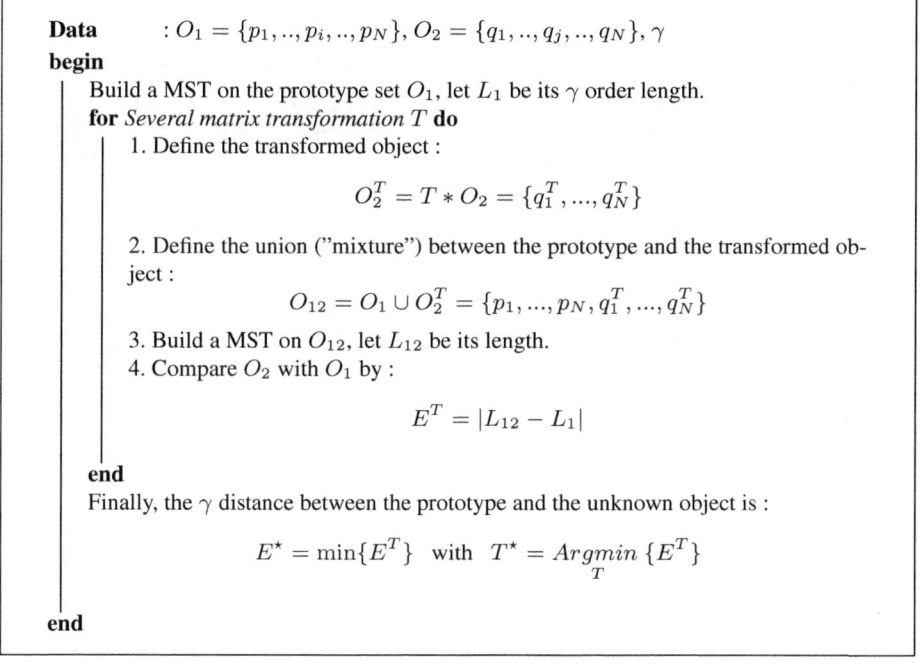

Data   : $O_1 = \{p_1, .., p_i, .., p_N\}$, $O_2 = \{q_1, .., q_j, .., q_N\}$, γ
begin

Build a MST on the prototype set O_1, let L_1 be its γ order length.
for *Several matrix transformation T* **do**
  1. Define the transformed object :

$$O_2^T = T * O_2 = \{q_1^T, ..., q_N^T\}$$

  2. Define the union ("mixture") between the prototype and the transformed object :
$$O_{12} = O_1 \cup O_2^T = \{p_1, ..., p_N, q_1^T, ..., q_N^T\}$$

  3. Build a MST on O_{12}, let L_{12} be its length.
  4. Compare O_2 with O_1 by :

$$E^T = |L_{12} - L_1|$$

end
Finally, the γ distance between the prototype and the unknown object is :

$$E^\star = \min\{E^T\} \quad \text{with} \quad T^\star = \underset{T}{Argmin}\{E^T\}$$

end

Algorithm 1. Objects matching by minimum spanning trees: previous algorithm

nearest neighbor points set. The mixture operation can restore the pixels which were randomly removed. By consequent the symbol topology is not irreversibly lost and the dominant skeleton is globally conserved. Outliers could resist to the preliminary filtering process and significantly damaged the skeleton by adding out of proportion edges length. A specific program against outliers can be provided by detecting and cutting edges which have too highest lengths. The classification results are given in the Table 1. The object's identity is that of the symbol for which the prototype minimizes the error. In this basic experiment all the objects were rightly recognized (minimal distances located on the diagonal) despite the degradation level of some objects.

Table 1. Results of previous algorithm : minimum spanning tree-based symbol recognition ($\gamma = 1$) operating in the image space. Matrix of minimal distances between symbol prototypes and unknown objects ($E^\star$).

	Objects					
	obj.1	obj.2	obj.3	obj.4	obj.5	obj.6
sink	**241.151**	813.667	796.598	788.656	763.928	661.071
television	771.110	**176.791**	338.359	915.355	918.071	805.424
washbasin	769.000	345.120	**60.968**	919.571	890.757	773.585
table	796.019	848.280	851.381	**616.001**	759.019	815.000
sofa	762.252	906.763	893.877	775.057	**63.083**	805.583
bed	689.019	784.095	765.703	806.026	828.034	**192.001**
Reality	sink	television	washbasin	table	sofa	bed

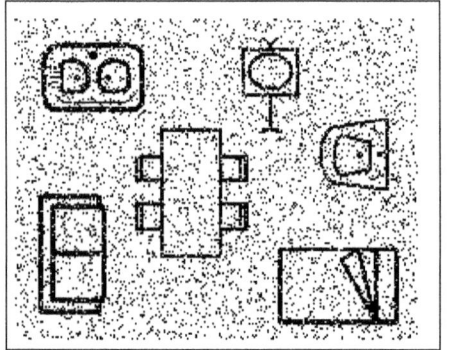

Fig. 2. Degraded symbols : 30% salt and pepper noise

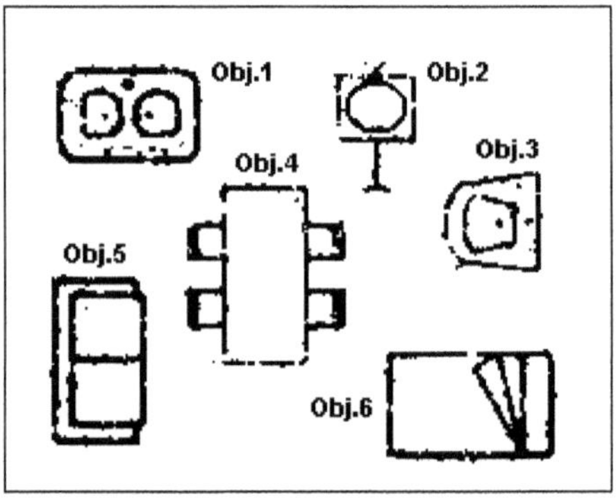

Fig. 3. After filtering and segmentation : unknown objects

If the mixture stage led to a significant recognition rate[3], it also introduces a growing time computing because twice more points must be connected by the trees. In the next Section, the algorithmic cost of our solution is estimated and ways to reduce it were pointed out.

3 Using Minimum Spanning Trees for Shape Description: New Algorithm

The time computing of our approach is a critical point. To give an idea, for a given object randomly rotated and degraded (binary images $d = 2$, $N = 1000$), less than $4mn^4$ are needed to establish the inter-objects distance ($E^\star$). In this section, the algorithm cost of our method is estimated and ways to reduce it are mentioned. The solution chosen is described and justified and the final algorithm gain is evaluated.

---

[3] That was confirmed on medium scale tests.

[4] It is the result of a blind and iterative research with the Matlab 6.5 implementation of the algorithm on a standard PC (Pentium 4 processor at 3.06 GHz with 512 MB RAM) with the input algorithm parameters ($T = 24$, $\gamma = 1$).

3.1   Algorithm Cost Estimation and Ways to Reduce It

The time computing is closely dependant of the algorithmic cost[5] of the MST[6] construction. For a given number of points to be connected N, evolving in dimension $\{2, 3\} \in d$, the MST algorithmic cost is increased by $N^2 \log N$. According to the algorithm provided in Section 2, $T + 1$ trees are required to determine the inter-object distance, where T, is the maximum number of iterations in the research. The first tree is calculated on N points, whereas the other T are calculated on the union of objects, i.e. $2N$ points[7]. The algorithmic cost therefore stands at $C_1 \approx N^2 \log(N) + T.\{(2N)^2 \log(2N)\}$.

Several solutions to algorithmic cost reduction exits, both algorithmic and from the image processing field. Thus the use of a faster algorithm (D. Karger and P. Klein in [13]) or even an incremental schema (M. Soss in [29]) are the methods currently being explored by the community. In practice, another solution consists in reducing the number of geometric transformations (rotations) applied, it is also possible to carry out a main component analysis in order to accelerate the research for the optimal transformation (identification of inertia axes). Image processing methods can also be used to reduce all the pixels to be taken into account, as, for example, in morphological erosion operations [28]. The extraction of the minimal skeleton of an image is in the center of some works in which the minimization of redundancy is tracked [19,15]. These morphological approaches, however, lead to a loss of structural information on the object to be classified. This could penalize our spatial description-based approach. The alternative that we propose consists in carrying out a global transformation of the image, whilst preserving the structural dimensions of the objects.

3.2   Cost Reduction via Discrete Cosine Transform

The main idea is that of working in an image representation mode in which the coefficients can be better decorrelated. The Discrete Cosine Transformation (DCT) was chosen for these "good" properties discussed below. Introduced in 1974 by N. Ahmed, T. Natarajan and K. Rao [2], it is expressed as follows :

$$F(u,v) = \frac{1}{\sqrt{2N}} C(u)C(v) \sum_{x=0}^{N-1} \sum_{y=0}^{N-1} I(x,y). \cos \frac{\pi u(2x+1)}{2N} \cos \frac{\pi v(2y+1)}{2N} \quad (1)$$

where: $I(x, y)$ is the original image and $F(u, v)$ is the transformed image; x, y is the spatial plane coordinates and u, v is the frequency plane coordinates;

$$C(u) = C(v) = \frac{1}{\sqrt{2}} \text{ for } u, v = 0 \quad C(u) = C(v) = 1 \text{ otherwise}$$

---

[5] By algorithmic *cost* we mean the number of operations required for a given number of points, as opposed to the *complexity* which refers more to the class of an algorithm (quadratic for the two algorithms proposed).

[6] The code chosen is that recently developed by A. Hero and O. Michel. Since the problem is a full NP, their algorithm provides a good estimate of the MST length in a reasonable time computed [12,10].

[7] Under the assumption that the objects are defined by the same number of points.

The term $\frac{1}{\sqrt{2}}$ enables the transformation to be orthogonal.

The DCT is very close to the Fourier Transformation (FT) and possesses interesting properties for our aim of minimizing the number of points that describe an object :

- strong *energy compaction*, the signal information tends to be concentrated in a few low-frequency components of the DCT;
- low *correlation* between the transformed coefficients, the DCT approaching the Karhunen-Love transform (which is optimal in the decorrelation sense);
- fast *computation*, only 11 multiplications are needed to perform the 1D DCT with the practical fast algorithm proposed by C. Loeffler, A. Ligtenberg, and G. Moschytz in [17].

What is more, K. Rao shows that the DCT provides a better approximation of an image with less coefficients than the FT [25]. Moreover, the capacity of the DCT to decorrelate data can be explained statistically. By considering the pixels of the images as realizations of a random stationary process, some works [8,25] show that the DCT is a good estimation of the Karhunen-Love transform; in addition, it does not come up against the diagonalization of the covariance matrix which is always delicate. These properties have been widely tested since the DCT was chosen for the JPEG compression norm[8] in 1992 [3,35].

3.3   Adaptation to Graphic Symbol Recognition Problem : Justifications

Some image compression concepts [25,7,24] are used in this section. The concepts surrounding the steps of transformation, quantization and coding have guided some of our choices. For example, the DCT is not applied to the whole image but to blocks of the image. For computation reasons, it is preferable to decompose the original image into K blocks of size $M \times M$ and to perform the DCT on each of the blocks. One should, however, ensure that the size of the blocks (directly related to the loss of some frequencies) is large enough to avoid the mosaic effect. Another argument, notably concerning our application, reinforces this choice - the construction of a tree in transformed space without partitioning the original image inevitably leads to the loss of spatial information in the image. After integration (on the whole image) it is difficult to link a frequency to a specific region of the object in the image. The spatial information is diluted in the set of frequencies. By partitioning the image and ordering the blocks, part of the spatial dimension is conserved, which should facilitate the tree-based comparison of objects.

The transformed matrices do not undergo any quantization process. The quantization is only to attenuate the amplitude of the coefficients in order to minimize the number of bits required to code them. On the contrary, it would appear to be interesting to maintain the greatest inter-coefficient variability possible, so as to discern the objects.

The DCT is very compact, in other words few coefficients are required for an estimation. What is more, the majority of the non null coefficients are concentrated around the continuous component. The further away you are from the low frequencies (in both directions) the closer the coefficients are to zero. This distribution can be used to omit

---

[8] Joint Photographic Expert Group.

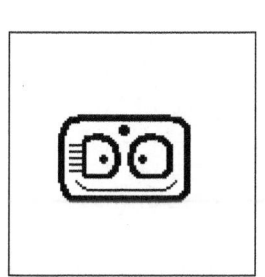
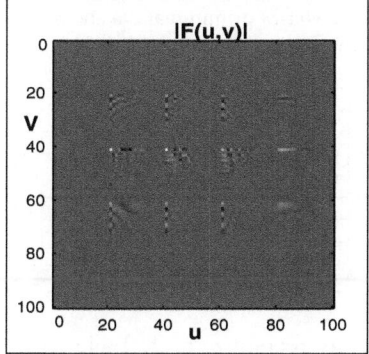

Fig. 4. From left to right respectively : the original image; the image expressed in DCT space ($K = 25$, $M = 20$), and the image reconstructed when only 19.5% of the original DCT coefficients are conserved (78 coefficients per blocks) via zig-zag algorithm

certain coefficients without risking alteration of the original image (cf. Figure 4). Furthermore, this trend is used by scanning techniques (coding theory) which aim to minimize the address of the last non-null coefficient to be transmitted. The zig-zag algorithm example is fairly well known [8,7]. The trees are therefore constructed on the sub-block DCT coefficients.

4 Algorithm Evaluation

The symbol recognition experiment carried out in Section 2.1 was reproduced under the same conditions. The algorithm presented in Section 2 was applied by integrating the stages described in Section 3.3. That is to say that the prototype object is firstly expressed into DCT space (K blocks of size $M \times M$). A selection of the coefficients is made using the zig-zag algorithm (cf. Figure 4). Only N' points per block are kept. In DCT space, a sub-block is a set of N' points evolving in dimension 3 ($u, v, F(u, v)$). An MST is constructed on each sub-block. Finally, K MSTs are performed and their lengths are the components of the features vector for the prototype processed. The same schema is applied to the image resulting of the mixture between the prototype and the input object. The same[9] DCT block partition is used in the two case. An Euclidean norm between features vectors is computed. This metric based on MST lengths built on sub-blocks DCT space is used as classification criterion. For multi-oriented objects, the inter-object distance is the minimum distance observed during the several geometric transformations (T). The results are given in Table 2. They are obtained with the set of parameters $\{K = 25, M = 20, N' = 78\}$, their values are guided by the analysis developed in Section 3.3. Complementary results observed on another sample test are given in the Appendix.

---

[9] The tree-based characterization of transformed coefficients only makes sense if the symbols are decomposed on the same image partition. For various symbols size, the highest image bounding box is used as reference.

Table 2. Results of new algorithm : minimum spanning tree-based symbol recognition ($\gamma = 1$) operating in reduced DCT space. Matrix of minimal distances between feature vectors related to symbol prototypes and unknown objects.

	Objects					
	obj.1	obj.2	obj.3	obj.4	obj.5	obj.6
sink	**16.752**	11031.616	11327.937	91.292	67.839	48.822
television	62.281	**7.885**	10561.515	110.458	82.309	78.956
washbasin	68.056	30.257	**13.196**	110.748	96.129	66.874
table	19217.062	11955.169	11892.729	**29.400**	23828.005	20298.654
sofa	19790.828	10893.603	11545.823	82.740	**17.976**	54.041
bed	19372.944	11006.482	11291.834	83.158	25060.909	**29.064**
Reality	sink	television	washbasin	table	sofa	bed

All the objects were rightly recognized (cf. Table 2). The recognition quality was measured using a criterion Δ, reflecting the discriminant nature of the decision. Δ is evaluated by comparing the distances associated with the two emerging symbols. For example, the *television* was identified (second column) with a discrimination index $\Delta = \frac{30.257 - 7.885}{7.885} = 283.7\%$. Within the framework of this experiment (model, noise level, etc.), the two approaches have a similar level of discrimination. The median levels were 244.3% vs. 271.7% when the trees operate in the image space and the DCT space, respectively. However, the algorithmic gain is undeniable when the comparison is carried out in transformed space, which we propose to estimate. The calculation cost of the DCT is negligible in comparison with that engendered by MST construction. This assumption seems acceptable given the fast algorithm class proposed by C. Loeffler, A. Ligtenberg and G. Moschytz in [17]. The cosine basis functions are pre-computed and stored and the separability property of the DCT is exploited. The cost (approximate) therefore evolves with $C_2 \approx K.N'^2 \log(N') + T.\{K.(N')^2 \log(N')\}$. The first term refers to the calculation of the features vector of the prototype object and the second refers to that of the feature vector of the mixture of objects to be compared. For $\{K = 25, T = 24, N' = 78, N = 1000\}$ the average algorithmic gain reaches $\frac{C_1}{C_2} = 44.463$. This gain is substantial and is mainly due to the second term of C_2 in relation with that of C_1. The complexity of the construction of trees is circumvented by constructing more trees (K more times) but on a smaller volume of data ($\frac{N'}{2N} = 3.9\%$). We therefore profit from the polynomial evolution of the cost in terms of number of points. Less than $3mn$ vs. $2h33mn12s$ is thus required to obtain the distance matrix synthesized in Table 2. From a decision making point of view, the time processing becomes reasonable even if it is not yet compatible with the time constraints of a real application related to a large database. However, this time could be shortened by the more in depth reduction of the number of DCT coefficient taking into account. Another hand, we know from image compression point of view that image restitution quality is linked to compression rate especially for loss techniques based on DCT (cf. Figure 5). So, the decision making speed up will be followed by a drop of recognition rate. Large scale tests have shown that a good trade-off between accuracy ($\sim 85.5\%$) and time processing ($\sim 2, 21s/sample$) is realized when 36 coefficients per block ($N' = 36$) are conserved. When less than 36 coefficients are selected, the part of the original image energy brought falls under 50%, causing deeply degradations on the reconstructed image from reduced DCT space (cf. Figure 5) then the link between the original image and the reduced DCT space is broken.

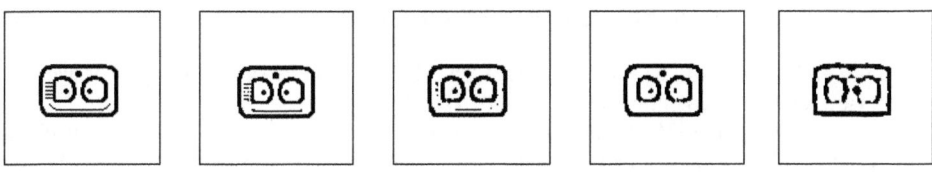

Fig. 5. The original image (on the left) and the reconstructed images from progressive reduction of DCT coefficients selected (resp. from 78, 55, 36, and 21 DCT coefficients per block). The part of the original image energy conserved is respectively 86.6%, 81.6%, 74% and 48%.

5   Conclusion

This article is focused on the way to reduce time computing on a previous algorithm dealing with the use of minimum spanning trees (and specific stages) for shape matching. Many ways leading to its reduction are described, such as the expression of the target object in Discrete Cosine Space (JPEG-1992 standard). We are therefore no longer operating in the image space but in a compact space in which the data is better decorrelated. If the expression of images in the DCT space appears to be adapted (energy compaction, noise resistance), the use of other transformations to improve the calculation time come rapidly to mind. We can cite the wavelet transform chosen in the new JPEG-2000 standard (sub-band coding [18,34]). However, it would be appropriate to verify that the final algorithmic gain is real. The cost estimated in Section 3.1 concerns points of dimension 3. In scale-position space the points are of a greater dimension, causing this estimation to be null and void. The algorithmic cost for trees construction should therefore be revised upwards. As a result a better approximation using scales should compensate the increased cost. Our proposition leads to a substantial reduction of the algorithm cost whilst conserving a competitive recognition rate. These results were confirmed by medium-scale tests (25 classes comprising 20 samples per class) involving varied and realistic sets of symbols from a reference database [1] to which increasing amounts of degradation have been added inducing loss of the connectedness of the shapes.

Acknowledgments

The authors would like to thank Pr. Alfred Hero (E.E.C.S. Department, University of Michigan, USA) and Pr. Olivier Michel (Astrophysical Laboratory, University of Nice Sophia-Antipolis, France) for the free use of their minimum spanning tree original software.

References

1. Fifth IAPR International Workshop on Graphics Recognition. Computer Vision Center, Barcelona, Catalonia, Spain, July 30-31 (2003)
2. Ahmed, N., Natarajan, T., Rao, K.: On image processing and a discrete cosine transform. IEEE Transactions on Computers C-23(1), 90–93 (1974)

3. Chen, W., Pratt, K.: Scene adaptive coder. IEEE Transactions on Communications COM-32, 225–232 (1984)
4. Egmont-Petersen, M., de Ridder, D., Handels, H.: Image processing with neural networks-a review. Pattern Recognition 35, 2279–2301 (2002)
5. Franco, P., Ogier, J.-M., Loonis, P., Mullot, R.: A topological measure for image object recognition. In: Lladós, J., Kwon, Y.-B. (eds.) GREC 2003. LNCS, vol. 3088, pp. 279–290. Springer, Heidelberg (2004)
6. Franco, P., Ogier, J.-M., Loonis, P., Mullot, R.: Template matching by minimum spanning trees. In: 5th IAPR International Conference on Graphics Recognition (GREC 2003), Barcelone, Spain, pp. 341–352 (2003)
7. Gersho, A., Gray, R.M.: Vector Quantization and Signal Compression. Kluwer Academic Publishers, Boston (1992) (6 edn., 1999)
8. Guichard, J., Nasse, D.: Traitement des images numériques pour la réduction du débit binaire. Le Traitement du Signal - Actes des Forums de France Télécom Recherche (2), 1–15 (1994)
9. Guillas, S., Bertet, K., Ogier, J.M.: Towards an iterative classification based on concept lattice. In: Yahia, S.B., Nguifo, E.M., Belohlavek, R. (eds.) CLA 2006. LNCS (LNAI), vol. 4923, pp. 256–262. Springer, Heidelberg (2008)
10. Hero, A., Michel, O.: Robust estimation of point process intensity features using k-minimal spanning trees. In: IEEE International Symposium on Information Theory, Germany, June 1997, p. 74 (1997)
11. Hero, A., Michel, O.: Robust entropy estimation strategies based on edge weighted random graphs. In: SPIE, International Symposium on Optical Science, Engineering and Instrumentation, San Diego (July 1998)
12. Hero, A., Michel, O.: Asymptotic theory of greedy approximations to minimal k-point random graphs. IEEE Transactions on Information Theory IT 45, 1921–1939 (1921)
13. Karger, D., Klein, P., Tarjan, R.: A randomized linear-time algorithm to find minimum spanning trees. Journal of the Association for Computing Machinery (ACM) 42(2), 321–328 (1995)
14. Khotanzad, A., Hong, Y.H.: Rotation invariant image recognition using features selected via a systematic method. Pattern Recognition (23), 1089–1101 (1990)
15. Kresch, R., Malah, D.: Morphological reduction of skeleton redundancy. Signal Processing 38, 143–151 (1994)
16. Lladós, J., Valveny, E., Sánchez, G., Martí, E.: Symbol recognition: Current advances and perspectives. In: Blostein, D., Kwon, Y.-B. (eds.) GREC 2001. LNCS, vol. 2390, pp. 104–128. Springer, Heidelberg (2002)
17. Loeffler, C., Ligtenberg, A., Moschytz, G.: Practical fast 1-d dct algorithms with 11 multiplications. In: International Conference on Acoustics, Speech, and Signal Processing (ICASSP 1989), pp. 988–991 (1989)
18. Mallat, S.G.: Theory for multiresolution signal decomposition: The wavelet representation. IEEE Transactions on PAMI 11(7), 674–693 (1989)
19. Maragos, P., Shafer, R.: Morphological skeleton representations and coding of binary images. IEEE Transactions on Accoustics, Speach and Signal Processing 34(5), 1228–1244 (1986)
20. Marchand-Maillet, S., Sharaiha, Y.M.: A minimum spanning tree approach to line image analysis. In: Proceedings of the 13th International Conference on Pattern Recognition, August 1996, vol. 2, pp. 225–230 (1996)
21. Oh, C., Ryu, Y.K.: Study on the center of rotation method based on minimum spanning tree matching for fingerprint recognition. Optical Engineering 43(4), 822–829 (2004)
22. Osowski, S., Dinh Nghia, D.: Fourier and wavelet descriptors for shape recognition using neural networks-a comparative study. Pattern Recognition 35, 1949–1957 (2002)
23. Pei, S., Lin, C.: Normalisation of rotationally symmetric shapes for pattern recognition. Pattern Recognition (25), 913–920 (1992)

24. Pennebaker, W.B., Mitchell, J.L.: The JPEG Still Image Data Compression Standard. Van Nostrand Reinhold, New York (1993)
25. Rao, K., Yip, P.: Discrete Cosine Transforms - Algorithms, Advantages, Applications. Academic Press, Boston (1990)
26. Redmond, C., Yukich, J.E.: Limit theorems and rates of convergence for euclidean functionals. Annals of Applied Probability 4(4), 1057–1073 (1994)
27. Rény, A.: On measures of entropy and information. In: Symposium on Mathematics Statistics and Probabilities, Berkeley, pp. 547–561 (1961)
28. Serra, J.: Image analysis and mathematical morphology. Theoretical Advances, vol. 2. Academic Press, London (1988)
29. Soss, M.: On the size of the sphere on influence graph. PhD thesis, Mc Gill University Scholl of Computer Science Montreal (1998)
30. Tabbone, S., Wendling, L.: Recognition of symbols in grey level line drawings from an adaptation of the radon transform. In: The 17th International Conference on Pattern Recognition, Cambridge, UK, pp. 570–573 (2004)
31. Tombre, K., Lamiroy, B.: Graphics recognition - from re-engineering to retrieval. In: 7th International Conference on Document Analysis and Recognition (ICDAR 2003), pp. 148–156. IEEE Computer Society, Los Alamitos (2003)
32. Tombre, K.: Graphics recognition: The last ten years and the next ten years. In: Liu, W., Lladós, J. (eds.) GREC 2005. LNCS, vol. 3926, pp. 422–426. Springer, Heidelberg (2006)
33. Toussaint, G.: The relative neighborhood graph of a finite planar set. Pattern Recognition, 261–268 (1980)
34. Vetterli, M., Kovacevic, J.: Wavelets and Subband Coding. Prentice Hall, Englewood Cliffs (1995)
35. Wallace, G.: The jpeg still picture compression standard. Communications of the Association for Computing Machinery 34(4), 30–44 (1991)
36. Xu, Y., Olman, V., Xu, D.: Minimum spanning trees for gene expression data clustering. Genome Informatics (12), 24–33 (2001)
37. Ying, X., Uberbacher, E.C.: 2d image segmentation using minimum spanning trees. Image and Vision Computing 15(1), 47–57 (1997)
38. Zahn, C.: Graph-theoretical method for detecting and describing gestalt clusters. IEEE Trans. on Computers, 68–86 (1971)

Appendix: Additional Results Lead on Reference GREC2003 Database

Another sample test : case of electrical symbols. Problem : Complex symbols recognition (i.e. shift, rotated, degraded and sometimes defined by disjoint regions).
Support : low resolution binary images (< 240 dpi).

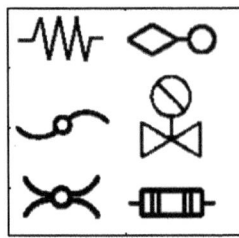 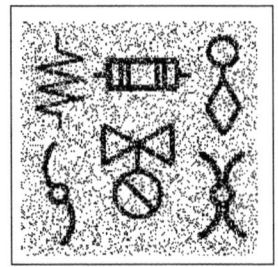 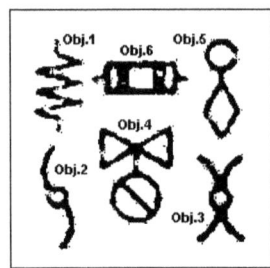

Fig. 6. From left to right respectively : original and degraded symbols ("salt and pepper" noise, level $NSR = 30\%$) and the objects to be classified (after blind filtering and segmentation)

Table 3. Results of first algorithm : minimum spanning tree-based symbol recognition ($\gamma = 1$) operating in the image space. Matrix of minimal distances between symbol prototypes and unknown objects ($E^\star$). Median discrimination level: $\Delta = 278.1\%$, time computing: $2h28mn34s$.

	Objects					
	obj.1	obj.2	obj.3	obj.4	obj.5	obj.6
symb.1	**134.656**	588.306	651.173	1146.886	564.485	749.122
symb.2	580.235	**138.171**	597.887	1238.154	522.313	935.886
symb.3	589.093	598.551	**90.892**	1150.123	633.514	770.414
symb.4	1087.915	1154.539	1126.148	**272.472**	1112.076	910.770
symb.5	564.656	494.585	643.242	1175.009	**91.656**	767.236
symb.6	679.679	858.757	737.521	991.402	703.514	**198.122**
Reality	symb.1	symb.2	symb.3	symb.4	symb.5	symb.6

Table 4. Results of new algorithm : minimum spanning tree-based symbol recognition ($\gamma = 1$) operating in reduced DCT space. Matrix of minimal distances between feature vectors related to symbol prototypes and unknown objects. Median discrimination level: $\Delta = 367.2\%$, time computing: $2mn3s$.

	Objects					
	obj.1	obj.2	obj.3	obj.4	obj.5	obj.6
symb.1	**8.762**	13564.972	20644.405	87.801	18334.453	24845.423
symb.2	54.728	**14.357**	38.627	107.445	47.365	63.031
symb.3	50.695	13593.552	**6.243**	100.729	18347.104	24884.850
symb.4	18280.536	13541.843	20588.324	**16.341**	18262.776	24870.032
symb.5	45.373	17.158	33.005	98.618	**7.271**	24876.376
symb.6	40.946	24.934	39.714	92.548	32.276	**8.507**
Reality	symb.1	symb.2	symb.3	symb.4	symb.5	symb.6

Unified Pairwise Spatial Relations: An Application to Graphical Symbol Retrieval

K.C. Santosh[1], Laurent Wendling[2], and Bart Lamiroy[3]

[1] INRIA Nancy-Grand Est
[2] Nancy Université Henri Poincaré
[3] Nancy Université LORIA - Campus Scientifique - BP 239 - 54506
Vandoeuvre-lés-Nancy Cedex, France
FirstName.LastName@loria.fr

Abstract. In this paper, we present a novel unifying concept of pairwise spatial relations. We develop two way directional relations with respect to a unique point set, based on topology of the studied objects and thus avoids problems related to erroneous choices of reference objects while preserving symmetry. The method is robust to any type of image configuration since the directional relations are topologically guided. An automatic prototype graphical symbol retrieval is presented in order to establish its expressiveness.

1 Introduction

Pairwise spatial relations can greatly ease image understanding, scene analysis and pattern recognition tasks. It has been widely used in many areas such as, GIS understanding [1,2] – where it is necessary to handle efficiently both inaccurate and vague spatial data –, analyzing architectural documents for automatic recognition [3], graphical drawing understanding from scanned color map documents [4] and defining efficient image retrieval methods [5,6,7]. However, it is still difficult to organise and obtain spatial relations in an automated way [8,9].

In general, there is no particular spatial reasoning approach that can adapt to any type of application. They can be either topological [10,11,12,13] or directional [14,15,16,17,18] in nature. Further, models are entirely depending on the characteristics of the studied objects as well as specific application driven needs for spatial relations, such as binary or metrical refinement: the level of detail in the expression of spatial predicates such as *Left, Right etc.*, varies widely from one application to another [19] as does to the precision of the quantised information. Moreover, the introduction of metric information often gives rise to asymmetry, rendering it subject to erroreneous choices of reference objects, which in turn affect the global positioning semantics.

It is possible however, to identify three main levels of information that are involved in spatial relations: topological (that describes neighborhood and incidence e.g. *Dis-Connected, Externally Connected...*), directional (that describes

J.-M. Ogier, W. Liu, and J. Lladós (Eds.): GREC 2009, LNCS 6020, pp. 163–174, 2010.
© Springer-Verlag Berlin Heidelberg 2010

order in space e.g. *Left, Right...*) and metric (e.g. *Near, Far...*). Unlike the existing models that separately treat topological and directional relations, this paper unifies topological and directional information into one descriptor as described in [8] i.e., *topologically* guided *quantised directional* relation with *symmetry*. This unification does not increase computational time. In addition, our method produces angular coverage over a cycle in $I\!R^2$ that avoids fluctuations of spatial predicates or other instabilities that may occur even with a small change in the quantised information. Further, we built upon the idea of semantic inverse theory [8] and preserved symmetry by using a unique reference point set instead of selecting an object from a pair. Moreover, this unique reference point set gives a very sound basis for determining metric relations. Currently, this aspect is beyond the scope of the paper.

We organise the rest of the paper as follows. Section 2 provides a literature review of existing methods with their strenghts and shortcomings. The proposed method appears in section 3, immediately followed by an example. Section 4 explores a series of tests. In section 5, a prototype application based on the proposed method is explored. Section 6 concludes the paper along with a few steps to go further.

2   Review

Topological relations are invariant to topological transformations [20]. These encompass, but are not restricted to rigid transforms as rotation, scaling, and translation. Since we are interested in developing topologically guided directional relations, we need to assess both topology and directional parts. We distinguish the following topological models: the 4-intersection model [10], the 9-intersection model [11], the Voronoi-based 9-intersection model [21], the general intersection model [22] and the calculus-based model [23]. In this paper, we will be considering the 9-intersection model instead of the 4-intersection [24]. The Voronoi-based 9-intersection model is found to be inappropriate in our context. As mentioned earlier, no existing model fully integrates topology. They rather have various degrees of sensitivty to or awareness of topological relations. The fact is that integrating both high level metrical directional and topologically sound descriptions is computationally expensive. Existing approaches present a trade-off between these factors.

The *cone-shaped* model reduces relative positionning to the discretised angle [17] of the sole centroids. It is robust to small variations of shape and size and separation. However, in cases where the centroids coincide it cannot produce any measure. It even leads to the computation of wrong directions, particularly in the case of concavity, where the centroid does not fall within the shape. Extensions like [25], do not lift such ambiguities, nor does it handle to overlapping regions.

Overlapping is a complex problem and approaches based on *angle histograms* are more efficient. Let two objects $\mathbb{A}$ and $\mathbb{B}$ be considered as the sets of their pixels: $\mathbb{A} = \{a_i\}_{i=1...m}$ and $\mathbb{B} = \{b_j\}_{j=1...n}$. The $m \times n$ pairs of points allow for the computation of a set of angles $\theta_{i,j}$ between each (a_i, b_j). The histogram H representing the frequency of occurrence of each angle f_θ can then be formulated as $H_\theta(\mathbb{A}, \mathbb{B}) = [\theta, f_\theta]$. Besides a higher time complexity, there is no significant difference between the *cone-shaped* and the *angle histogram* model when objects are separated by a relatively large distance. The approach was thoroughly studied from its accuracy point of view, and its ambiguity for describing different pairs of objects, resulting in identical histograms [15]. The work was also extended to include metric information in [26], but cannot handle complex objects with holes.

Approaches based on the *Minimum Bounding Rectangle* (MBR) [18,27,5,28,12] give more interesting relations as they also approximate shape and size of the object. The quality of the bounding rectangle depends on compactness[1] of the tile. The sole information used in the MBR approaches is derived from the geometry of the bounding rectangle from which externally aligned orientations: *Left*, *Right*, *Top* and *Bottom etc.* are straightforwardly derived. There are 36 possible configurations of pairwise spatial relations with MBRs and 218 possible spatial relations between non-empty and connected regions [29]). Further, MBR approximates topological relations, which in turn may express false connection/overlapping.

The *F-Histogram* model gives coherent results [16] at the risk of high processing time. It is generic and depends on a sound mathematical framework. It considers pairs of longitudinal sections instead of pairs of points. It does not cover basic topological relations such as, *Inside* and *Overlap* nor does it integrate metric information. Another well-known approach uses *fuzzy landscapes* [14], and is based on fuzzy morphological operators.

3   Proposed Method

In addition to the shortcomings mentioned earlier, proper reference is always a primary factor to organise spatial relations between the objects. It is to be reminded that a change of reference object implies a change in spatial predicates. This, in its turn, may eventually affect overall spatial reasoning (if reference is not given).

In our method, we propose to unify topology and directional relations between the objects $\mathbb{A}$ and $\mathbb{B}$. The proposed method is summarised in two steps. We first extract a unique reference point set $\mathbb{R}$ based on their MBR $(\hat{\mathbb{A}}, \hat{\mathbb{B}})$ topology. This $\mathbb{R}$, thus avoids problems related to erroneous choices of reference entities and will guarantee that subsequent computations of spatial relations $\mathfrak{R}$. In addition, it preserves symmetry.

3.1   Unique Reference Point Set $\mathbb{R}$ Based on Topology

Fig. 1 shows examples of topological configurations and the corresponding reference region $\mathbb{R}$ that they define. $\mathbb{R}$ is derived from the topological relation between

---

[1] Compactness $= \frac{Area(\mathbb{A})}{Area(MBR(\mathbb{A}))}$.

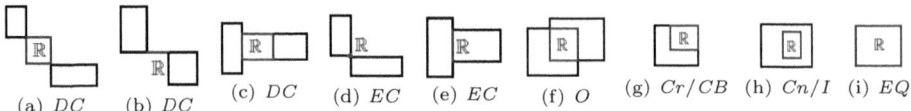

Fig. 1. $\mathbb{R}$ via topological relations

$\hat{\mathbb{A}}$ and $\hat{\mathbb{B}}$, as either the common region of two neighbouring sides in the case of disconnected components or the intersection in the case of overlapping, equal or otherwise connected components. In what follows we shall use the characteristic points $\mathbb{R}p_i$ (extrema and centroid) of $\mathbb{R}$ as,

$$\mathbb{R} = \{\mathbb{R}p_i\}_{i=1\ldots 2n+1}$$

where n is the dimensionality of the region. The dimension of $\mathbb{R}$ changes with the topological relations (Fig. 1). In this illustration, $\mathbb{R}$ becomes both 1D (b) and 2D (a,c) when two MBRs are *Dis-Connected* (*DC*) while, 0D (d) and 1D (e) when they are *Externally Connected* (*EC*). Similarly, only 2D (f, g, h, i) when *Overlapping* (*O*), *Cover/Covered By* (*Cr/CB*), *Contain/Inside* (*Cn/I*), and *Equal* (*EQ*) occur. These are the basic topological predicates closely related to human understanding in conncetion with the Region Connection Calculus-8 (RCC-8) [13]. We express the topological relations in a 9-dimensional binary space based on the 9-intersection model [11]. It uses on the intersections of the boundaries ($\partial*$), interiors ($*^o$) and exteriors ($*^-$) of two shapes $\mathbb{A}$ and $\mathbb{B}$. The topological configuration $Topo.(\mathbb{A}, \mathbb{B})$ is a vector in this space in which componetns equal 0 if the corresponding intersection is empty, and 1 otherwise, as shown here:

$$Topo.(\mathbb{A}, \mathbb{B}) = \begin{bmatrix} \mathbb{A}^o \cap \mathbb{B}^o & \mathbb{A}^o \cap \partial\mathbb{B} & \mathbb{A}^o \cap \mathbb{B}^- \\ \partial\mathbb{A} \cap \mathbb{B}^o & \partial\mathbb{A} \cap \partial\mathbb{B} & \partial\mathbb{A} \cap \mathbb{B}^- \\ \mathbb{A}^- \cap \mathbb{B}^o & \mathbb{A}^- \cap \partial\mathbb{B} & \mathbb{A}^- \cap \mathbb{B}^- \end{bmatrix}$$

Therefore 3×3 binary signature for $DC(\mathbb{A}, \mathbb{B}) = \begin{bmatrix} 0 & 0 & 1 \\ 0 & 0 & 1 \\ 1 & 1 & 1 \end{bmatrix}$, $EC(\mathbb{A}, \mathbb{B}) = \begin{bmatrix} 0 & 0 & 1 \\ 0 & 1 & 1 \\ 1 & 1 & 1 \end{bmatrix}$, $\ldots$, $EQ(\mathbb{A}, \mathbb{B}) = \begin{bmatrix} 1 & 0 & 0 \\ 0 & 1 & 0 \\ 0 & 0 & 1 \end{bmatrix}$.

3.2   Directional Relations - Radial Line Model (RLM)

The model precisely yields angular coverage over a cycle in $I\!\!R^2$ and thus avoids the use of spatial predicates as in the existing models. It is to remind that the level of expresion of spatial predicates is sensitive to every small change in quantised information. The model further, explores both qualitative and quantitative process.

Binary Relations. Let $\mathbb{X}$ be one of the initial objects $\mathbb{A}$ or $\mathbb{B}$ and let their reference region be $\mathbb{R}$. At every $\mathbb{R}p_i$, we cover the surrounding space at regular radial intervals of $\Theta = 2\pi/m$, such that $\theta_j = j\Theta$. It rotates over a cycle and intersecting with $\mathbb{X}$, and generates binary values at every step of its rotation (Fig. 2). This gives a boolean histogram of angular coverage,

$$\mathcal{H}(\mathbb{X}, \mathbb{R}p_i) = [I(\mathbb{R}p_i, j\Theta)]_{j=0..m} \text{ where } I(\mathbb{R}p_i, \theta_j) = \begin{cases} 1 \text{ if } line(\mathbb{R}p_i, \theta_j) \cap \mathbb{X} \neq \emptyset \\ 0 \text{ otherwise} \end{cases}$$

This is extended *wlog* to the sector defined by two successive angle values: $Cone(\mathbb{R}p_i, \theta_j, \theta_{j+1})$. The process is repeated for every $\mathbb{R}p_i$.

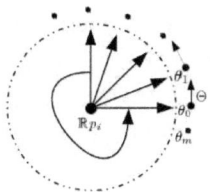

Fig. 2. Radial line $line(\mathbb{R}p_i, \theta_j)$ rotation

Refined Relations. In order to be robust to noise and to border conditions due to discritization, we extend the boolean description by partially building on the cloud model [30]. We normalise the coverage with respect to the total area of the object under consideration with respect to every $\mathbb{R}p_i$, $\frac{Area(Cone(\mathbb{R}p_i, \theta_j, \theta_{j+1}))}{Area(\mathbb{X})}$ such that $\sum \mathcal{H}(.) = 1$. This goes without loss of generality and it is robust to any type of point set (either a point, a line or a region). It is not only convey information about the presence of objects in a given direction, but also infores about the proportion of the object that is lying there.

We average the resulting histograms (from every $\mathbb{R}p_i$) to produce $\Re(\mathbb{X}, \mathbb{R})$.

Remarks

- Spatial Relations: We use $\Re_{Bin}(.)$ and $\Re_{Ref}(.)$ for binary and refined relations respectively.
- Symmetry: Due to $\mathbb{R}$, RLM yields two way directional relations as well as it preserves symmetry. For symmetry reasons, we use $\Re(\star, *) = \{\mathcal{H}(\star, \mathbb{R}), \mathcal{H}(*, \mathbb{R})\}$. This guarantees that, $\Re(\star, *) = \Re(*, \star)$.
- Resolution: Θ determines a trade-off between precision and time complexity, determining *resolution* of $\mathcal{H}$. Smaller the *resolution*, better the information exploitation.

3.3   An Example

Fig. 3 shows an example illustrating our method for a pair of truly overlapping objects. We first show how to determine $\mathbb{R}$ from $Topo.(\hat{\mathbb{A}},\hat{\mathbb{B}})$. Fig. 4 shows how both boolean and metrical refinement histograms are produced from $Topo.(\mathbb{R}, \mathbb{X})$.

As an example, we use $\Theta = \pi/20$ to produce $\mathcal{H}$ for $\Re_{Bin}(.)$ and $\Re_{Ref}(.)$. For every $\mathbb{R}p_i$, the visual representations of binary (blue) and refined (red in blue mask – zoomed ×3) histograms are shown for object $\mathbb{A}$ and $\mathbb{B}$ in Fig. 4. For easier understanding, the directional relation signatures with respect to the reference centroid point $\mathbb{R}p_c$ are:

$$\mathcal{H}_{Bin}(\mathbb{A}, \mathbb{R}p_c) = [0\ 0\ 0\ 0\ 0\ 0\ 1\ 1\ 1\ 1\ 1\ 1\ 1\ 1\ 1\ 1\ 0\ 0\ 0\ 0\ 0\ 0\ 0\ 0\ 0\ 0\ 0\ 0]$$

$$\mathcal{H}_{Ref}(\mathbb{A}, \mathbb{R}p_c) = [0\ 0\ 0\ 0\ 0\ 0\ 0.0261\ 0.0722\ 0.0746\ 0.0775\ 0.0708\ 0.0746\ 0.0841$$
$$0.0675\ 0.0433\ 0.0299\ 0.0328\ 0.0323\ 0.0328\ 0.0352\ 0.0299\ 0.0328$$
$$0.0323\ 0.0328\ 0.0352\ 0.0375\ 0.0361\ 0.0095\ 0\ 0\ 0\ 0\ 0\ 0\ 0\ 0\ 0\ 0\ 0\ 0]$$

After averaging, it is found that $\Re(\mathbb{A}, \mathbb{R}) \neq \emptyset$ while $\Re(\mathbb{B}, \mathbb{R})$ is. It is due to the fact that $Cr(\mathbb{R}, \mathbb{X})$ or $CB(\mathbb{X}, \mathbb{R})$.

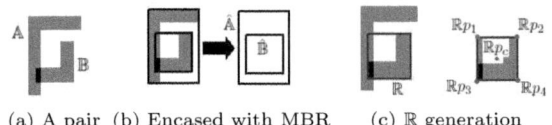

(a) A pair   (b) Encased with MBR   (c) $\mathbb{R}$ generation

Fig. 3. An example to illustrate the proposed method (a truly overlapping case)

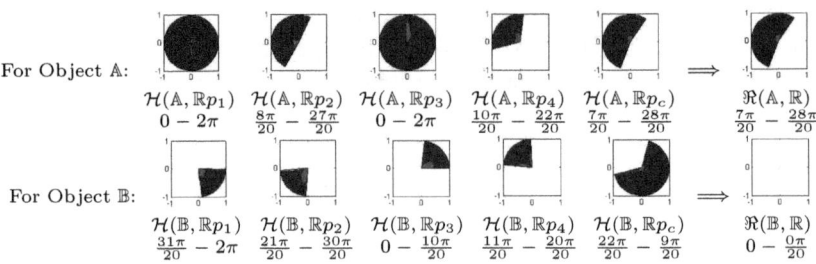

Fig. 4. $\frac{7\pi}{20} - \frac{28\pi}{20}(\mathbb{A}, \mathbb{R})$ (1st row) and $0 - \frac{0\pi}{20}(\mathbb{B}, \mathbb{R})$ (2nd row)

4   Experiments

In this experiment, we use segmented and labeled objects in order just to explore the expressive power of the method. Table. 1 shows the behaviour of our method

Table 1. A series of tests $\left(\Theta = \frac{\pi}{20}\right)$

Image with $\mathbb{R}$	Topology (A,B)	$(\hat{A},\hat{B})$	$\mathfrak{R}_{Bin.}(.) + \mathfrak{R}_{Ref.}(.)$ $\mathfrak{R}(A,R)$	$\mathfrak{R}(B,R)$	Image with $\mathbb{R}$	Topology (A,B)	$(\hat{A},\hat{B})$	$\mathfrak{R}_{Bin.}(.) + \mathfrak{R}_{Ref.}(.)$ $\mathfrak{R}(A,R)$	$\mathfrak{R}(B,R)$
Experiment I									
(a)	DC	DC	$\frac{11\pi}{20} - \frac{28\pi}{20}$	$\frac{33\pi}{20} - \frac{8\pi}{20}$	(d)	DC	DC	$\frac{11\pi}{20} - \frac{28\pi}{20}$	$\frac{33\pi}{20} - \frac{8\pi}{20}$
(b)	DC	DC	$\frac{11\pi}{20} - \frac{28\pi}{20}$	$\frac{33\pi}{20} - \frac{8\pi}{20}$	(e)	DC	DC	$\frac{14\pi}{20} - \frac{28\pi}{20}$	$\frac{33\pi}{20} - \frac{2\pi}{20}$
(c)	DC	DC	$\frac{11\pi}{20} - \frac{28\pi}{20}$	$\frac{33\pi}{20} - \frac{8\pi}{20}$	(f)	DC	DC	$\frac{14\pi}{20} - \frac{28\pi}{20}$	$\frac{33\pi}{20} - \frac{2\pi}{20}$
Experiment II									
(i)	DC	DC	$\frac{11\pi}{20} - \frac{30\pi}{20}$	$\frac{33\pi}{20} - \frac{7\pi}{20}$	(iv)	DC	Cn	$\frac{04\pi}{20} - \frac{37\pi}{20}$	$0 - \frac{0\pi}{20}$
(ii)	DC	O	$\frac{11\pi}{10} - \frac{31\pi}{20}$	$\frac{31\pi}{20} - \frac{10\pi}{20}$	(v)	DC	Cn	$\frac{02\pi}{20} - \frac{39\pi}{20}$	$0 - \frac{0\pi}{20}$
(iii)	DC	Cn	$\frac{09\pi}{20} - \frac{32\pi}{20}$	$0 - \frac{0\pi}{20}$	(vi)	DC	Cn	$0 - \frac{40\pi}{20}$	$0 - \frac{0\pi}{20}$

on a series of objects, ranging from simple, solid and regular pairs of objects to concave, as well as complex ones, covering all possible topologies. The overall result shows a comparison between the topology of the shapes themselves (A, B) as well as their MBR $(\hat{A}, \hat{B})$. Comparison made with the topology between them determines the qualtity of the MBR tile. The difference in topological relations is due to MBR false connection/overlapping. It is however, not a problem in our method since it uses the initial objects (A, B) to produce $\mathfrak{R}$ after the discovery of $\mathbb{R}$. Further, $\mathfrak{R}_{Bin}$ and $\mathfrak{R}_{Ref}$ of both objects with respect to $\mathbb{R}$ are also shown in Table 1.

4.1   Discussions

In a few congurations (Table 1), RLM yields identical $\mathfrak{R}_{Bin}$ but different $\mathfrak{R}_{Ref}$ between different pairs of objects. This behaviour can be observed in Experiment I for (a), (b), (c) and (d) as well as (e) and (f). It is to be noted that $\mathfrak{R}_{Ref}$ is only used to cross validate when $\mathfrak{R}_{Bin}$ is found to be non discriminant. Experiment II shows the behaviour of our method on progressive coverage of one object by another. In this illustration, a progressive angular coverage as well as effect of *inclusion* topological relations on directional relations are clearly demonstrated. As in Fig. 4 $\Re(\mathbb{B}, \mathbb{R}) = \emptyset$ because $Cn(\mathbb{R}, \mathbb{B})$.

Overall, directional relations are topologically guided. For DC, EC and O relations, it is straightforward. But for all *inclusion* relations like Cr/CB, Cn/I, and EQ, $\mathcal{H}(\mathbb{X}, \mathbb{R}) = \emptyset$. Therefore, only one part of $\Re$ needs to be computed. This eventually reduces time complexity as RLM gives no measure.

4.2   Time Complexity

In this section we analyse the time complexity behaviour both with respect to the precision (Θ) and the size of the objects. Unlike the existing models described in section 2, processing time does not increase exponentially with the size of the images but has only a little effect. Fig. 5 shows time complexity measure by increasing the size of the image (scaling step +0.2). Overall, the RLM takes almost the same time to make a complete rotation over a cycle in all size of images. This is the main reason for the time complexity graph being approximately level. In order to increase the speed, one can use boundaries of objects to compute the boolean histogram.

It has no doubt that cone shaped and classical projection models run faster than the RLM and histogram of angles (and its variant). In our method, resolution Θ determines which one to trade off: either quality or computational load. It is to be noted that the RLM resolution should be chosen based on the size of objects under consideration. Further, time complexity is compensated to $\frac{n(n-1)}{2}$ for n objects due to the symmetry relations.

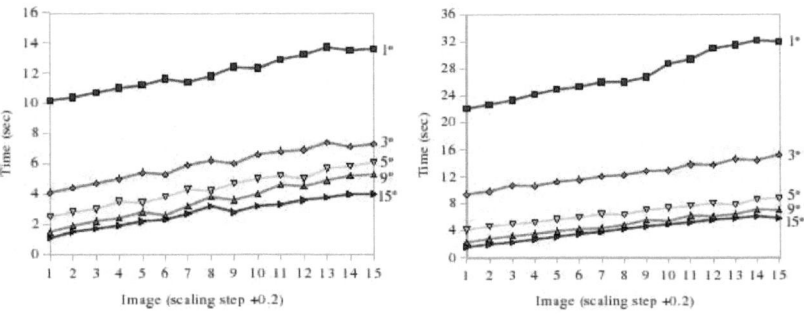

Fig. 5. Time complexity for $\Re_{Bin}$ (left) and $\Re_{Ref}$ (right) for a number of different resolutions

5   An Application

5.1   Symbol Description via Spatial Relations

We use the visual vocabulary presented in [31,32] to organise spatial relations and use them for symbol description. In our case, the vocabulary consists of:

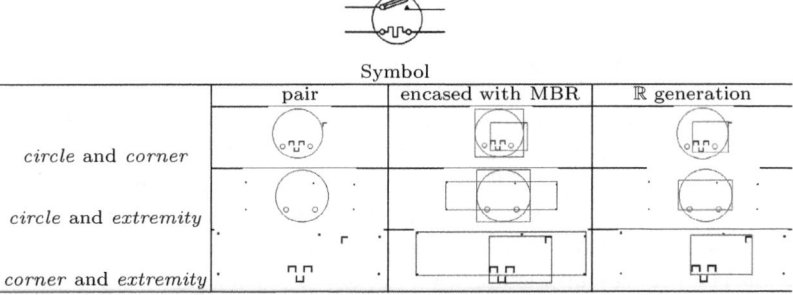

Fig. 6. Reference point set $\mathbb{R}$ generation for all possible pairs of classes

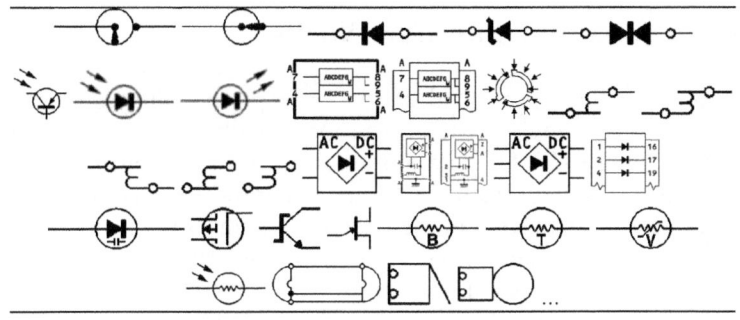

Fig. 7. A small set of electrical symbols

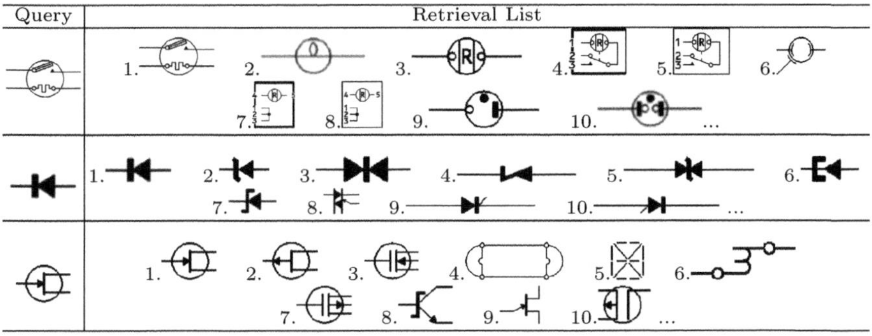

Fig. 8. Retrieval lists based on similarity for a few chosen queries $\left(\Theta = \frac{\pi}{180}\right)$

circles, corners, loose ends and *thick* (filled) components. To handle arbitrary number of vocabulary elements, we group them together into 'classes' having the same type, as shown in Fig. 6. The symbol is then modeled as a graph in which each group is a typed node, and the arcs, representing the spatial relations $\mathfrak{R}$.

5.2   Symbol Retrieval

We use straightforward graph matching to retrieve similar symbols with respect to the chosen query. We manually choose query symbols which are matched with symbols in the database. We employ similarity ranking based on the geometric distance. We take manhattan distance metric between the corresponding relations in the graphs, $\sum_i^n |(\Re_i(.) - \Re'_i(.))|$. Fusion of matching scores from individual relations reflects how similar the symbol in the database with query symbol. Based on the similarity value, we rank retrieval symbols.

Since it is a protype application, we use small database and a few test queries. A sample of the database is shown in Fig. 7. Ranking retrieval lists for a few choosen queries are shown in Fig. 8.

6   Conclusions and Further Works

In this paper, we have presented a new concept of unifying pairwise spatial relations. Since directional relations are topologically guided, one does not need to model them separately. The method provides accurate spatial organisation for any type of image configuration. In addition, it produces symmetric directional relations thanks to the use of a unique reference point set. These two way spatial relations are developed at one pass, while this is not the case in existing models.

One of the possible applications – prototype symbol retrieval – is reported, using a small database. Further work consists of using intra-class spatial relations as well as pre-filtering techniques to establish precision and recall in symbol retrieval. We will further develop the use of this method for scene matching and image analysis tasks which will ultimately bring it into the context of full image recognition.

References

1. Worboys, M.: GIS - A computing perspective. Taylor and Francis, Abington (1995)
2. Goodchild, M., Gopal, S.: The Accuracy of Spatial Databases. Taylor and Francis, Basingstoke (1990)
3. Vandenbrande, J.H., Requicha, A.A.G.: Spatial Reasoning for the Automatic Recognition of Machinable Features in Solid Models. IEEE PAMI 15(12), 1269–1285 (1993)
4. Centeno, J.S.: Segmentation of Thematic Maps Using Colour and Spatial Attributes. In: Chhabra, A.K., Tombre, K. (eds.) GREC 1997. LNCS, vol. 1389, pp. 221–230. Springer, Heidelberg (1998)
5. Lee, S.H., Hsu, F.J.: Spatial Reasoning and Similarity Retrieval of Images Using 2D C-string Knowledge Representation. PR 25(3), 305–318 (1992)
6. Heidemann, G.: Combining spatial and colour information for content based image retrieval. CVIU 94, 234–270 (2004)
7. Medasani, S., Krishnapuram, R.: A fuzzy approach to content-based image retrieval. In: Proc. of FUZZ-IEEE, Seoul, Korea, pp. 1251–1260 (1997)
8. Freeman, J.: The modelling of spatial relations. CGIP 4, 156–171 (1975)

9. Rosenfeld, A., Kak, A.: Digital picture processing, vol. 2. Academic Press, London (1982)
10. Egenhofer, M., Franzosa, R.: Point-set Topological Spatial Relations. Intl. Journal of GIS 5(2), 161–174 (1991)
11. Egenhofer, M., Herring, J.R.: Categorizing Binary Topological Relations Between Regions, Lines, and Points in Geographic Databases. University of Maine, Research Report (1991)
12. Papadias, D., Sellis, T., Theodoridis, Y., Egenhofer, M.J.: Topological Relations in the world of Minimum Bounding Rectangles: a Study with R-trees. In: Intl. Conf. on Managament Data, pp. 92–103 (1995)
13. Renz, J., Nebel, B.: Spatial Reasoning with Topological Information. In: Freksa, C., Habel, C., Wender, K.F. (eds.) Spatial Cognition 1998. LNCS (LNAI), vol. 1404, pp. 351–372. Springer, Heidelberg (1998)
14. Bloch, I.: Fuzzy relative position between objects in image processing: a morphological approach. IEEE PAMI 21(7), 657–664 (1999)
15. Wang, X., Keller, J.: Human-Based Spatial Relationship Generalization Through Neural/Fuzzy Approaches. Fuzzy Sets and Systems 101, 5–20 (1999)
16. Matsakis, P., Wendling, L.: A New Way to Represent the Relative Position Between Areal Objects. IEEE PAMI 21(7), 634–643 (1999)
17. Mitra, D.: A Class of Star-Algebras for Point-Based Qualitative Reasoning in Two-Dimensional Space. In: 15th Intl. Florida AI Research Society Conf., pp. 486–491 (2002)
18. Peuquet, D., CI-Xiang, Z.: An algorithm to determine the directional relationship between arbitrarily-shaped polygons in the plane. PR 20(1), 65–74 (1987)
19. Retz-Schmidt, G.: Various Views on Spatial Prepositions. AI Magazine, 95–104 (1988)
20. Egenhofer, M.J.: A Formal Definition of Binary Topological Relationships. In: Litwin, W., Schek, H.-J. (eds.) FODO 1989. LNCS, vol. 367, pp. 457–472. Springer, Heidelberg (1989)
21. Chen, J., Li, C., Li, Z., Gold, C.: A Voronoi-based 9-intersection Model for Spatial Relations. Intl. Journal of GIS 15(3), 201–220 (2001)
22. Abdelmoty, A., El-Geresy, B.: A General Method for Spatial Reasoning in Spatial Databases. In: The Fourth International Conference on Information and Knowledge Management, pp. 312–317 (1995)
23. Clementini, E., Felice, P.D., van Oosterom, P.: A small set of formal topological relationships suitable for end-user interaction. In: Abel, D.J., Ooi, B.-C. (eds.) SSD 1993. LNCS, vol. 692, pp. 277–336. Springer, Heidelberg (1993)
24. Egenhofer, M., Sharma, J., Mark, D.: A Critical Comparison of the 4-Intersection and 9-Intersection Models for Spatial Relations: Formal Analysis. In: McMaster, R., Armstrong, M. (eds.), pp. 56–69 (1993)
25. Miyajima, K., Ralescu, A.: Spatial Organization in 2D Segmented Images: Representation and Recognition of Primitive Spatial Relations. Fuzzy Sets and Systems 2(65), 225–236 (1994)
26. Wang, Y., Makedon, F.: R-histogram: Quantitative representation of spatial relations for similarity-based image retrieval. In: The 11th Annual ACM International Conf. on Multimedia, pp. 323–326 (2003)
27. Dutta, S.: Approximate spatial reasoning: integrating qualitative and quantitative constraints. Intl. Journal of Approximate Reasoning 5, 307–331 (1991)
28. Jungert, E.: Qualitative spatial reasoning for determination of object relations using symbolic interval projections. In: IEEE Sympo. on Visual Lang., pp. 24–27 (1993)

29. Sun, H., Chen, X.: Research on Technologies of Spatial Configuration Information Retrieval. In: IEEE 8th ACIS Int. Conf. on Software Engineering, AI, Networking, and Parallel/Distributed Computing, pp. 396–401 (2007)
30. Xuehua, T., Lingkui, M., Kun, Q.: Study on the Uncertain Directional Relations Model based on Cloud Model. In: Intl. Archives of the Photogrammetry, Remote Sensing and Spatial Infor. Sc., pp. 345–350 (2008)
31. Santosh, K.C., Lamiroy, B., Ropers, J.P.: Utilisation de Programmation Logique Inductive pour la Reconnaissance de Symboles. In: 9èmes Journées Francophones Extraction et Gestion des Connaissances, pp. 35–42 (2009)
32. Santosh, K.C., Lamiroy, B., Ropers, J.P.: Inductive logic programming for symbol recognition. In: 10th ICDAR, Barcelona, Spain, pp. 1330–1334 (2009)

Real Scene Sign Recognition

Linlin Li and Chew Lim Tan

Computer Science, National University of Singapore, Singapore
{lilinlin,tancl}@comp.nus.edu.sg

Abstract. A common problem encountered in recognizing signs in real-scene images is the perspective deformation. In this paper, we employ a descriptor named Cross Ratio Spectrum for recognizing real scene signs. Particularly, this method will be applied in two different ways: recognizing a multi-component sign as an whole entity or recognizing individual components separately. For the second strategy, a graph matching is used to finally decide the identify of the query sign.

Keywords: Graphics Recognition, Real Scene Recognition, Perspective Deformation.

1 Introduction

With the advancement of camera technology, many techniques are developed for real scene symbol/character recognition. Traffic sign recognition [3,4] is implemented in Driver Support Systems to recognize the traffic signs put on the road e.g. "slow", "school ahead", or "turn ahead". Another application is license plate recognition [9], which is practically useful in parking lot billing, toll collecting monitoring, road law enforcement, and security management. Cargo container code recognition systems [5] are used in ports to automatically read cargo container codes for cargo tracking and allocation. Signboard recognition systems or translation cameras recognize signs captured by a portable camera, helping international tourists to overcome language barrier.

Many difficulties are encountered in real scene symbol/character recognition, including uneven illumination, occlusion, blur, low resolution as well as perspective deformation. For traffic sign recognition, license plate recognition, and cargo container code recognition, the recognition target is far away from the camera and moving, and thus issues needed to be resolved are blur and low resolution. For translation cameras, because the recognition target is often near the camera, the perspective distortion and uneven illumination become the main obstacles.

We are particularly interested in signboard recognition in this papers. Besides perspective distortion and uneven illumination, another difficulty of signboard recognition is in the concise nature of signs: a sign often comprises of only a few words/characters and some graphic symbols displaying a certain format. It will cause problems in both detection and recognition. An approach to address the perspective issue is to use Affine invariant detectors and Affine invariant descriptors. However, existing Affine invariant descriptors, like SIFT, work well

J.-M. Ogier, W. Liu, and J. Lladós (Eds.): GREC 2009, LNCS 6020, pp. 175–186, 2010.

on complex objects with great variation in intensity. However, the simplicity and symmetry of symbols make the Affine invariant descriptors not discriminative enough.

In our early paper [6], a real-scene character recognition method was proposed, based on a descriptor named cross ratio spectrum. The main contribution of this paper is to propose two strategies to recognize multi-component signboards. In Section 2, we will show the performance of our recognition method, treating multi-component signboards as whole entities. Since this strategy is only useful when the boundary of a signboard is known, we will discuss a more general case when the such condition is satisfied in Section 3. In particular, a graph matching method is proposed to assemble the individual component recognition results gotten by our previous method [6]. With this strategy, the recognition can be conducted for real scene images without prior knowledge about the boundary of a signboard.

2   Recognize Perspectively Deformed Symbols

In this section, a brief review of the recognition method proposed in [6] will be made. The experimental result of applying it on whole signboards will be also presented.

2.1   Comparing Two Cross Ratio Spectra

Cross Ratio is a fundamental invariant for projective transformation [7]. The cross ratio of four collinear points (P_1, P_2, P_3, P_4) displayed in order is defined as:

$$cross\_ratio(P_1, P_2, P_3, P_4) = \frac{P_1 P_3}{P_2 P_3} / \frac{P_1 P_4}{P_2 P_4} \qquad (1)$$

where $P_i P_j$ denotes the distance between P_i and P_j. $cross\_ratio(P_1, P_2, P_3, P_4)$ remains constant under any projective transformation.

Suppose there are two sample points P_1 and P_k on the convex contour of a symbol H, as shown in Fig. 1. I_1 and I_2 are the intersections between the line $P_1 P_k$ and the symbol contour. The cross ratio, defined by P_1, I_1, I_2, and P_k, is denoted by $CR(P_1, P_k)$. When there are more than two intersections between two points, only the first two intersections (near P_1) are used. If the number of intersections is 0 or 1, and thus no cross ratio value can be computed, the pseudo-cross ratio value is assigned as -1 and 0 respectively.

A cross ratio spectrum is a sequence of cross ratios. Suppose the sample point sequence of the convex hull of P is $\{P_s, s = [1 : S]\}$, where P_2 is the anti-clock-wise neighbor pixels of P_1, etc. The Cross Ratio Spectrum (CRS) of a pixel P_i is defined as:

$$CRS(P_i) = \{CR(P_i, P_{i+1}), ..., CR(P_i, P_n), CR(P_i, P_1), ..., CR(P_i, P_{i-1})\}$$

An example of a cross ratio spectrum is shown in Fig. 1. An important hypothesis about the cross ratio spectrum is that:

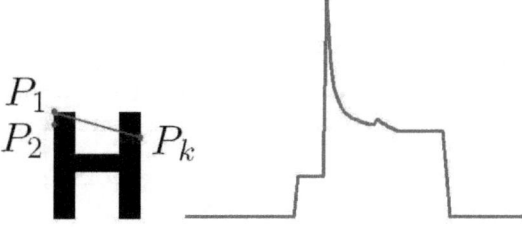

Fig. 1. The Cross Ratio Spectrum of point P_1 at the left top corner of symbol 'H'

If P_i and P'_i are two mapping points in a symbol P and its perspective version P', spectrum $CRS(P_i)$ is an uneven stretching version of spectrum $CRS(P'_i)$.

Hence, we use Dynamic Time Warping (DTW) to compare the similarity between two spectra. In the following sections, Q refers to an unknown symbol with M sample points on the convex hull, and T refers to a template symbol with N sample points. The notation of $CRS(Q_i)$ is rewritten as $CRS(Q_i) = \{q_u, u = 1 : M-1\}$ for simplicity. Similarly, $CRS(T_j) = \{t_v, v = 1 : N-1\}$. The comparison between two sample points Q_i and T_j is formulated as:

$$DTW(u,v) = min \begin{cases} DTW(u-1, v-1) + c(u,v) \\ DTW(u-1, v) + c(u,v) \\ DTW(u, v-1) + c(u,v) \end{cases} \quad (2)$$

$$c(u,v) = \frac{abs(log(CR(Q_i, Q_u)) - log(CR(T_j, T_v)))}{log(CR(Q_i, Q_u)) + log(CR(T_j, T_v))} \quad (3)$$

If $CR(.,.)$ is -1 or 0, $log(CR(.,.))$ is assigned as -1 and -0.5 respectively. The distance between points Q_i and T_j is given by the last item:

$$DTW\_dist(Q_i, T_j) = DTW(M-1, N-1) \quad (4)$$

2.2   Comparing Two Symbols

In order to compare two symbols Q and T. Two steps are followed:

- DTW comparisons are conducted between each pair of Q_i and T_j, and a DTW-distance-table is constructed as the table showed in Fig. 2(a). Cells in the table denote the distances of corresponding pixel pairs.
- Each time, a DTW is applied to a sub-table comprising of column $\{\hbar, \hbar + 1, ..., \hbar + M - 1\}$ of the table, to align T_1 with $Q_\hbar$ and T_N with $Q_{\hbar+M-1}$ as the boundary condition. The comparison is formulated as follows:

$$DTW(i,j) = min \begin{cases} DTW(i-1, j-1) + c(i,j) \\ DTW(i-1, j) + c(i,j) \\ DTW(i, j-1) + c(i,j) \end{cases} \quad (5)$$

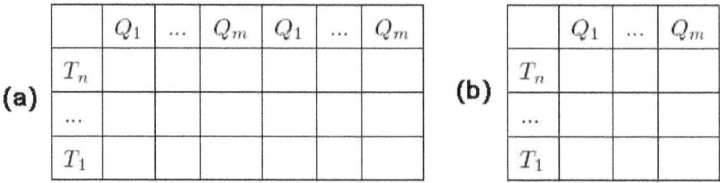

	Q_1	...	Q_m	Q_1	...	Q_m
T_n						
...						
T_1						

	Q_1	...	Q_m
T_n			
...			
T_1			

(a)   (b)

Fig. 2. (a)DTW distance table. (b)Searching in a sub-table.

Fig. 3. Samples of testing symbols

$$c(i, j) = DTW\_dist\_table(\hbar + i - 1, j) \qquad (6)$$

where $i = 1 : M$ and $j = 1 : N$. A sub-table is shown in Fig. 2 (b) when $\hbar = 1$. A candidate distance between Q and T is given by $DTW(M, N)$. M DTW comparisons are conducted. Among M candidate distances, the smallest one gives the desirable global distance.

The comparison algorithm has a bi-quadratic time complexity of $O(M^2 * N^2)$. This will be solved by the indexing step in Section 3.1. We take a 1NN recognition strategy int the experiment: a query is compared with all templates, and the template which has the smallest distance with the query gives the identity of the query.

2.3   Synthetic Symbol Testing

In this section, the ability of handling perspective deformation of the proposed method will be illustrated with a well defined synthetic image set. Scale Invariant Feature Transforms (SIFT) with Harris-Affine detector [1], Shape Context [2], are employed as comparative methods.

---

[1] http://www.robots.ox.ac.uk/~vgg/research/affine/index.html
[2] http://www.eecs.berkeley.edu/Research/Projects/CS/vision/shape/sc_digits.html

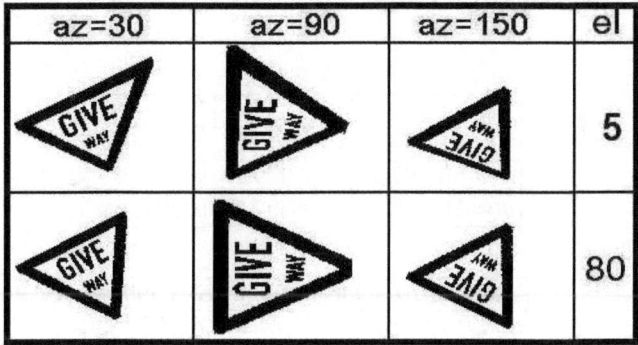

Fig. 4. Deformed versions of a symbol

Shape context is a global descriptor, in which each sample point on the shape contour is represented by the distribution of the remaining points relative to it, and a point-to-point correspondence between the query and a template is solved by a bipartite graph matching. After that, a Thin Plate Spline model-based transformation is estimated for a better alignment between two shapes. The distance between two shapes is given by a sum of shape context distances. Iterations are employed for better recognition result. Our experiment follows the same process as introduced in [2].

SIFT is a local Affine invariant descriptor which describes a local region around a key point. SIFT descriptor is robust to occlusion, and does not require segmentation. However, a foreseeable problem of applying SIFT descriptor to symbols is the lack of discriminating power, because of the simple and symmetrical structure of symbols. In order to solve the structural ambiguity and maximize the recognition strength of SIFT descriptor, the recognition process is designed as follows. A Harris-Affine detector is used to detect Affine invariant key points. For each key point of Q, its first 20 nearest neighbors are found in the training set. If the distance is less than a threshold (200 in the experiment), the neighbor is kept, otherwise is thrown away. RANSAC fitting algorithm is then used to further filter false matches. False matches (outliers) are removed by checking for agreement, between each match and the perspective transformation model (8 degrees of freedom) generated by RANSAC. The identity of Q is given by the template which has the maximum number of correct matches with Q.

In our experiment, the convex hull of a symbol is extracted by [1], and the points are sampled in an equal distance manner. A subset of a standard traffic sign database[3] (45 signs with red and blue frames) are employed as the template set. Some symbols are shown in Fig. 3. 12 testing datasets are generated by Matlab using various perspective parameters. The perspective images are generated by setting the target point at a specific point o', and setting the perspective viewing angle as 25° (to model a general camera lens), while changing the

---

[3] http://en.wikipedia.org/wiki/Road_signs_in_Singapore

Table 1. Recognition accuracy of synthetic images

(a) Our method

el=	10°	30°	50°	70°	90°
$n = 0$	99.25	100	100	100	100
$n = 50$	97.77	97.77	99.25	100	100
$n = 100$	95.92	97.77	97.77	100	100

(b) SIFT

el=	10°	30°	50°	70°	90°
$n = 0$	54.07	60.74	74.44	74.44	74.44
$n = 50$	52.22	53.33	68.88	73.33	74.44
$n = 100$	38.14	44.44	51.48	51.48	74.44

(c) Shape Context

el=	10°	30°	50°	70°	90°
$n = 0$	71.85	82.96	94.07	100	100
$n = 50$	51.48	74.81	88.88	100	100
$n = 100$	45.18	60.74	74.81	88.88	100

azimuth (az) and elevation (el) angles gradually. Point o' is at the same horizontal line as the mass center of a symbol, denoted by o, with a distance of $n \times h$, where n is a positive integer and h is the height of the symbol. Generally, the larger the n is, the greater the deformation is. For each testing set, n and el are predefined, and az is set as $\{30°, 90°, 150°, 210°, 270°, 330°\}$ respectively. Therefore, each testing set comprises of $6 \times 45 = 270$ symbols. Deformed versions of a symbol with different perspective parameters are shown in Fig. 4.

Tables 1(a), (b), and (c) show the recognition accuracy using our method, SIFT, and Shape Context methods respectively, where accuracy is the number of correctly recognized symbols over the number of total query symbols. The accuracy in each cell is based on a testing set comprising of 270 symbols generated with corresponding perspective parameters. It is easy to see that when symbols are deformed by perspective projection, our method has a better recognition accuracy than other methods. Table 1(a) shows that the performance of our method degrades only a little with increasing deformation. For the performance of SIFT descriptor shown in Table 1(b), when the perspective deformation is moderate, such as $n = \{0, 50\}$ and $el \geq 50°$, errors are mainly caused by the structural similarity of symbols. However, when the deformation is more severe, the descriptor is not resistant to the deformation any longer. Table 1(c) shows that when the deformation is moderate, Shape Context has a very good recognition accuracy. However, when the perspective becomes more severe, it is not able to work well. Under a perspective deformation, some parts of a symbol expand, while some parts shrink, which affects the statistics calculated from the symbol. Therefore, statistic-based methods like SIFT and Shape Context will not work under severe perspective deformation.

Fig. 5. Rectify photos by the correspondence given by different methods, rectified images are scaled for better viewing purpose. (a) real-scene symbols (b) by our method (c)by SIFT (d)by Shape Context (e) template.

The alignment information is useful for perspective rectification. Fig. 5 shows the results of rectifying two symbols by ours method, SIFT, and Shape Context respectively, using Least Square method to evaluate a transformation model based on correspondences between a real-scene symbol and the template achieved by the three methods.

3   Identifying Signboards in Real Scene

In Section 2, we take a signboard as a whole entity, assuming that its boundary is already known. However, it is difficult to detect the boundary of a signboard with disjointed components in a real scene image with presence of many irrelevant objects. It is even more difficult when several signboards gather together or incomplete signboards exist. In these cases, the strategy introduced in Section 2 cannot be applied directly.

(a)   (b)

Fig. 6. (a)Locating a signboard out of a real scene image. (b)The identity of the signboard.

The task of this section is to identify template signboards in real scene images, as shown in Fig. 6. First, regions likely to contain signboards are found out, and then are decomposed into components. A component is a homogenous area with uniform color. Second, components are recognized with the method proposed in [6]. Finally, a graph matching process is employed to find the identity of signboards present in the image.

3.1   Indexing Templates

The training set is the same as used in Section 2. These signboards are indexed in three layers: sign, component and point. The index structure is shown in Fig. 7. In the sign layer, topology information of signs is kept. Details can be found in Section 3.3. In the component layer, the point index information for each component is maintained. The point layer stores actual CRS descriptors of points.

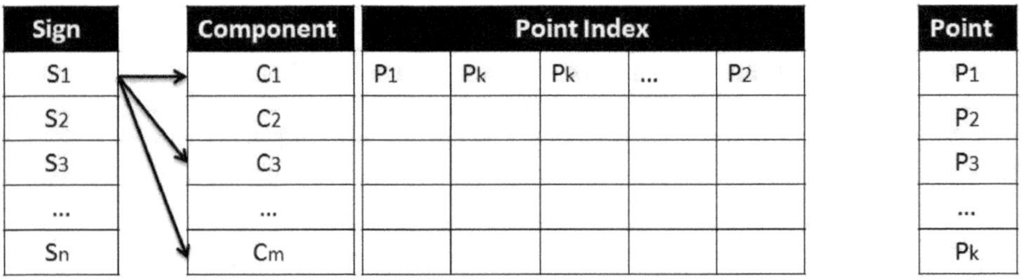

Sign	Component	Point Index					Point
S1	C1	P1	Pk	Pk	...	P2	P1
S2	C2						P2
S3	C3						P3
...	...						...
Sn	Cm						Pk

Fig. 7. The index structure

In order to build the component layer, we first dismantle template signboards into components by Color Structure Code segmentation[4] [8]. All foreground components, namely red, blue, black components, and white components surrounding by blue or red components are indexed. Duplicate components are removed as follows. The template component set Γ is initialized as $\Gamma = \emptyset$. If a component cannot be recognized with Γ correctly, it is added to Γ, otherwise thrown away. We got 138 template components from the training set.

For the point layer, CRSs of all points in Γ are extracted. Based on an important observation that many neighboring points have similar spectra, we will further reduce the number of points needed to be indexed by KNN clustering. In particular, CRSs of all 11040 points extracted from Γ (80 points from each of 138 template signboard) are obtained. Pairwise DTW distances are computed for these points. KNN clustering method is applied on these distances. 400 clusters are formed. The centroid of a cluster is defined as the CRS which has the minimum sum of distances to the other CRSs in the cluster. The centroid

---

[4] http://www.uni-koblenz.de/~lb/lb_research/research.csc.html

and member CRSs for each cluster are recorded. When a query comes in, it is compared to the centroid of each cluster. The results are used to fill up the DTW-distance-table (Fig. 2(a)), referring to the member list of clusters. Details about the indexing and searching process can be found in [10].

3.2   Searching Index

When a query image Q comes in, it is dismantled into components by Color Structure Code segmentation method [8], as shown in Fig. 8. Components which are too small are thrown away. The nearest neighbor for each remaining query component is found in Γ with the method proposed in [10]. If the distance between the query component and its nearest neighbor is larger than a certain threshold, the match fails. Note that the segmentation algorithm tends to over-segment due to uneven illumination. Therefore, in this case, the query component is merged with its adjacent components to form a new query, as shown in Fig. 9, if the hue difference between the query component and its adjacent neighbor is less than 5%. Then the index searching is run again with the new query component.

(a)   (b)

(c)

Fig. 8. Preprocessing: (a)Original image. (b)Segmentation results. (c)Examples of components obtained from the original image.

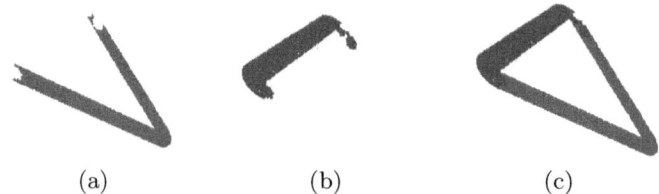

Fig. 9. (a)A component cannot be matched to any template. (b)An adjacent component of (a) which has similar hue. (c)A new component is formed by merging (a) with (b).

3.3   Template Model and Query Graph

Signboards may have identical components but different layout, such as signboards in Fig. 10(a) and (c). In order to differentiate them, directed graphs are built to represent their topology information, with components as vertices, spatial relationships as edges.

For template signboards, if component V_i is encompassed by component V_j, there is an arc from V_j to V_i. A dummy vertex is added for each template model, represented by $\otimes$. It has an arc to each vertex whose component is not encompassed by any other component. This dummy vertex actually refers to the background of a signboard. Template models for signboards in Fig. 10 (a) and (c) are shown in Fig. 10 (b) and (d), respectively.

For a query image, a dummy vertex is assigned to a component if it has not been assigned with any identity. Arcs are added as follows. An arc is added from one vertex to another vertex, if the corresponding component encompass another, as for template processing. An arc is added from a dummy vertex to a vertex if two corresponding components are adjacent and the dummy component is not encompassed by the other. Then, we obtain all subgraphs, starting at a dummy vertex and comprising of all vertices to which there are paths from the starting vertex. They are denoted as $SG = \{SG_i, i = 1 : K\}$. If subgraph

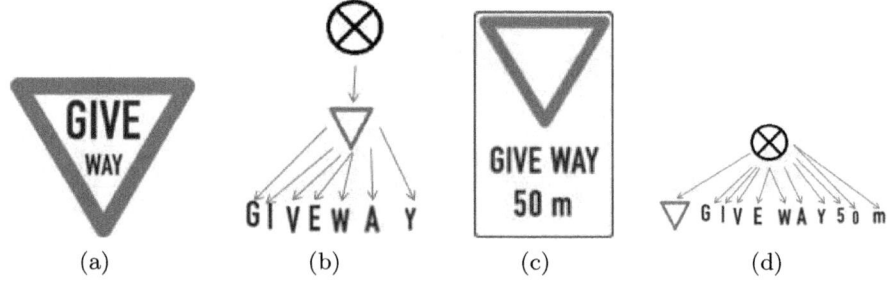

Fig. 10. (a)Signboard. (b)The graph model for signboard (a). (c)Signboard. (d)The graph model for signboard (c).

Fig. 11. Samples of testing data

SG_1 is a subgraph of subgraph SG_2, SG_1 will be removed from SG. For example, $\{\{\bigotimes^1, V_j\}, \{\bigotimes^1 \longrightarrow V_j\}\}$ will be removed, if $\{\{\bigotimes^2, V_j, V_i\}, \{\bigotimes^2 \longrightarrow V_j, \bigotimes^2 \longrightarrow V_i\}\}$ exists.

3.4   Graph Matching

The remaining elements in SG will be matched against all template models. We define that dummy nodes can be matched to each other without any cost. If SG_i is a subgraph of a template model, the match is successful. The final identity of SG_i is given by the model which has the maximum number of matches with it. Finally, all subgraphs of the query image which share the same identity are grouped together. This matching processing is able to handle both gathering signboards and incomplete signboards.

3.5   Experiment Results

Our testing data comprises of 100 real scene images. Examples are shown in Fig. 11. Many of them have elevation angles smaller than 20°, leading to severe perspective distortion.

In the experiment, we first use a simple yet effective color thresholding method proposed in [3] to detect possible regions of signboards. A loose threshold is set to avoid loss of signboards in this step. Overlapping regions are merged together to form a larger region. Then we apply our recognition method introduced in Section 3 on these regions. 203 regions are extracted in total, within which there are 142 target signboards. Our method identifies 137 signboards, out of which 129 is correct, leading to recognition precision and recall at 94.16% and 90.84%.

4   Conclusion

In this paper, we proposed two strategies to apply a symbol recognition method to recognize real scene signboards, namely holistic and dismantling/assembling strategies. We do recommend using holistic recognition for better performance, if good detection and segmentation algorithms are available, because this increases the distinctiveness of symbols. However, the dismantling/assembling strategy will give more flexibility. For example, speed limit signs have the same format: a circle with a number in it, representing the speed limit. With different numbers, the sign may have many different variants. In the dismantling/assembling strategy, all these variants can be represented by a circle and 10 digits.

References

1. Barber, C.B., Dobkin, D.P., Huhdanpaa, H.T.: The quickhull algorithm for convex hulls. ACM Transactions on Mathematical Software 22(4), 469–483 (1996)
2. Belongie, S., Malik, J., Puzicha, J.: Shape matching and object recognition using shape contexts. IEEE Transactions on Pattern Analysis and Machine Intelligence 24, 509–522 (2002)
3. de la Escalera, A., Moreno, L., Salichs, M., Armingol, J.: Road traffic sign detection and classification. IEEE Transactions on Industrial Electronics 44(6) (1997)
4. Lalondeand, M., Li, Y.: Road signs recognition - survey of the state of the art. Technique Report, CRIM-IIT (1995)
5. Lee, S.W., Kim, J.S.: Multi-lingual, multi-font, multi-size large-set character recognition using self-organizing neural network. In: Proceedings of the 3rd International Conference on Document Analysis and Recognition, vol. 1, pp. 23–33 (1995)
6. Li, L., Tan, C.L.: Character recognition under severe perspective distortion. In: Proceedings of the 19th International Conference on Pattern Recognition (2008)
7. Mundy, J.L., Zisserman, A.P.: Geometric invariance in computer vision. MIT Press, Cambridge (1992)
8. Rehrmann, V., Priese, L.: Fast and robust segmentation of natural color scenes. In: Chin, R., Pong, T.-C. (eds.) ACCV 1998. LNCS, vol. 1351. Springer, Heidelberg (1997)
9. Yamaguchi, T., Maruyama, M., Miyao, H., Nakano, Y.: Digit recognition in a natural scene with skew and slant normalization. International Journal of Document Analysis and Recognition 7(2-3), 168–177 (2005)
10. Zhou, P., Li, L., Tan, C.L.: Character recognition under severe perspective distortion. In: Proceedings of the 10th International Conference on Document Analysis and Recognitionn (2009)

Symbol Recognition Using a Concept Lattice of Graphical Patterns

Marçal Rusiñol[1], Karell Bertet[2], Jean-Marc Ogier[2], and Josep Lladós[1]

[1] Computer Vision Center, Dept. Ciències de la Computació
Edifici O, UAB, 08193 Bellaterra, Spain
{marcal,josep}@cvc.uab.cat
[2] L3I, University of La Rochelle
Av. M. Crépeau, 17042 La Rochelle Cédex 1, France
{kbertet,jmogier}@univ-lr.fr

Abstract. In this paper we propose a new approach to recognize symbols by the use of a concept lattice. We propose to build a concept lattice in terms of graphical patterns. Each model symbol is decomposed in a set of composing graphical patterns taken as primitives. Each one of these primitives is described by boundary moment invariants. The obtained concept lattice relates which symbolic patterns compose a given graphical symbol. A Hasse diagram is derived from the context and is used to recognize symbols affected by noise. We present some preliminary results over a variation of the dataset of symbols from the *GREC 2005* symbol recognition contest.

Keywords: Graphics Recognition, Symbol Classification, Concept Lattices, Shape Descriptors.

1 Introduction

In order to tackle the problem of recognizing graphic symbols, a wide variety of symbol descriptors have been proposed in the literature. In most cases the applications have to cope with large corpora of graphical entities. In such conditions, the final performance of the systems not only depends on the description technique but also on which kind of data structure is used to provide efficient access and organize the feature descriptors.

In other cases, data structures are not used to provide efficient access to the data but also convey themselves some kind of information. In the field of Graphics Recognition the most clear example of such structures is the use of graphs, which have been applied over the years in structural pattern recognition problems. As other examples of data structures which have also been used in the symbol recognition domain due to its inherent codification of information we can cite for instance trees, dendrograms, or concept lattices.

Concept lattices are used as knowledge representation and its application to the symbol recognition domain was first proposed by Bertet and Ogier in [1]. The followed symbol recognition scheme by using concept lattices is to first build a

J.-M. Ogier, W. Liu, and J. Lladós (Eds.): GREC 2009, LNCS 6020, pp. 187–198, 2010.
© Springer-Verlag Berlin Heidelberg 2010

binary table where the symbols correspond to the rows of the table and features from the descriptor vector correspond to the columns. A boolean value in a given cell indicates whether if a given symbol has a certain feature. From this table a Hasse diagram is derived and the classification of symbols is done in terms of traversal of this diagram. Guillas et al. presented in [5] a symbol recognition approach using concept lattices of pixel-based descriptors, whereas Coustaty et al. proposed in [3] the use of a structural description technique as features.

In this paper we propose a new approach to recognize symbols by the use of a concept lattice. Instead of using a numeric values arising from the feature vector, we propose to build the lattice in terms of symbolic patterns. Each symbol we want to recognize is represented by a set of composing primitives. Each one of these primitives is described by a well-known shape descriptor. The concept lattice relates which primitives compose a given graphical symbol. The obtained concept lattice from the context is then used to recognize symbols affected by noise. We present some preliminary results over a variation of the dataset of symbols from the *GREC 2005* Symbol Recognition Contest.

The remainder of this paper is structured as follows: the next section presents the followed steps to extract the primitives from the graphical symbols and how they are described. In section 3, we detail how the concept lattice is build from the set of model symbols and the sets of primitives. Section 4 presents the experimental setup by using a large dataset of distorted graphical symbols. Finally, the conclusions and future research lines can be found in Section 6.

2   Primitive Extraction and Symbol Description

Let us detail in this section how a graphical symbol is decomposed in a set of primitives representing simple graphical patterns, and how these primitives are described by the use of the well-known boundary moment invariants.

2.1   Extracting Primitives from Symbols

Our research work is mainly focused on the management of graphical data appearing in line-drawing images. Since these documents are mainly composed by lines, we choose to work with a vectorial representation of the symbols rather than at pixel level. In order to convert the symbol images to the vector domain, we use the raster-to-vector process proposed by Rosin and West in [9]. Instead of polygonally approximate the skeleton of the symbols, in our method, we approximate the contour of the closed loops conforming a symbol and its external contour in order to tackle with symbols which do not contain any loop at all. However, line segments are not suitable to be used as primitives due to its instability in terms of artifacts, fragmentation, errors in junctions, etc. A higher level entity has to be used as primitive. Adjacent vectors are merged together into a polyline instance. These polylines represent the graphical patterns conforming a given graphical symbol.

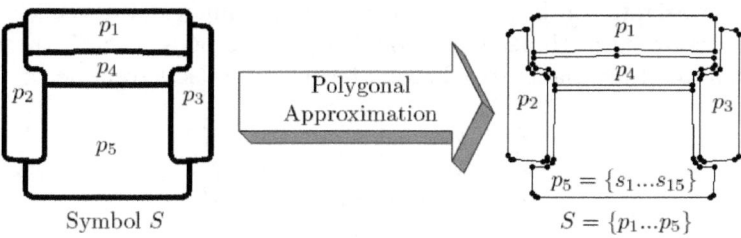

Fig. 1. Symbol representation in terms of a polygonal approximation of the contours of closed regions

Formally, let $p = \{s_1...s_n\}$ be a polyline consisting of n segments s_i. A symbol is represented in terms of its polylines representing loops and denoted as $S = \{p_1...p_m\}$. We can appreciate in Fig. 1 how the different parts of a symbol are detached making the regions meaningful primitives. Let us review in the next section how can we coarsely describe these primitives by the use of boundary moment invariants.

2.2 Primitive Description by Boundary Moment Invariants

The central $(p + q)$th order moment for a digital image $I(x, y)$ is expressed by

$$\mu_{pq} = \sum_x \sum_y (x - \bar{x})^p (y - \bar{y})^q I(x, y) \tag{1}$$

The use of the centroid $c = (\bar{x}, \bar{y})$ allows to be invariant to translation. The geometric moments can also be computed among the contour of the object as introduced by Chen in [2] and by Sardana et al. in [10] by using eq. 1 only for the pixels of the boundary of the object. A normalization by the object perimeter is used to achieve invariance to scale by using the following equation:

$$\eta_{pq} = \frac{\mu_{pq}}{\mu_{00}^\gamma} \quad \text{where} \quad \gamma = p + q + 1 \tag{2}$$

By sampling the polygonal approximation we can use the boundary moments as geometric descriptors of the primitives. In order to obtain invariance to rotation we use the set of seven functions proposed by Hu in [6] involving moments up to third order.

$$
\begin{aligned}
\phi_1 &= \eta_{20} + \eta_{02} \\
\phi_2 &= (\eta_{20} - \eta_{02})^2 + (2\eta_{11})^2 \\
\phi_3 &= (\eta_{30} - 3\eta_{12})^2 + (3\eta_{21} - \eta_{03})^2 \\
\phi_4 &= (\eta_{30} + \eta_{12})^2 + (\eta_{21} + \eta_{03})^2 \\
\phi_5 &= (\eta_{30} - 3\eta_{12})(\eta_{30} + \eta_{12})[(\eta_{30} + \eta_{12})^2 - 3(\eta_{21} + \eta_{03})^2] + \\
&\quad (3\eta_{21} - \eta_{03})(\eta_{21} + \eta_{03})[3(\eta_{30} + \eta_{12})^2 - (\eta_{21} + \eta_{03})^2] \\
\phi_6 &= (\eta_{20} - \eta_{02})[(\eta_{30} + \eta_{12})^2 - (\eta_{21} + \eta_{03})^2] + 4\eta_{11}(\eta_{30} + \eta_{12})(\eta_{21} + \eta_{03}) \\
\phi_7 &= (3\eta_{21} - \eta_{03})(\eta_{30} + \eta_{12})[(\eta_{30} + \eta_{12})^2 - 3(\eta_{21} + \eta_{03})^2] - \\
&\quad (\eta_{30} - 3\eta_{12})(\eta_{21} + \eta_{03})[3(\eta_{30} + \eta_{12})^2 - (\eta_{21} + \eta_{03})^2]
\end{aligned}
\tag{3}
$$

Moment invariants can be normalized to get the different invariants into similar numerical ranges. Hupkens and de Clippeleir proposed in [7] the following normalization of invariants to achieve a better robustness to noise.

$$
\begin{aligned}
\phi_1' &= \phi_1, & \phi_4' &= \phi_4 \, / \, \phi_1^3, \\
\phi_2' &= \phi_2 \, / \, \phi_1^2, & \phi_5' &= \phi_5 \, / \, \phi_1^6, \\
\phi_3' &= \phi_3 \, / \, \phi_1^3, & \phi_6' &= \phi_6 \, / \, \phi_1^4, \\
& & \phi_7' &= \phi_7 \, / \, \phi_1^6
\end{aligned}
\tag{4}
$$

Formally, each primitive p_i of a symbol S is described by a seven-dimensional feature vector

$$
f_i = [\phi_1', \phi_2', \phi_3', \phi_4', \phi_5', \phi_6', \phi_7']
$$

arising from the boundary moment invariant descriptors. The description space is quantized to transform this continuous set of values into a discrete set of symbolic graphical patterns. Let us detail in the next section how the concept lattice is build from the set of model symbols and the corresponding sets of primitives.

3   Concept Lattice of Graphical Patterns

Let us begin by reviewing the mathematical foundation of the concept lattices. We then focus on its application to the particular problem of symbol recognition by the traversal of the concept lattice.

3.1   Foundations of the Concept Lattice

We formally define a concept lattice by the formal concept analysis theory [4]. A concept lattice is a representation of a *formal context* $C = (G, M, R)$ where G is a set of *objects* and M is a set of *attributes*. R is a *relation* between these two sets. The fact that a certain object o has the attribute a is denoted as oRa.

From an object set $O \subseteq G$ we define as $f(O)$ the set of attributes in relation R with the objects from O.

$$
f(O) = \{a \in M \mid oRa \;\forall o \in O\}
\tag{5}
$$

We analogously define $g(A)$ as being the set of objects in relation with the attributes from a set $A \subseteq M$.

$$
g(A) = \{o \in G \mid oRa \;\forall a \in A\}
\tag{6}
$$

A *formal concept* for the context C is defined as a pair of objects and attributes (O, A) in relation according to R. The objects $O \subseteq G$ and the attributes $A \subseteq M$ must verify that $f(O) = A$ and $g(A) = O$. We denote as $\beta(C)$ all the concepts of the context C.

Formally, a concept lattice is defined as the set of concepts ordered by the *order relation*[1] $\leq$ defined for two concepts (O_1, A_1) and (O_2, A_2), as:

$$(O_1, A_1) \leq (O_2, A_2) \iff O_1 \subseteq O_2 \tag{7}$$

the set of concepts and the order relation form the *concept lattice* $(\beta(C), \leq)$ of the context $C = (G, M, R)$.

By defining a *cover relation* $\prec$ as:

$$(O_1, A_1) \prec (O_2, A_2) \iff \begin{cases} (O_1, A_1) < (O_2, A_2) \\ \nexists (O_3, A_3) \in \beta(C) \mid (O_1, A_1) < (O_3, A_3) < (O_2, A_2) \end{cases} \tag{8}$$

the *Hasse diagram* $(\beta(C), \prec)$ of a concept lattice $(\beta(C), \leq)$ is obtained.

A context may be seen as a table, where the objects correspond to the rows of the table and the attributes correspond to the columns. A boolean value in cell (o, a) indicates whether if a given object o has the attribute a.

Let us see in the next section how the concept lattices can be applied to the symbol recognition problem.

3.2   On the Use of Concept Lattices for Symbol Description

Concept lattices are used as knowledge representation and its application to the symbol recognition domain was first proposed by Bertet and Ogier in [1]. A classical symbol recognition scheme can define a context C where G corresponds to the set of graphical symbols we want to recognize and M corresponds to a set of attributes arising from the symbol descriptors. Guillas et al. presented in [5] a symbol recognition approach using concept lattices of pixel-based descriptors, whereas Coustaty et al. proposed in [3] the use of a structural description technique to define the context C.

In all these previous approaches, graphical symbols are represented by a numerical descriptor. Each value of this feature vector is discretized in a number of intervals following a cutting criterion. Given a cutting value dividing a feature in two intervals, each model symbol has a membership relation with one of the intervals, that enables to differentiate the two subsets of symbols. The process of cutting the description space in intervals is repeated until each class can be distinguished. A concept lattice is build from the binary relationship between intervals and symbol families. When a symbol has to be recognized its description vector is also cut into intervals and the traversal of the Hasse diagram results in the class where the symbol belongs to.

However, these approaches may be very sensitive to noise, occlusions or even non-perfect symbol segmentations. If a single value of the feature vector is assigned to an incorrect interval, then the symbol can not be correctly recognized. In this paper we propose to describe symbols by graphical patterns taken as primitives and to build a concept lattice representing that a symbol family contain

---

[1] An order relation is a reflexive, antisymmetric and transitive binary relation.

Objects	Attributes				
	composite	even	odd	prime	square
	c	e	o	p	s
1			1		1
2		1		1	
3			1	1	
4	1	1			1
5			1	1	
6	1	1			
7			1	1	
8	1	1			
9	1		1		1
10	1	1			

(a) Context $C = (G, M, R)$

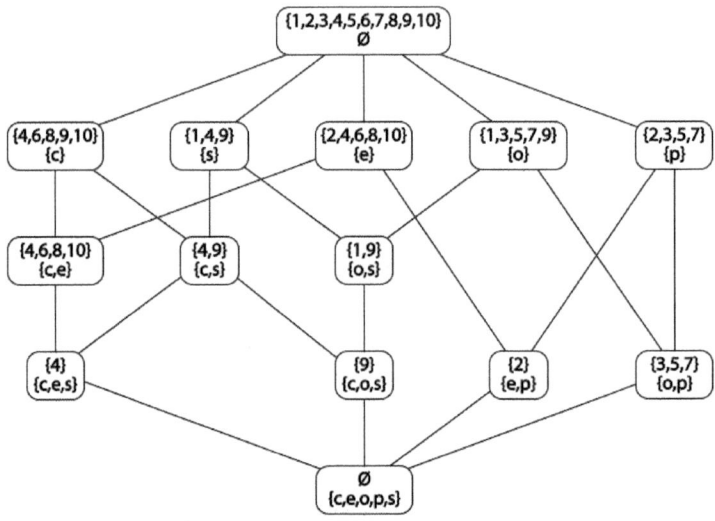

(b) Hasse diagram

Fig. 2. A concept lattice represented by a Hasse diagram for integers from 1 to 10 and several number attributes

or not a given simple shape. We can see an example of the proposed approach in Fig. 3. From the model symbols, we construct a set of attributes being graphical patterns. This set of attributes is constructed by clustering by similarity the space formed by all the feature vectors f_i describing the graphical primitives p_i composing the symbols in the database. The context C defines then a relationship between symbol classes and composing primitives. Let us see in the next section how we can use this context and the Hasse diagram derived from the concept lattice to recognize distorted symbols.

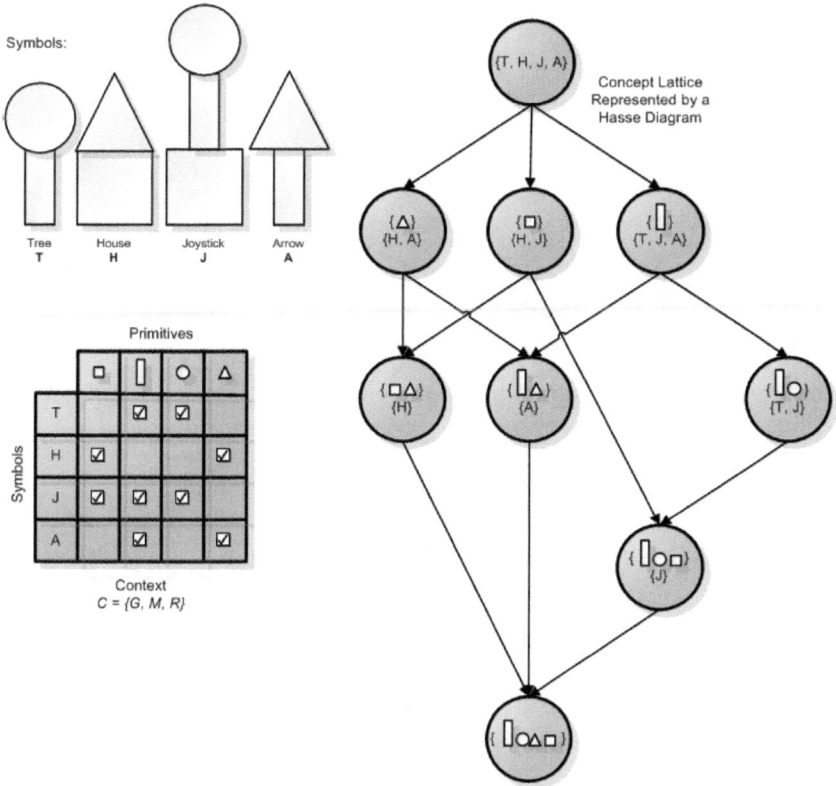

Fig. 3. Intuitive idea of the proposed approach

3.3 Traversing the Hasse Diagram for Symbol Recognition

Given a query symbol S^q and its corresponding feature vectors f_i^q describing each of the primitives p_i^q which compose the symbol, the Hasse diagram is traversed in order to recognize the query symbol. Starting from the topmost concept of the Hasse diagram, all the concepts of the poset containing a given set of attributes (primitives) of the query symbol are visited. A voting scheme accumulates evidences of the hypothetic symbols which may be the query symbol. These hypothetic symbols are the ones found in each poset concept.

Let us see the example in Fig 4. A noisy instance of a symbol from the family *joystick* has been taken as query symbol to recognize. The square and the circle primitives can be correctly identified despite the noise, however, the rectangle is not correctly recognized. When traversing the Hasse diagram, at each poset concept, we accumulate evidences of the plausible symbols. At the end, the symbol family accumulating more votes is taken as the class where the symbol belongs to.

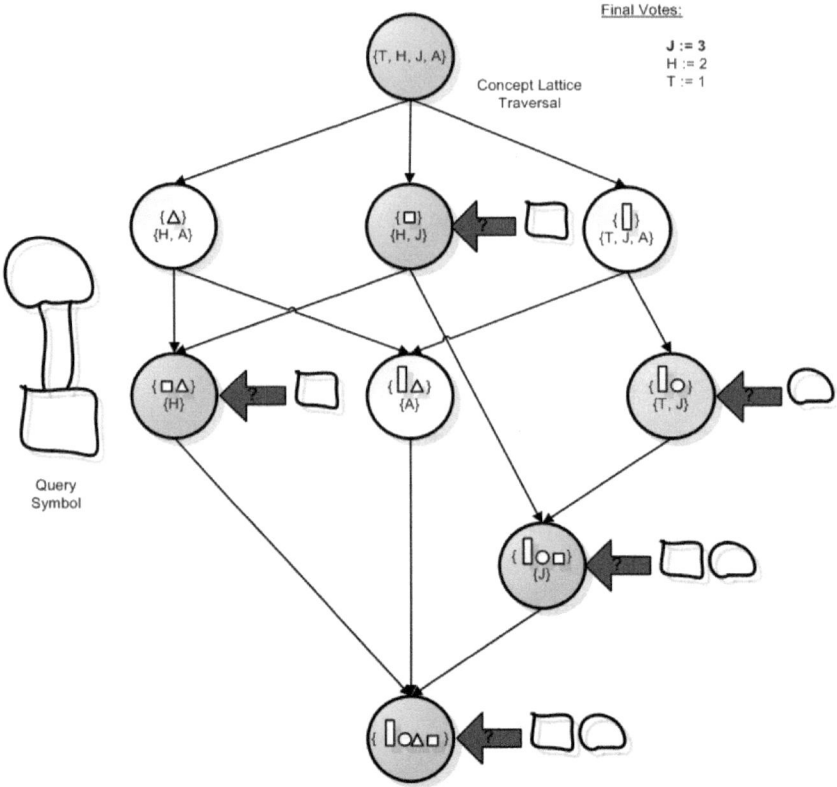

Fig. 4. Traversing the Hasse diagram for symbol recognition

In order to recognize symbols having different number of primitives a normalization of the voting space is done according to the theoretical number of votes which a given symbol should have obtained. A symbol having m primitives should receive $2^m - 1$ votes if all its primitives have been correctly identified. Let us see in the next section the obtained experimental results.

4   Experimental Results

We present in this section the experimental results for a symbol recognition problem. Let us first detail the symbol dataset we use and then present the obtained results.

4.1   Symbol Dataset

In order to carry out our experiments we have build a database of symbols in vectorial format with vectorial distortions. We have used all the 150 symbols of the original GREC2005 symbol database [11] as models. In order to generate realistic vectorial deformations, we first applied a degradation model to the bitmap images, and then applied a raster-to-vector process to these degraded images.

The bitmap images are degraded using the method presented by Kanungo et al. in [8] to simulate the noise introduced by the scanning process. Three different parameter configurations are used to obtain three different degradation levels. Some simple morphological operations are applied to these degraded images to get rid of the background noise. A connected component analysis is applied to label the closed regions and to extract the internal and external contours composing a symbol. These distorted contours are then polygonally approximated by using the Rosin and West algorithm introduced in [9]. In this dataset, the graphical symbols are composed by several polylines each one composed by a set of adjacent segments. The number of polylines which composes a symbol is constant for a given class, but the number of segments of these polylines is affected by the distortion model and varies from an instance to another. Fig. 5 shows an example of this distortion as well as some complementary characteristics of this dataset.[2]

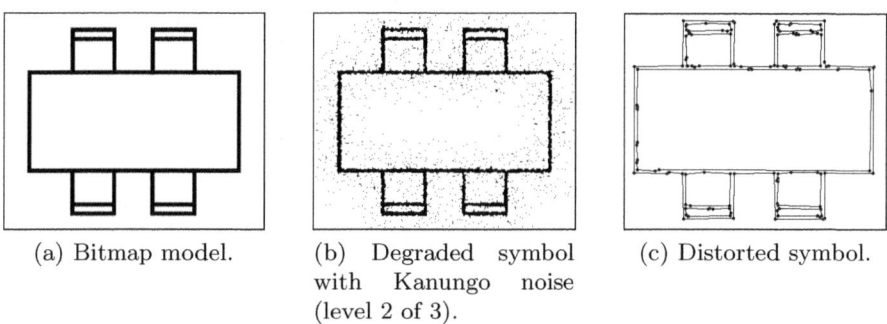

(a) Bitmap model.   (b) Degraded symbol with Kanungo noise (level 2 of 3).   (c) Distorted symbol.

Property	Value
Number of classes	150
Total number of elements	45,000 (300 elements/class)
Max. number of polylines in a symbol	16
Min. number of polylines in a symbol	1
Mean number of polylines in a symbol	3.9
Max. number of segments in a symbol	264
Min. number of segments in a symbol	11
Mean number of segments in a symbol	73.7

(d) Details on the GREC-POLY database.

Fig. 5. Example and characteristics for the GREC-POLY database

---

[2] The vectorial symbol dataset is public available and can be downloaded through the following website http://www.cvc.uab.cat/~marcal/GREC-POLY/

4.2   Evaluation

We present in Table 1 the obtained recognition results for the whole recognition experiment. Each one of the 45,000 degraded symbols is classified into the most likely symbol class depending on the value of the votes. However, we show in Table 1 the obtained recognition rates when considering just the topmost element, the two highest classes or the first three classes.

As we can appreciate, the noise introduced by the lowest distortion level is quite well tolerated. However, the deformation of the medium and higher level really impairs the overall performance of the method. Nevertheless, the box plot shown in Fig. 6, indicates that the performance is also highly dependent on the symbol design. Even in the highest level of distortion, the upper quartile attains good recognition rates whereas in the lowest level of distortion some symbol designs are badly recognized provoking some outliers in the box plot.

Table 1. Recognition results

Considered Results	Recognition rates(%)		
	Distortion levels		
	1 (low)	2 (medium)	3 (high)
top 1	78.93	64.61	53.92
top 2	80.41	66.03	55.24
top 3	80.58	66.13	55.38

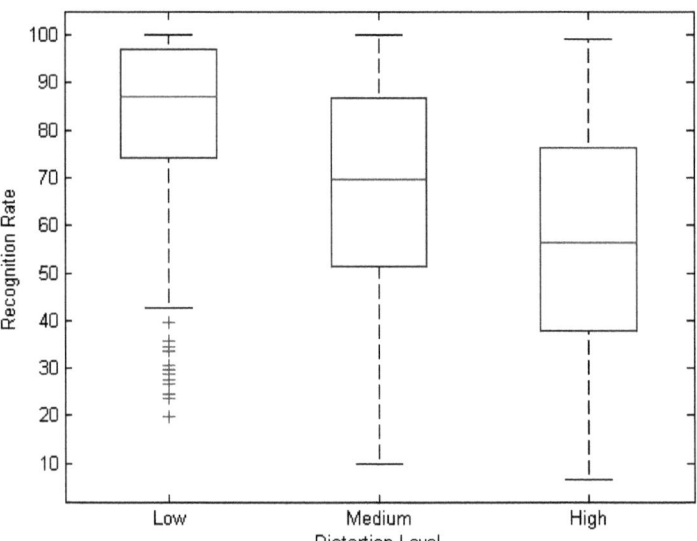

Fig. 6. Box plot of the recognition rates depending on the distortion levels

5   Conclusions and Future Work

In this paper we have proposed a new approach to recognize symbols by the use of a concept lattice of graphical patterns. Each model symbol has been decomposed in a set of graphical patterns taken as primitives. Each one of these primitives has been described by boundary moment invariants. The obtained concept lattice relates which symbolic patterns compose a given graphical symbol. The concept lattice was then used to recognize symbols affected by noise. We have presented some preliminary results over a variation of the dataset of symbols from the *GREC 2005* symbol recognition contest.

Despite the simplicity of the used descriptor, the obtained results are encouraging. The use of concept lattices as knowledge representation and its combination with voting approaches accumulating evidences to validate symbol class hypotheses seems a promising approach. The main novelty of this paper is the use of concept lattices from a symbolic set of attributes instead of numeric ones used in the previous approaches. The use of symbolic description of graphical symbols has been proven to be a powerful tool. We believe that this research line has to be further studied since there is still room for improvements.

The remaining challenge is to try to apply such kind of approaches to recognize non-segmented graphical symbols which may appear within its real context. The combined use of the knowledge representation given by the concept lattices and some spatial coherence rules may be envisaged in order to tackle with the problem of recognizing graphical symbols appearing in complete documents.

Acknowledgments

This work has been partially supported by the Spanish projects TIN2006-15694-C02-02, TIN2009-14633-C03-03 and CONSOLIDER-INGENIO 2010 (CSD2007-00018).

References

1. Bertet, K., Ogier, J.M.: Graphic Recognition: The Concept Lattice Approach. In: Lladós, J., Kwon, Y.-B. (eds.) GREC 2003. LNCS, vol. 3088, pp. 265–278. Springer, Heidelberg (2004)
2. Chen, C.C.: Improved Moment Invariants for Shape Discrimination. Pattern Recognition 26(5), 683–686 (1993)
3. Coustaty, M., Guillas, S., Visani, M., Bertet, K., Ogier, J.M.: On the Joint Use of a Structural Signature and a Galois Lattice Classifier for Symbol Recognition. In: Liu, W., Lladós, J., Ogier, J.-M. (eds.) GREC 2007. LNCS, vol. 5046, pp. 61–70. Springer, Heidelberg (2008)
4. Ganter, B., Wille, R.: Formal Concept Analysis – Mathematical Foundations. Springer, Heidelberg (1997)
5. Guillas, S., Bertet, K., Ogier, J.M.: A Generic Description of the Concept Lattices' Classifier: Application to Symbol Recognition. In: Liu, W., Lladós, J. (eds.) GREC 2005. LNCS, vol. 3926, pp. 47–60. Springer, Heidelberg (2006)

6. Hu, M.: Visual Pattern Recognition by Moment Invariants. IRE Transactions on Information Theory 8, 179–187 (1962)
7. Hupkens, T.M., de Clippeleir, J.: Noise and intensity invariant moments. Pattern Recognition Letters 16(4), 371–376 (1995)
8. Kanungo, T., Haralick, R.M., Philips, I.: Global and Local Document Degradation Models. In: Proceedings of the Second International Conference on Document Analysis and Recognition, ICDAR 1993, pp. 730–734 (1993)
9. Rosin, P.L., West, G.A.W.: Segmentation of Edges into Lines and Arcs. Image and Vision Computing 7(2), 109–114 (1989)
10. Sardana, H.K., Daemi, M.F., Ibrahim, M.K.: Global Description of Edge Patterns Using Moments. Pattern Recognition 27(1), 109–118 (1994)
11. Valveny, E., Dosch, P.: Report on the Second Symbol Recognition Contest. In: Liu, W., Lladós, J. (eds.) GREC 2005. LNCS, vol. 3926, pp. 381–397. Springer, Heidelberg (2006)

Touching Text Character Localization in Graphical Documents Using SIFT

Partha Pratim Roy[1], Umapada Pal[2], and Josep Lladós[1]

[1] Computer Vision Center, Universitat Autònoma de Barcelona, 08193, Bellaterra
(Barcelona), Spain
[2] Computer Vision and Pattern Recognition Unit, Indian Statistical Institute,
Kolkata - 108, India

Abstract. Interpretation of graphical document images is a challenging
task as it requires proper understanding of text/graphics symbols present
in such documents. Difficulties arise in graphical document recognition
when text and symbol overlapped/touched. Intersection of text and symbols with graphical lines and curves occur frequently in graphical documents and hence separation of such symbols is very difficult.

Several pattern recognition and classification techniques exist to recognize isolated text/symbol. But, the touching/overlapping text and
symbol recognition has not yet been dealt successfully. An interesting
technique, Scale Invariant Feature Transform (SIFT), originally devised
for object recognition can take care of overlapping problems. Even if
SIFT features have emerged as a very powerful object descriptors, their
employment in graphical documents context has not been investigated
much. In this paper we present the adaptation of the SIFT approach in
the context of text character localization (spotting) in graphical documents. We evaluate the applicability of this technique in such documents
and discuss the scope of improvement by combining some state-of-the-art
approaches.

1   Introduction

With the rapid progress of research in document image analysis and document
image understanding many applications are coming up to manage the paper
documents in electronic form to facilitates indexing, viewing, extracting the in-
tended portions, etc. These applications include documents that to be digitized
and stored in a database.

Due to the emergence of Geographical Information Systems (GIS), map acqui-
sition and recognition have become a pursued topics, both by the industry and
the academy. The interpretation of graphical documents does not only require
the recognition of graphical parts but the detection and recognition of multi-
oriented text. The problem for detection and recognition of such text characters
is many-folded. Text/symbols many times touch/overlap with long graphical
lines. Sometimes, the text lines are curvi-linear to annotate graphical objects.
Thus the recognition of such document is more difficult due to the usage of
multi-oriented and multi-scale environment.

J.-M. Ogier, W. Liu, and J. Lladós (Eds.): GREC 2009, LNCS 6020, pp. 199–211, 2010.
© Springer-Verlag Berlin Heidelberg 2010

Fig. 1. Locations of isolated (bounded by blue box) and touching (bounded by red box) character ('R') are shown in a part of graphical image

Separation of text/graphics in document image is one of the fundamental aims in graphics recognition. It requires proper discrimination of text/graphics [1], [2], [3]. Here, the aim is to segment the document into two layers: a layer assumed to contain text and symbols, and the other one containing the rest of graphical objects representing street, river, border of the regions, etc. The problem has received a great deal of attention in the literature because of the different processing approach of text and graphics. At the component level the problem is not too intensed. The spatial distribution of the components and their sizes can be measured in a number of ways, and fairly reliable classification can be obtained. Difficulties arise however, when either there is text and symbol overlapped in the graphics components, or text and symbol touched with graphics. See Fig.1, where some characters are touched/overlapped with graphical lines, and segmentation of such documents is very difficult. Text/Symbol identification in complex document is done in two ways. Majority of the methods use segmentation approach of these text/symbols and then recognize them. On the other hand, a few methods work on recognizing the symbols before approaching segmentation. There exists many pieces of published work on text/graphics separation. Algorithm due to Fletcher and Kasturi [2] uses simple heuristics based on the characteristics of text characters. The method is insensitive in text font style, size and orientation. One of the assumptions was that the text characters do not touch with graphics or other characters and each text character forms an isolated component. Directional mathematical morphology approach has been used by Luo et al. [4] for separation of character strings from maps. The idea is to separate large linear segments by directional morphology and histogram analysis of these segments. Large segments are considered as part of graphics; effectively leaving small text character segments. Tan et al. [5] illustrates a system using *Pyramid* structure. Multi-resolution representations of such a pyramid structure help to select different regions for segmentation. Cao and Tan [1] proposed a method of detecting and extracting text characters that are touched to graphics. It is based on the interpretation of intersection of lines in the overlapped region on the vectorized image of text and graphics. A consolidated method, proposed by Tombre et al. [3] used connected components analysis to make it stable for graphics-rich documents. Although there exist many research work [1], [2], [3], the separation of text and graphics when text and graphics portion intersects has not yet been dealt with successfully because of the complexity of the problem. The shape of non-analytic curves found from text character is difficult to analyze where background noise or overlapping/touching of graphical lines exist.

Recently, the Scale Invariant Feature Transform (SIFT) [6] has emerged as a cut edge methodology in general object recognition as well as for other machine vision applications [7]. One of the interesting features of the SIFT approach is the capability to capture the main features of an object by means of local patterns extracted from a scale-space decomposition of the image. The main advantage of SIFT is that, this approach works in invariant to image scale, rotation, addition of noise/occlusion. With the wide applicability and potential of this technique, for the classification of 2D objects, recently, this approach is also investigated in graphical symbol recognition [8].

In this paper we present the evaluation of the SIFT approach in the context of text character spotting in graphical documents which deals with localization and detection of multiple instances of text character. It is applied to take care of segmentation of text/symbol components from graphical component. We evaluate the potentiality and applicability of this technique in such documents and discuss the scope of improvement with combination of state-of-the-art approaches.

The rest of the paper is organized as follows. The system overview is discussed in Section 2. In Section 3, we explain briefly the SIFT approach for object recognition. We describe the isolated component extraction and recognition procedure in Section 4. In Section 5, we present the adaptation of SIFT approach to detect text characters in graphical documents. The experimental results are demonstrated in Section 6. Finally conclusion is given in Section 7.

2   System Overview

Graphical documents normally contain text characters printed in different fonts. Sometimes, in a single document more than one font exist to describe the entities used. Also, the shapes of text images are different from one font to another and they do not comprise analytical curves always. Thus, when these characters touch with graphical lines, it is not easy to segment them.

To take care of it, our system works in a combination of bottom-up and top-down approaches to separate and locate text characters. We extract the knowledge from bottom-up approach and use them in top-bottom approach. First, the isolated characters are extracted from graphical document. Next, these characters are labelled using a rotation invariant character recognition system. Given a query text character, the system learns the different fonts of that character from these recognized (labelled) character sets. Thus, the different shapes of each character are learnt dynamically using this bottom-up approach. Next these different shapes of the character are used as query images to search other instances of characters of similar shapes in touching/overlapped graphical regions. In Fig.1, we show some isolated characters (bounded by blue box). The isolated characters

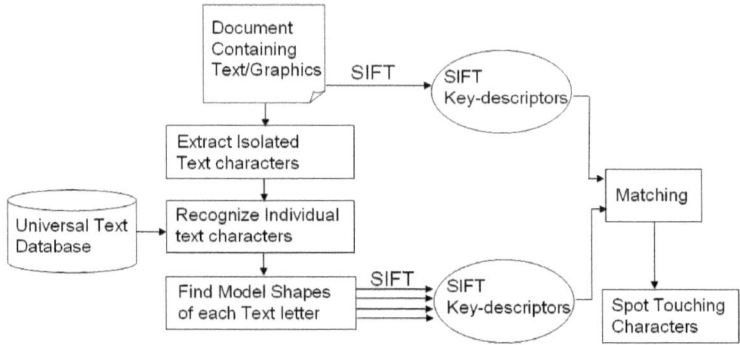

Fig. 2. Touching Text Character Spotting System

are learnt online to have the knowledge of shape of the query character. Next, other similar characters which are touching (bounded by red box in Fig.1) are searched using this knowledge. Thus, if a character is touched with graphical lines, a top-bottom approach is used to locate them.

The flow chart of the proposed scheme is shown in Fig.2. We use a connected component analysis to separate isolated characters from graphical components. These characters are recognized using a Support Vector Machine (SVM) classifier. The SVM is trained before to build the character shape models from corresponding feature of different text characters of the database. Finally, the recognized models are queried using SIFT to locate text character in touching/overlapping zones.

3   SIFT Approach for Object Recognition

The SIFT approach [6] to object recognition is a combination of selecting "local-features" and their "matching" method. SIFT features are invariant to image scale and rotation, and are demonstrated to provide robust matching in both the spatial and frequency domains, reducing the probability of disruption by occlusion, clutter, or noise. In addition, the features are highly distinctive, which allows a single feature to be correctly matched against a large database of features. SIFT feature usage for object recognition involves two steps - (1) SIFT feature extraction and (2) Object recognition.

3.1   Feature Extraction

Following are the major stages of computation used to generate the set of image features.

Scale space extrema detection: Distinctive points are selected by identifying maxima/minima of the document image after applying the image with a

Difference-Of-Gaussian (DOG) filter. This is done by convolving the image with Gaussian filters at different scales and taking differences of the resultant images.

Key-point Localization: At each candidate location, a detailed model is fit to determine location and scale. Keypoints are selected based on measures of their stability. Once a keypoint candidate has been found by comparing a pixel to its neighbors, a detailed step is performed using the neighbor data for location, scale, and ratio of principal curvatures.

Orientation assignment: One or more orientations are assigned to each key-point based on local image gradient directions. This method is used to incorporate rotation invariance to the key-point. To determine a key-point orientation, a gradient orientation histogram is computed in the neighborhood of the key-point.

Generation of key-point descriptor: The descriptor is meant to encode the key-point and information about the neighboring points. The information is encoded based on the local image gradients at these points. Once a keypoint orientation has been selected, the feature descriptor is computed as a set of orientation histograms on 4×4 pixel neighborhoods. Each histogram contains 8 bins. Thus the descriptor obtained is a 128 ($4 \times 4 \times 8$) element descriptor.

3.2   Object Recognition

Recognizing an object using SIFT can be performed using the following steps. *Key-point matching:* Key-point matching is done by matching the key-point descriptors from the test image with those of a template query image, using a nearest neighbor approach. The nearest neighbor match is compared with the next (second nearest) closest one to ensure that a match is only accepted if it is distinctive enough. *Hough Transform:* The Generalized Hough Transform (GHT) [9] is used to cluster key-point matches that are consistent with a single object hypothesis. Each key-point is characterized by a 2D location, a scale and an orientation. Thus a 4 dimensional hough space is used for this purpose. Finally, a set of potential hypothesis of the object in the test image is obtained using HT.

4   Text Character Extraction and Labeling

Here, we present the approach to extract the text character images automatically. It works in 3 stages namely isolated text character extraction, labeling these text characters and finally selecting shape/font of each text character label.

4.1   Text Component Extraction Using CC Analysis

In map, text and graphics appear simultaneously. We used the connected component analysis [3] for initial segmentation of isolated text components. The information using geometrical and statistical features [2] of the connected component perform well to group a component into one between text or graphics layer. For each connected component, we use a minimum enclosing bounding

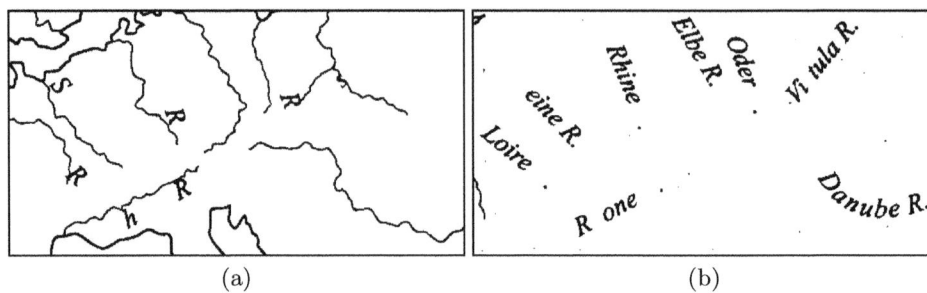

$$(a) \qquad\qquad\qquad\qquad (b)$$

Fig. 3. (a)Long graphical regions and (b) text regions of Fig.1

box which describes the height and width of the character shape. These rotated bounding boxes are better adjusted to the character components in rotation invariant environment. The components are filtered to be as a member of a text component based on its attributes (bounding box, pixel density, ratio of dimensions, area). A histogram on the size of components is analyzed for this purpose. By a correct threshold selection obtained dynamically from the histogram, the large graphical components are discarded, leaving the smaller graphics and text components. In our experiment, the threshold T is considered as,

$$T = n \times max(A_{mp}, A_{avg}) \tag{1}$$

where, A_{mp} and A_{avg} are frequency of most populated area and average area respectively. The value of n was set to 3 from the experiment [10].

4.2   Text Character Modeling

We generate universal character model for each character shapes using training of data which are extracted from different graphical documents. The isolated characters are recognized using the change of angle of characters' contour pixels configuration. We describe the feature extraction and character model generation method as follows.

Feature Extraction: We proposed a zone-based signature for character recognition in our earlier paper [11]. Circular ring and convex hull ring based concept have been used to divide a character into several zones to compute features. To make the system rotation invariant, the features are mainly based on angular information of the external and internal contour pixels of the characters. Given a sequence of contour pixels, the change of angle of each pixel is calculated from the neighbour pixels. The angles obtained from all the contour pixels of a character are grouped into 8 bins corresponding to eight angular intervals of 45°. For a character, frequency of the angles of 8 bins will be similar even if the character is rotated at any angle in any direction.

Circular and convex hull rings are constructed on a character as follows. A set of circular rings is defined as the concentric circles considering their center as the center of minimum enclosing circle (MEC) of the character and the minimum enclosing circle is the outer ring of the set. Similarly, convex hull rings are also constructed from the convex hull shape of the character.

To get more local feature, we compute angular information (slope) of the contour pixels with respect to the center of the MEC. We grouped the contour pixels of a character into 8 bins based on their angular information with respect to the neighbor pixels. Now, angular information (slope) with respect to the centre of MEC of the bin-wise contour pixels is computed. The angle obtained from all the contour pixels of a bin are grouped into 8 sets corresponding to 8 angular information. Thus, for 8 bins of contour pixels, we have $8 \times 8 = 64$ dimensional slope features.

Finally, considering 7 circular rings and 7 convex hull rings, we have 56 (8×7) feature from convex hull ring, 56 (8×7) features from circular ring and 64 (8×8) features from angular information with respect to center of MEC. As a result, we have 176 (56+56+64) dimensional feature vector for the classification. This feature has been selected based on experiment. Normalization of the feature is done to obtain scale invariance [11].

Character Classification: For recognition, we feed the features in a Support Vector Machine (SVM) classifier. SVM classifier has been used to build the character shape model from corresponding feature of our training data. Given a connected component, we compute the recognition confidence to obtain the corresponding class and use the label to describe the component as the text character.

The SVM is defined for two-class problem and it looks for the optimal hyperplane which maximizes the distance, the margin, between the nearest examples of both classes, named support vectors (SV_s). Given a training database of M data: $x_m \| m=1,\ldots,M$, the linear SVM classifier is then defined as:

$$f(x) = \sum_j \alpha_j x_j \cdot x + b$$

Where, x_j is the set of support vectors and the parameters α_j and b have been determined by solving a quadratic problem [12]. The linear SVM can be extended to a non-linear classifier by replacing the inner product between the input vector x and the SV_s x_j, to a kernel function k defined as:

$$k(x,y) = \phi(x) \cdot \phi(y)$$

This kernel function should satisfy the Mercer's Condition [12]. Some examples of kernel functions are polynomial kernels $(x \cdot y)^p$ and Gaussian kernels $k(x,y) = exp(-\frac{\|x-y\|^2}{\sigma^2})$. Gaussian kernel has been chosen in our experiments to recognize multi-oriented text character. Details of SVM can be found in [12].

Identification of Text Components: Both English uppercase and lowercase alpha-numeric characters were considered for our experiment, so we should have 62 classes (26 for uppercase, 26 for lowercase and 10 for digit). But because of

Fig. 4. A part of Fig.3(b) with their character label shown in Red color

shape similarity due to orientation of some of the characters ('d', 'p'; 'b', 'q'; etc) are grouped together.

The isolated components which were included in text components group using CC analysis as described in Section 4.1 are recognized and labelled as text character using SVM. For each component, we compute the recognition confidence for all character class models using our SVM classification process and rank the confidence scores in descending order. If we recognize a component with a very high accuracy, we accept it as a good-shaped character. If the difference between top two recognition scores of a component is high, it is also considered as good-shaped character. The score difference is selected based on experimental result. In Fig.4, we show a portion of the text layer of Fig.3(b) containing isolated characters and their recognition label. Thus, given a character ascii value, we find different font styles used in the document from these isolated character shapes. It is to be noticed that, the character 'n' is identified as 'u' because of its shape similarity nature in rotation invariant environment. Thus, given an ascii character to search in the document, we find all the shapes used in that document corresponding to the ascii value using this appraoch. The rest of the components are not considered as isolated components. Sometimes due to background noise the recognition confidence is low. Also, due to noise, two or three characters may touch. These touching characters may be filtered in text character layer using CC analysis due to their small size. Using the recognition confidence we reject these components and add them in graphical layer of the document. Next, the good shapes of isolated characters found in the document are clustered according to different font style which may present in the document. This is explained as follows.

4.3 Font/Style Adaptability

Graphical documents may contain text characters of different fonts (for e.g. "Arial", "Times New Roman", etc.) or style (normal, italics) to annotate and give importance to the location names present in the document. To learn online the font used in the document, different fonts are trained in SVM before. It is to recognize variation of style or font of characters. Thus, for each alpha-numeric text character value, the system finds component shapes according to their font style. We may use all the components to search other instances present in the mixed graphics layer. We reduce the time complexity by finding unique representative shape for each font and style of each character. To select the unique model shape of each font style, a clustering algorithm is employed. The different classes will represent different font shapes of a character. The representative text character images are used as template (query image) to find out other text character images in the document which were not extracted or recognize well due to touching or noise.

5   Locating Similar Text Characters Using SIFT

We use character shape model as query model to locate overlapped/touching characters from graphical documents. SIFT is used to detect and describe the local features in this context. Before proceeding directly to locate text characters, we reduce the search regions of probable text characters in the document using skeleton analysis.

5.1   Potential Text Regions in Graphical Layer

According to text and graphics feature, it is assumed that the length of segments of the characters are smaller compared to that of graphics. When a line touches a symbol or text of blob like shape (dense pixels), the arrangement of segments in thinned (skeleton) image is not easy always to separate text components perfectly. It needs post-processing, which is a difficult task. Here, we use the segment length of skeleton for separating the long lines. The segments of skeleton are decomposed at the intersection point of the skeleton. Intersection points refer to those points where more than one segment intersect in skeleton image. Next, the length of these segments are computed. Based on the bounding box (BB) information of a segment, the length of a segment (L_s) is calculated as:

$$L_s = Max(Height_{BB}, Width_{BB}) \qquad (2)$$

The skeleton segments having length L_s larger than T (T is computed from Eq.1) are chosen for elimination. The remaining portion after removal of long segments are considered for potential regions of touching text characters according to their feature. For example, Fig.3(a) demonstrates an initial mixed component with touching characters. The components after removing long graphical segments are shown in Fig.5.

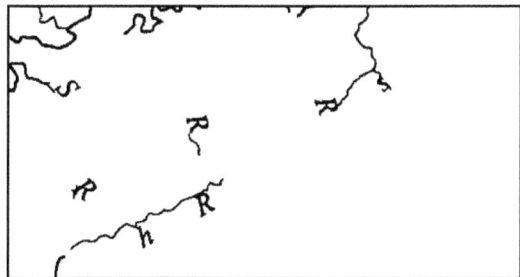

Fig. 5. Potential Regions of Text Characters

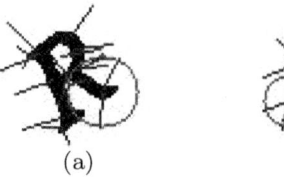

(a)   (b)

Fig. 6. SIFT features of text character. Circle denotes the zone of corresponding descriptor.

5.2   Text Character Spotting

For each text font model we find the local SIFT features. As explained in Section 3, each feature is composed by four parts: the locus (location in which the feature has been found), the scale, the orientation and the descriptor. The descriptor is a vector of 128 dimension. Some examples of the keypoints using SIFT approach to two different text characters are shown in Fig.6. Here, in the figure the line denotes the direction of dominant direction. The figure explains the different keypoint-feature descriptors corresponding to different text characters. SIFT is also applied in the document of graphical layer containing touching characters.

If we search for all text character using the approach described in Section 4 only, we find many false alarms due to curvature nature of graphical lines. Since, we estimated before the size of character using connected components analysis, different size criteria can be made to reduce more false positives.

As text character images are very small, they produce a set of SIFT keypoint descriptors, out of which many of them are for corner points. These corner points are not always distinctive to produce the location of the character in the graphical document, specifically for map images. To reject the keypoint descriptors which consist of only local corner point regions, a criteria is set to reject them. This filtering is done by comparing the region of key-descriptor to the size of query character shape. A size threshold is chosen for this purpose. The value of threshold is selected as $T_{K1} \geq S/2$, where S is the size of the character template. Thus, only the keypoints which captures major part of the text character are only accepted.

The graphical document also contain many corners due to its nature which leads to have many false positive. We remove the SIFT keypoints whose size are small or large. It is assumed that the size of text character will not be very small. A size threshold T_{K2} is chosen to discard these small descriptors. The value of T_{K2} is set to 6 from experimental result. Again, the maximum size (T) of the text components is computed in our earlier stage. This parameter is used to restrict the size of keypoint which are found larger from graphical document.

Finally, the local SIFT features of each text characters are matched with each features to the character SIFT features. The corresponding matching locations are detected as probable zones of text characters. See Fig.7, where we have shown

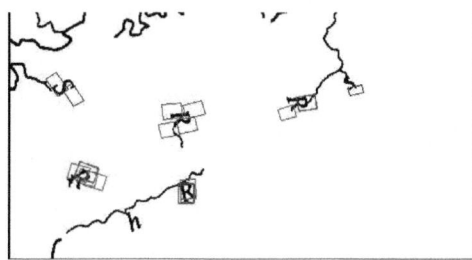

Fig. 7. Matching result of SIFT features of text characters 'R' and 'S' in the document image. Corresponding SIFT features are marked by different colors. Here, red color indicates matching of character 'R' and blue color indicates matching of character 'S'.

probable location of two specific characters 'R' and 'S' in the document. Here, we have shown detection of two characters only. It is to be noticed that we have located the characters 'R' and 'S' by our approach which were intersected by curve lines.

6   Experimental Result

We have considered 10 different real geographical maps to test our method. Images were digitized by a flatbed scanner at 300 DPI. They contain text characters of different scale and orientation. Sometimes text characters are broken into different components due to printing or noise issue. Graphical long lines are touched/overlapped with text in many places. We show some qualitative examples in Fig.8 and Fig.9. In these image potions, the characters 'g', 'E' and 'B' are touched with graphical lines. We have tested these images in two ways. Fig.8 is the result of spotting of the character 'g' after removing all other isolated characters. In Fig.9 the spotting is done without removing other isolated characters. It is to be noticed that, here, we get many false alarms due to similar shape characters ('R' due to 'B' and 'T' due to 'E').

From the experimental result, it is found that this approach works better to take care of locating characters of arbitrary shape and orientation in graphical documents. Also, it is observed that, it finds the locations of corresponding

Fig. 8. Character 'g' is localized after removing isolated characters

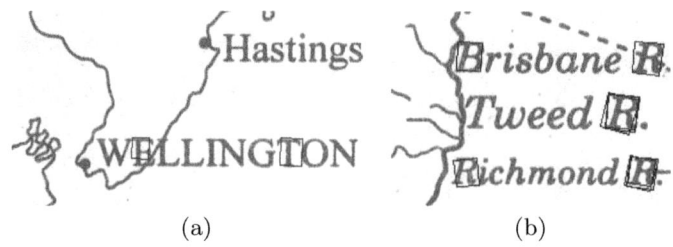

(a)   (b)

Fig. 9. Character 'E' and 'B' are localized in two different images. Here the spotting is done without removing other isolated characters.

characters with some false positive which are taken care according to different approaches we described earlier.

7   Conclusion

In this paper a combination of bottom-up and top-down approach has been investigated in the context of text character localization (spotting) in graphical documents. We have adapted SIFT in this application to locate text characters which are touched/overlapped. As our text extraction system is based on connected component analysis, when there is white noise in the text characters, we miss such text characters using our algorithm. Here, SIFT is useful to give us the probable location of character which are not extracted properly due to white noise or overlapping with graphical lines. The local text character model shapes are identified dynamically for each text character. Next, we used them to locate the similar images in other areas of the graphical image. This enables us to detect the text characters which are touched with graphical lines on-the-fly. SIFT descriptor finds the possible locations of text characters. It is appreciable that, using SIFT the other instances of text characters in different pose such as scale and rotation are detected using character model.

As explained earlier, our system learns the character shapes online from the isolated characters present in the document. Thus, if it does not find any isolated chacters from the document, then we can not seach the isolated characters in the rest of the document. Another problem is that, SVM is learnt using different character shapes before in offline. If some shapes are not learnt properly, our system may not recognize characters properly. Also, as SIFT features depends only on local neighbor area, in graphical documents due to curvature nature of graphical lines it generates many false positives. We have combined some state-of-the-algorithm using size criterion to reduce the volume of false alarms.

Acknowledgments

This work has been partially supported by the Spanish projects CONSOLIDER-INGENIO 2010 (CSD2007-00018), TIN2008-04998 and TIN2009-14633-C03-03.

References

1. Cao, R., Tan, C.: Text/graphics separation in maps. In: Blostein, D., Kwon, Y.-B. (eds.) GREC 2001. LNCS, vol. 2390, p. 167. Springer, Heidelberg (2002)
2. Fletcher, L.A., Kasturi, R.: A robust algorithm for text string separation from mixed text/graphics images. IEEE Transactions on PAMI 10(6), 910–918 (1988)
3. Tombre, K., Tabbone, S., Peissier, L., Lamiroy, B., Dosch, P.: Text/Graphics separation revisited. In: Lopresti, D.P., Hu, J., Kashi, R.S. (eds.) DAS 2002. LNCS, vol. 2423, pp. 200–211. Springer, Heidelberg (2002)
4. Luo, H., Agam, G., Dinstein, I.: Directional mathematical morphology approach for line thinning and extraction of character strings from maps and line drawings. In: Proceedings of the ICDAR, Washington, DC, USA, p. 257 (1995)
5. Tan, C.L., Ng, P.O.: Text extraction using pyramid. Pattern Recognition 31(1), 63–72 (1998)
6. Lowe, D.G.: Distinctive image features from scale-invariant keypoints. International Journal of Computer Vision 60, 91–110 (2004)
7. Bicego, M., Lagorio, A., Grosso, E., Tistarelli, M.: On the use of SIFT features for face authentication. In: Proceedings of CVPRW, USA, p. 35 (2006)
8. Rusiñol, M., Lladós, J.: Word and Symbol Spotting Using Spatial Organization of Local Descriptors. In: Proceedings of IAPR Workshop on DAS, pp. 489–496 (2008)
9. Ballard, D.: Generalizing the Hough transform to detect arbitrary shapes. Pattern Recognition 13(2), 111–122 (1981)
10. Tombre, K., Lamiroy, B.: Graphics recognition - from re-engineering to retrieval. In: Proceedings of the ICDAR, pp. 148–155 (2003)
11. Roy, P.P., Pal, U., Lladós, J., Delalandre, M.: Multi-oriented and multi-sized touching character segmentation using dynamic programming. In: Proceedings of ICDAR, Barcelona, Spain, pp. 11–15 (2009)
12. Vapnik, V.: The Nature of Statistical Learning Theory. Springer, Heidelberg (1995)

Graphical Drop Caps Indexing

Hassan Chouaib, Florence Cloppet, and Nicole Vincent

Laboratoire CRIP5, Paris Descartes University – Paris 5
45, rue des Saints Pères 75006 Paris, France
{hassan.chouaib,florence.cloppet,
nicole.vincent}@mi.parisdescartes.fr

Abstract. This paper presents a method for graphical drop caps indexing. Drop caps are extracted from old books. Finding a method classifying them according to styles defined by the historian is of considerable interest. The developed method is a statistical approach, where all possible patterns included in a pixel mask are processed in order to extract indexes that characterize the image. Then these indexes are used to classify a query drop cap by searching its most similar drop caps in the indexed base.

Keywords: Graphics Recognition, Graphics indexing, Zipf law, tf-idf, Drop caps, law mixture.

1 Introduction

Drop caps are specific graphics; they are neither images, nor writing but they are really both considering on the one hand the letter and on the other hand the background that is used to ornament the book. Several problems can be associated with them, such as recognition of the letter, analysis of the background scenery or analysis of the far background. For example the far background can be black and leaf can be drawn as background of the letter. The background contains as well as the letter font can be different according to the book manufacturer. It is of interest for the historians to find the identity, and the location of the manufacturer or the printing period of a document to better understand, that communication could be real in those days. In this paper, the far background analysis is the core of the study. The Historian of the CESR (**C**entre d'**E**tude **S**upérieur de la **R**enaissance in Tours) have at their disposal large bases of drop caps and one of their aim is to index them. They want to analyse the drawings contained in the background, maybe flowers, leafs, or scenery involving small children. But they are also interested by the far background, which has been classified in four types. The simplest one is white; non ink contributes to this background. On the contrary others are black and the scenery has to be carved in the log of wood used for printing. Two other types are more sophisticated. One is spotted that is to say some dots are carved in the wood and the printed result is a black layer where small white dots appear. Finally the last one comprises many small parallel lines that build different textures giving the feeling of a real painting. Then one possible index can have four instances: white, black, spotted and hatched. The four styles are shown in Fig. 1. Drop caps have already been studied [3] from the style point of view using

J.-M. Ogier, W. Liu, and J. Lladós (Eds.): GREC 2009, LNCS 6020, pp. 212–219, 2010.
© Springer-Verlag Berlin Heidelberg 2010

Fig. 1. Drop caps styles

some statistical local information. Here, the method relies on the same approach but instead of a global point of view we are paying more attention to local shapes for their own sake.

Our goal is to make a first indexing of the drop caps according to the description of far background styles that we have just mentioned. Then we have to achieve four classes recognition system. It will be performed in a two-step process. First of all, some indexes are to be extracted from the image in order to build descriptors that characterize each image. Then, these indexes are used to compare drop caps. A similarity measure has to be defined and we assume very similar images belong to the same style. After a learning phase, a drop cap image whose index is not yet known, is considered as a query image and compared to the most similar images contained in the reference database.

In the next section will be presented the features we have made use of and in section 3 we define a similarity measure inspired of the textual information retrieval. In section 4 some results are presented and discussed.

2   Representation Space

Before we can define some descriptor the representation space has to be chosen. Some references have to be highlighted, most significant for own purpose. In the first subsection we justify our choices based on the use in the field of text analysis.

2.1   From Text to Image Modeling

In this study we have been inspired by method used for text understanding. Texts are mad of a sequence of characters that are gathered in words of different lengths. In images the primary elements are pixels, they are not ordered but they are displayed in 2D space. What the eye perceives are spots of neighboring pixels, these spots are assumed to have in the image a role similar to word in a text. To simplify the model we limit the spots to a fix number of pixels. How to choose the spot shape is a problem. Some possible shapes are shown in Fig. 2.

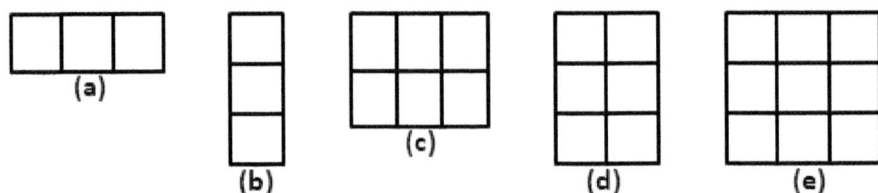

Fig. 2. Shape examples

If considering an image as a set of horizantal pixel lines, shape (a) in Fig. 2 is the most natural. Similary when considering a set of vertical pixel lines, shape (b) in Fig.2 is more appropriate. But images have more complex topology. Then shapes (c), (d) and (e) respect this 2D aspect. Shape (e) corresponds to the most commun definition of a neighbour in the field of discret image processing, the most isotropic neighbour. Indeed when reading, we are looking to the text with a privileged direction in the movement of the eye. The neighbourhood of a dot is longer on the horizantal direction rather than in the verticale one leading to shape (c). Finally an image is seen as a set of spots that are organised in the plane.

Salton [5] space representation is not relying on the order of the text words but only on their frequencies, much information is lost but enough is left to index texts, we want to do the same in the case of images.

Texts are always referring to a dictionary. In fact not all sequences of character have a meaning but only a rather small number of them. The number of words is the dimension of the text representation space.

2.2 Set of Reference Patterns

The drop caps images to be studied are grey level images with 256 grey levels. In our case words are any possible pattern contained in a shape as shown in Fig. 2. A shape is a mask of size nxm. Then the number of possible words, which are patterns, would be 256^{nxm}.

Then the representation space has a dimension equal to the number of possible patterns. With shape (e) this would lead to more than 10^{21} different words. This is a far too large dictionary and information would be too much dispatched. Then the dictionary has to be reduced. This can be done in two different ways: either the number of pixels in a mask or the number of grey levels characterizing the pixels can be reduced.

In our case, drop caps images are graphics, thus they are fundamentally binary images. In fact, when studying closely the images, we can see three levels that can be described as black, grey and white. Thus, a quantization was applied on the original image to obtain a new 3 grey levels image. With 3 grey levels the number of possible patterns is reduced to 3^{nxm}. With shape (e) this would lead to 3^9 different patterns that is to say 19683 patterns, with 2x3 shape ((c) in Fig. 2) we obtain only 729 different patterns. This is more reasonable as the images we are studying, are rather small ones. The dimension of the representation space is 729.

2.3 Descriptors

As in text representation an image is associated with a frequency vector. The frequencies are computed while scanning the image with a 2x3 mask. The occurrence frequency for each different pattern within the mask is computed.

Another piece of information related to the frequency through zipf law [8] in text is the rank of each pattern when frequencies are sorted in a decreasing order. Then we have also computed this information for the 729 patterns.

A drop caps image can be represented in two ways in a 729 dimension space.

3 Similarity Measurement

In order to take into account the property of the set of images we are studying, some new type of information is introduced. Let us consider each pattern in an image as a term in a document. Then the tf-idf model used in classical textual Information Retrieval [2,5] methods can be applied for comparison with ornamental letters contained in the database.

The aim of using tf-idf (**term f**requency- **i**nverse **d**ocument **f**requency) is to reduce the importance of a pattern present in the image and in the same time present in all images. The tf-idf helps us to find the most relevant patterns that represent an image.

Tf-idf [5,6] is the most common weighting method used to describe documents in the Vector Space Model, particularly in Information Retrieval problems. The term frequency component (tf) of a term t_i for a document d_j is calculated according to:

$$tf_{i,j} = \frac{frequency_{i,j}}{Max(frequency_{i,j})}$$

In addition, idf *(inverse document frequency)* measures how infrequent a word is in the collection. This value is estimated using the whole training text collection at hand. Accordingly, if a word is very frequent in the text collection, it is not considered to be particularly representative of this document. The idf, is normally computed as follows:

$$idf_i = \frac{N}{n_i}$$

where N is the total number of documents in the collection, and n_i is the number of documents in which the term t_i appears.

In *tf.idf* weighting schemes, the component of the weight for term t_i in a document d_j at position is of the form is:

$$W_{i,j} = tf * idf$$

Thus, tf-idf is a scheme to weight each value according to the use of the feature in all the patterns. It is strongly based on the frequency use.

To quantify the similarity between a document d and a query q, several measures of similarity are proposed. The most popular measures are shown in Table 1.

Let us note: $|d| = \sum_{i=1}^{n} d_i^2$ and $d.\,q = \sum_{i=1}^{n} d_i * q_i$.

Table 1. Some similiraty measures

Cosine	Jaccard
$Sim(d, q) = \dfrac{d.q}{\|d\|.\|q\|}$	$Sim(d, q) = \dfrac{d.q}{\|d\| + \|q\| - d.q}$

4   Experiments and Results

Whereas our method is quite general, we are testing it on a specific problem presented at the beginning of the paper. The database we use contains 2670 drop caps. The different styles are not represented in an equal way and in table2; the contribution of each class in the database is indicated.This database is provided by the **CESR**.

Table 2. Drop caps number for each style

Spotted	Hatched	Black	White	Total
258	674	258	1480	2670

The simplest classifier, using a vector space model, is a Knn classifier. This avoids the difficulty of a learning phase specially when the number of training data is low. The efficiency of the descriptors we introduced is assessed on this database using the leave one out method.

As we have introduced three descriptors (rank, frequency and tf-idf associated with frequency) several systems can be compared.

The dimension of the representation space is quite high and their right distance to compare elements is difficult to choose. Nevertheless we have made use of the Euclidian distance in the three cases. Moreover in the case of tf-idf of frequency we also introduced classical similarity measures and the distance associated with, by the formula:

$$d(d, q) = 1 - sim(d, q)$$

We tested two similarity measures, that are the cosine and Jaccard measures.

One parameter has to be fixed in the knn approach. We have compared results using four values of k: 1, 3, 5 and 10.Table 3 resumes some results we obtained.

Table 3. Recognition rate using each descriptor

K	Rank	Frequency	tf-idf		
			Euclidian	Jaccard	Cosine
1	71.9	79.843	**83**	**83.03**	83.3
3	73.7	80.7	82.84	82.58	84.31
5	75.73	**80.855**	82.43	82.32	**84.89**
10	**77.27**	79.67	81.57	82.13	83.7

We see the rank information gives lower results than the frequencies. This proves some information is lost when only rank is considered instead of frequency. Indeed it has been shown in [3] that distribution of patterns in an image, when drop caps are considered, cannot be modelled by a single zipf law but rather by a mixture of zipf laws. Then the rank and frequency are not equivalent.

As expected introducing the tf-idf weighting scheme the result are even better, they have been improved by up to 4%. In this case the choice of the similarity measure has some significant influence on the recognition rate. The high dimension of our representation space may justify these better results. That reach 84.89% using k=5 and the cosine similarity.

Now lets us look at the results in more details. Table 4 shows the recognition rate for each style using pattern frequency and tf-idf weighting scheme.

Table 4. Contribution of using tf-idf

	Spotted	**Hatched**	**Black**	**White**
Frequency	60.85	81.3	43.58	90.61
t-idf (Cosine)	75.58	79.38	72.375	91.22

We can notice that the recognition rate increased for most of styles and especially for the Black style. This improvement shows the success of tf-idf model to find the most representative pattern of each style. The fewer examples we have in a class the more important is the improvement obtained by this approach.

Furthermore we have investigated in the tf-idf model and looked for discriminating power of each feature. To do this we computed the Fisher score [7] associated with them. We show in Fig. 3 the score obtained on 729 frequency pattern considering two classes of drop caps the spotted and the hatched. The second graph of Fig. 3 contains tf-idf weighting associated with 729 patterns.

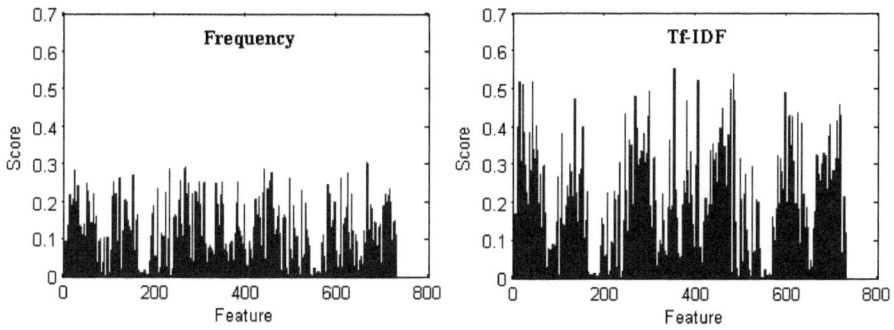

Fig. 3. Fisher score of frequency and tf-idf associated features

We see Fisher score of frequency never exceeds 0.3. Whereas in the second case more than 12% have a discriminating power higher than 0.3. It is usually assumed [7] that features with discriminating power lower than 0.01 are not interesting in the recognition process. These are 24.5% of them in the frequency case and 17.8% in tf-idf case. That is an increase of efficient features of 9%. The discriminating power mean moves from 0.06 to 0.12.

Our results were compared to those obtained with another method, the Z-method; proposed in [4].This method uses seven features to classify the drop caps. These features are based on the Zipf law [8]. Indeed the characteristics are factors built from the whole set of initial features. The first six features are shown in **Fig**.4.

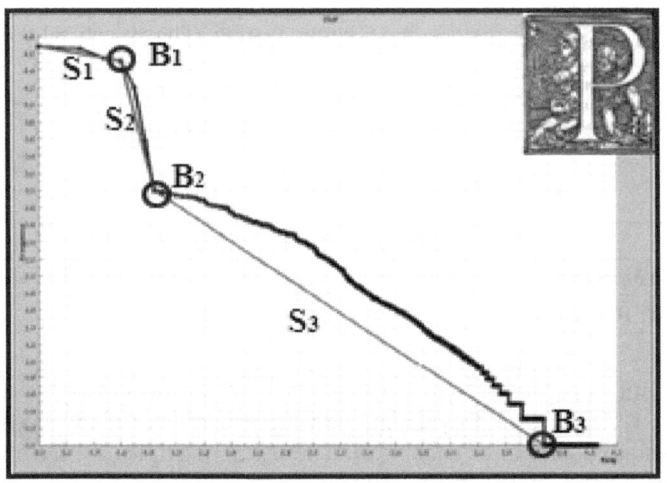

Fig. 4. Features of Z-method extracted using Zipf grapf

Features S_1, S_2 and S_3 correspond to 3 slopes. B_1, B_2 and B_3 represent the abscissa of these three points. The seventh is the slope of the inverse Zipf graph's line. The results of this method are illustrated in the table 5; they show that when a too global approach is used as in [4] some discriminating information about images is lost.

Table 5. Comparison with Z-method

K	tf-idf (cosine)	Z-method
1	83.3	74
3	84.31	77.27
5	**84.89**	**77.95**
10	83.7	77.44

5 Conclusion

Drop caps are particular graphics containing several layers. In this paper, we have proposed a method to index drop caps according to the far background. Statistical information extracted by scanning an image using a 2x3 mask helps us to define a set of descriptors that represents drop caps. Two descriptors that we have defined are Frequency pattern and its rank which are used with a Knn classifier to define our recognition system. We improved the recognition rate introducing a tf-idf weighting scheme. This model helps us to enhance the discrimination power of feature in our context. In our experiments we have shown that keeping all pattern information gives better results than factorizing information in seven features as in [4]. However, the space dimension is quite high, and other feature selection methods can be incorporated to define a new representation space. Other perspectives are to use other similarity measures and to test our approach on other graphic databases such as the GREC one.

References

1. Jiang, Y.-G., Ngo, C.-W., Yang, J.: Towards Optimal Bag-of-Features for Object Categorization and Semantic Video Retrieval. In: International Conference on Image and Video Retrieval (CIVR 2007) (2007)
2. Jones, K.S.: Experiments in relevance weighting of search terms. Information Processing and Management 15, 133–144 (1979)
3. Coustaty, M., Ogier, J.-M., Pareti, R., Vincent, N.: Drop Caps Decomposition For Indexing A New Letter Extraction Method. In: ICDAR 2009 (2009)
4. Pareti, R., Vincent, N.: Ancient letters indexing. In: International Conference on Patteri Recognition ICPR (2006)
5. Salton, G., Wong, A., Yang, C.S.: A Vector Space Model for Auto- Page 46, 50, 74 matic Indexing. Communications of the ACM 18, 613–620 (1975)
6. Salton, G., Allan, J.: Automatic Text Decomposition and Structuring. In: Proceedings RIAO 1994 Conference, CID, Paris, October 1994, pp. 6–20 (1994)
7. Gadat, S., Younes, L.: Stochastic Algorithm for Feature Selection in Pattern Recognition. Journal of Machine Learning Research 8, 509–547 (2007)
8. Caron, Y., Makris, P., Vincent, N.: Use of power law models in detecting region of interest. Pattern Recognition 40, 2521–2529 (2007)

Content Recognition and Indexing in the LiveMemory Platform

Rafael Dueire Lins, Gabriel Torreão, and Gabriel Pereira e Silva

Universidade Federal de Pernambuco, Recife - PE, Brazil
rdl@ufpe.br, gabrieltorreao@gmail.com, gfps@cin.ufpe.br

Abstract. The proceedings of many technical events in different areas of knowledge witness the history of the development of that area. LiveMemory is a user friendly tool developed to generate digital libraries of event proceedings. This paper describes the module designed to perform content recognition in LiveMemory.

Keywords: Digital libraries, image indexing, content extraction.

1 Introduction

LiveMemory is a software platform designed to generate digital libraries from proceedings of technical events. Until today, only very few prestigious events have proceedings printed and widely distributed by international publishing houses. Thus, copies of the proceedings are restricted to those who attended the event. In this case, past proceedings are difficult to obtain and very often disappear; bringing gaps into the history of the evolution of events and even research areas. The digital version of proceedings, which started to appear at the end of the 1990's, possibly made things even worse. Only conference attendees were able to obtain copies of the CDs of the proceedings. LiveMemory was used to generate a digital library released in a DVD containing the whole history of the 25 years of the proceedings of the Symposium of the Brazilian Telecommunications Society, the most relevant academic event in the area in Latin America. The problems faced in the generation of the SBrT digital library ranged from compensating paper aging effects, filtering back-to-front noise [5], correcting page orientation and skew during scanning, to image binarization and compression. LiveMemory merges together proceedings that were scanned and volumes that were already in digital form. The SBrT'2008 digital library was organized per year of the event.

This paper outlines the functionality of the LiveMemory platform in general and addresses the way it recognizes the contents of the pages, making possible general indexing of documents and better access to the information in the library. This module works by getting information from two different sources. The first one is the image of the pages of the "Table of Contents" of the volume. The second one is each paper page image. Besides those pages there are introductory pages such as the history of the event, the address of the volume editor, etc. There may also be track or session separation pages, remissive index, etc. Pages are segmented to find the block areas which correspond to the information and then transcribed via OCR. The

J.-M. Ogier, W. Liu, and J. Lladós (Eds.): GREC 2009, LNCS 6020, pp. 220–230, 2010.

transcription of the blocks of the Table of Contents and headings of papers are cross analyzed to generate the entries of the navigation index (hyperlinks) in the digital library. It is important to remark that the volumes of SBrT varied widely in layout from one year to another, or even within the same volume, as most of those volumes were typewritten according to "loose" requirements stated that each year editor at a time there were no word processors. Even the page numbering systems adopted varied from one year to another. Some volumes are numbered with Indo-Arabic numerals throughout, some others use Roman numerals in introductory pages, there are volumes that are split into "sessions" or "tracks" and each paper gets a numbering according to its position in there. The title and page number segmentation process was developed in MatLab© and correctly spotted the required information in almost 100% of times. In the cross reference system, that information was checked against the transcription of the pages of the Table of Contents and in case of inconsistent information the priority is given to the index in the calculus of page attributes.

This paper is organized as follows. In the next section one provides a brief overview of the features of the LiveMemory platform. Section 3 details the page content functionality of the platform. The information cross-reference modulo is described in Section 4. The concluding section details the results obtained for the content detection module in LiveMemory in the development of the SBrT Digital Library, presents the conclusions and draws lines for further work.

2 LiveMemory Image Pre-processing Routines

The top-level interface of the LiveMemory platform allows the user to generate the opening screen of the proceedings to be generated. In that screen, the user provides the information of the number of volumes to be inserted. The LiveMemory environment automatically builds the hierarchy of directories for the different volumes. The user may also provide a wallpaper image to the screen and an opening soundtrack to be played when the library is accessed. The user must provide information of which volumes are already in digital form and which volumes are originally in paper. In the previous version of LiveMemory the only entry to the library is through the top menu that provides buttons to volumes. To improve that situation a few difficulties need to be overcome. The volumes that were originally in digital form use several different technologies. Some volumes are one large pdf file where all pages/articles appear one after another. Some others are structured/browsable pdf files where each article has an entry in the index. Some volumes have some search and indexing software that point at pdf files. Some other volumes are encapsulated Flash or database protected files. Being able to "unstructured" all the available data to generate a global library index or re-index by author or keywords them is far from being a trivial task, which is considered out of the scope of this paper.

This section outlines the image processing functionalities in LiveMemory. All printed proceedings are scanned in true color with a resolution of 200 dpi and stored in uncompressed bmp file format. The scanned images are loaded in a directory that corresponds to the year of the event. LiveMemory is targeted at non-experts in image processing, thus the image processing part is as automated as possible and asks for no parameter input. The set of tools to suitably filter images encompasses the following routines:

- content identification,
- image binarization,
- noise border removal,
- orientation and skew correction,
- page size normalization,
- salt-and-pepper filtering, and
- image compression in Tiff_G4 file format.

Content identification for index generation is explained in the next section. The most important image processing routines are outlined below. LiveMemory makes use of some of the functionalities of BigBatch [4] a platform to process monochromatic documents. Similarly, to BigBatch, the document process interface may work in user driven or batch modes.

2.1 Image Binarization

Monochromatic images claim much less space than their color equivalent, are much faster loaded for visualization, need less toner for printing, etc. Most proceedings were printed in black-and-white. Thus, it is advantageous to have the pages in their monochromatic version, whenever possible. One phenomenon observed in several of the proceedings digitized by the authors to the SBrT Digital Library is that several volumes exhibit a light back-to-front interference [5], also known as bleeding or showtrough. Fig.1 zooms into a part of a page of a volume of SBrT with such noise. To minimize such phenomenon, LiveMemory successfully uses an entropy based binarization algorithm that was designed to remove back-to-front interference in historical documents [5].

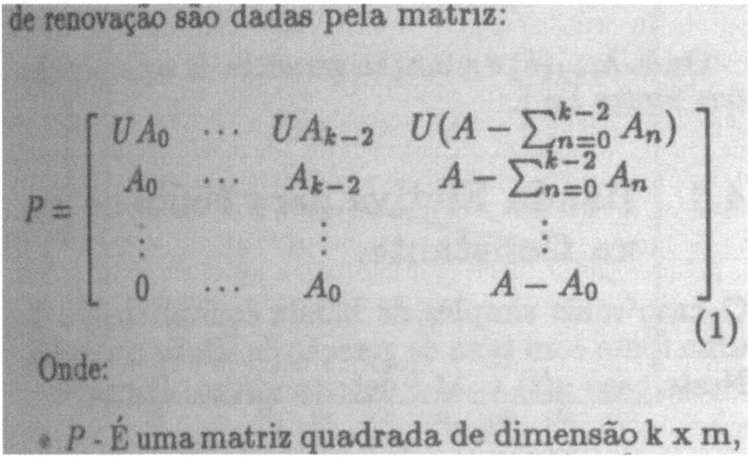

de renovação são dadas pela matriz:

$$P = \begin{bmatrix} UA_0 & \cdots & UA_{k-2} & U(A - \sum_{n=0}^{k-2} A_n) \\ A_0 & \cdots & A_{k-2} & A - \sum_{n=0}^{k-2} A_n \\ \vdots & & \vdots & \vdots \\ 0 & \cdots & A_0 & A - A_0 \end{bmatrix} \tag{1}$$

Onde:

• P - É uma matriz quadrada de dimensão k x m,

Fig. 1. Part of a document with light back-to-front noise

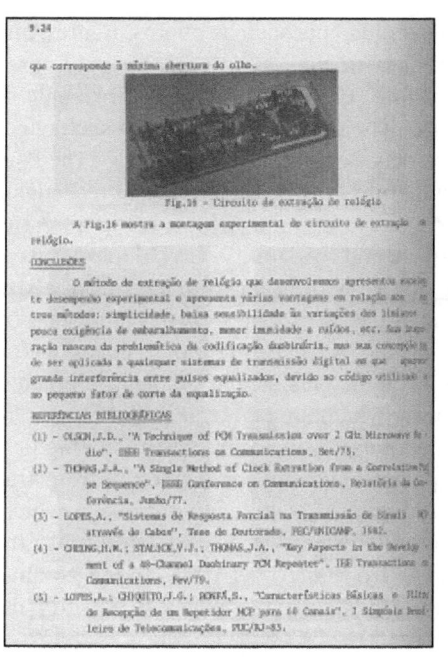

Fig. 2. Proceedings page with photo in true-color Size: 431 kB - JPG, 435kB - pdf

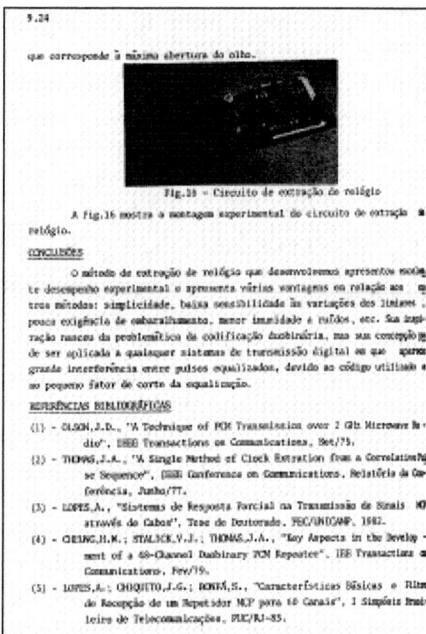

Fig. 3. Monochromatic version of Figure 02 Size: 122kB - Tiff, 351kB-pdf

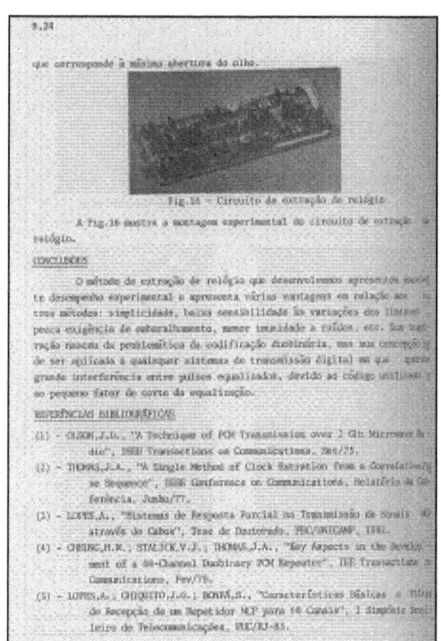

Fig. 4. Versions of **Fig.3.** Size: 3.03 kB - Tiff and 230 kB – pdf.

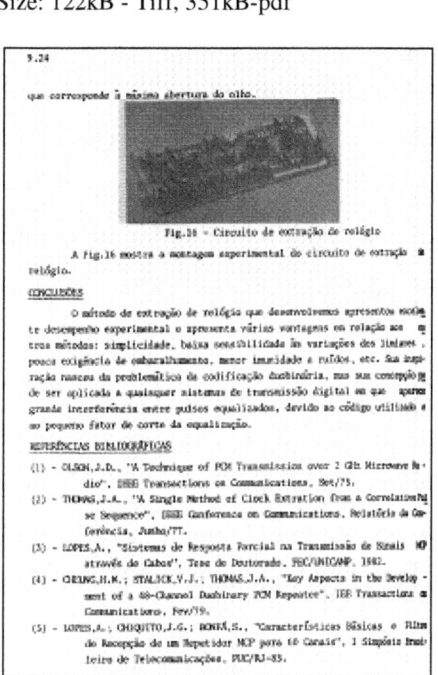

Fig. 5. LiveMemory Versions of **Fig.5.** Size: 3.03 kB - Tiff and 343 kB - pdf.

224   R.D. Lins, G. Torreão, and G. Pereira e Silva

Very often, paper pages incorporate graphical elements such as photos, figures, and graphs that are printed using dithering techniques in such a way that resemble gray scale images, although printed in black and white. Figure 2 provides an example of such a page, also with some back-to-front noise. The direct binarization of such pages does not yield satisfactory graphical results as may be observed in Figure 3. The conversion of page with photos, figures and graphs into gray scale provides a reasonable alternative in size, but introduces non-uniform pages into the volume as the majority of pages are monochromatic for the sake of space and readability. LiveMemory image processing module automatically sweeps the directory of scanned images from a volume looking for pages that encompass graphical elements. These pages are found by using projection profile both in the horizontal and vertical directions. Pages whose projection presents large contiguous areas indicate the presence of graphical elements. The projections allow splitting pages into blocks, which are tagged. Similar blocks are merged together. In such way, LiveMemory decomposes pages into text and graphical elements. Text areas are binarized. The graphical elements are converted from true color into gray scale. Figure 5 provides an example of such synthetic image which, although it brings no gain in space, if compared with gray scale, it is uniform to the reader as there is no difference in the text areas from the other pages in the volume. Layout analysis is performed in the different kinds of paper pages to identify the fields of interest with the aid of an OCR platform.

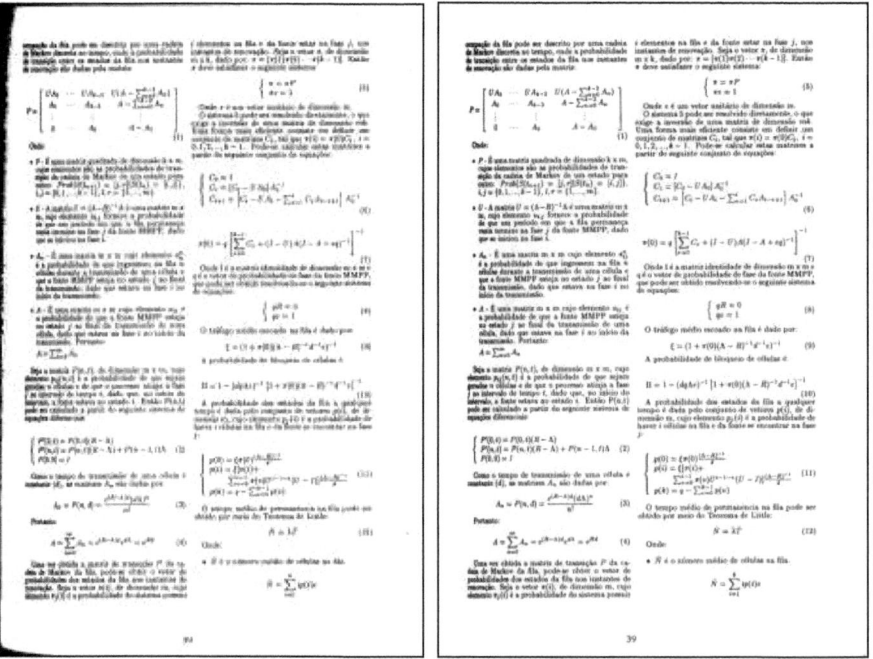

Fig. 6. Page with and without black border

2.2 Black Border Removal

As one may observe in the case of the page shown on the left hand side of Figure 6, the monochromatic version of the document exhibits a black border on its left margin. This border is the result of the uneven illumination of the scanning process due to volume binding. The same phenomenon appears, for different reasons whenever the volume of the proceedings is unbound and the loose pages are scanned using a production line automatically fed monochromatic scanner. The difference between the two cases aforementioned is that in the former the black border is within the document area, while in the latter case of automatically fed monochromatic scanners the noise surrounds the document. The right hand side of Figure 6 presents the same document of the left hand side with the black noisy border removed. The algorithm used in LiveMemory for black border removal is described in reference [1].

3 Paper Preparation

The experience with the digital volumes integrated into the SBrT digital library showed that, in general, there are standard layouts in the articles in one proceeding volume and that editors were careful enough to include headings with title and data of the authors. This information may be used for indexing articles and volumes in a similar way to the one proposed in reference [6].

A volume of proceedings has a somehow standard format that may be split into four parts:

- Volume presentation.
- Table of Contents.
- Papers.
- Remissive Index (optional).

The volume presentation frequently encompasses a title page, a forward (or preface) by the conference chairperson, the list of people on the program committee and other optional items. The Table of Contents is a list of authors, paper title, and page numbers. In general, roman numerals are used for page numbering the Volume presentation and the Table of Contents parts. Some conferences that use the Track format structure their proceedings differently, as:

- Volume presentation.
- Table of Contents.
- Track *1* (Track presentation+Papers)...Track *n* (Track presentation+Papers).
- Remissive Index (optional).

In this version of LiveMemory, the user provides information of the kind of structuring used in each volume. The papers themelves encompass front or title and content pages. The front pages of papers include:

- Paper Title.
- List of authors and affiliation.
- Abstract (or summary).
- Abstract in a foreign language (optional).
- List of keywords (optional).
- Classification indices (optional).

Identifying all these elements allows a complete navigation in the contents of papers.

3.1 Block Image Segmentation and Classification

LiveMemory segmentation algorithm uses projection profile to iteratively split the page in blocks in a top-down fashion. At first, one takes the projection profile of the whole page as the example shown in Figure 7, where one finds information blocks.

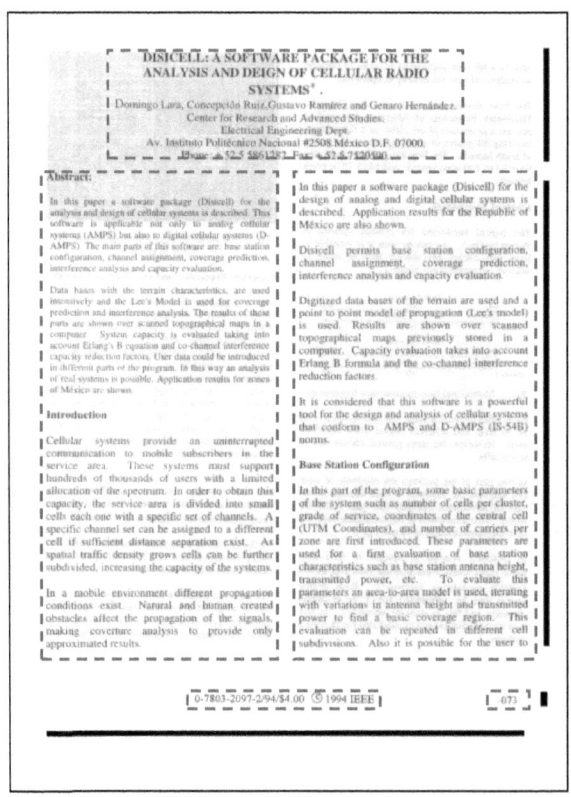

Fig. 7. – Front page block decomposition

Each of those blocks are then recursively split into sub-blocks until reaching blocks that envelope each line. If 25% of blocks are aligned vertically then that alignment point is a column margin. The analysis of the height, width and the position within the page and in relation to the other blocks in it allows one to re-merge similar blocks and to decide about their nature. For instance, in general the block that contains the title of an article is on the top-center of a page, not aligned with the column margins. A minimum width parameter is also used, to distinguish between a title and a page numbering. In the case of the paper proceedings of SBrT, no page heading had any other information besides page number.

Similarly, the blocks on the bottom of a page generally contain information about the paper and page number. In the case of the paper shown in Figure 7 the block on the top of the page contains the paper title, authors' names and affiliations. On the bottom of the page, one finds two blocks. The central one contains "0-7803-2097-2/94/$4.00 ©1994 IEEE" (every four years the SBrT annual event is called the International Telecommunications Symposium, an IEEE event) and the rightmost one presents "073", the page number in the proceedings volume. One may also observe that the page of Figure 7 also has back-to-front interference, which is more pronounced on the left hand side of the title block.

The segmentation and classification of the information on the Table of Contents is aided by user provided information on its general layout.

4 Automatic Categorization

Automatic categorization (or classification) of textual information in pre-defined classes is a research area of rising importance due to the ever growing availability of documents in digital format, thus a greater need of organizing them. A common approach to address such problem is based on machine learning: frequently an inductive process automatically generates a classifier having as starting point a set of pre-classified documents in each of the categories of interest. The advantages of such strategy in relation to the knowledge engineering one (which consists in experts in the subject manually defining a classifier) is due to its performance once it brings a considerable economy of efforts of the experts in the field, besides providing systematic grounds for extensibility, allowing more easily to address new domains in a much faster way. The LiveMemory platform follows the machine learning approach for text categorization, but also makes use of some artifices that yield an efficiency rise of such a process.

4.1 Document Organization

Indexing documents with a controlled dictionary is an example of the more general problem of database organization. Often, several other problems relative to document filing either for personal organization or in structuring a corporative database may be solved using text categorization techniques. An example of such may be found in the interesting paper in reference [9], which addresses the problem of efficiently finding patents of different categories with a high accuracy rate. Another example is the automatic fulfilment of columns in a newspaper (such as Economics, Health, International Politics, etc.). LiveMemory attempts to group together the papers in proceedings according to its concentration area. In the case of a telecommunications conference, for instance, one has mobile communications, satellite communications, computer networks, cryptography, etc. To make searches in a proceedings database in which the paper subject has been previously annotated is far more precise and faster. Very often, papers have no keywords and have to be swept to find them in their abstract or even the paper body, with the aid of a subject dictionary.

4.2 Document Organization

Text categorization dates back to 1961 with the work reported in [10] about the probabilistic text classification. Since then it has been used in a large number of different applications. LiveMemory is a tool that aims to work with proceedings independly of the language they are written. In the specific case of the proceedings of SBrT, there are papers in English, Portuguese and Spanish. Thus one has two different outlooks to cluster the papers into: one is in the language the paper is written in and another is its subject area. In the case of the SBrT digital library, priority is given to the paper concentration area, using as guidance the information within the volume such as tables of contents, indices, separation pages, etc. Even section ordering and conference schedule that often appears in some volumes are used to infer information about the content of papers, as chairpersons tend to organize sections with related papers. All information gathered must be cross-checked with abstracts and sometimes introductory and/or concluding sections. A different research concern, but somehow related to the one reported, focuses in finding document authorship and detecting plagiarism.

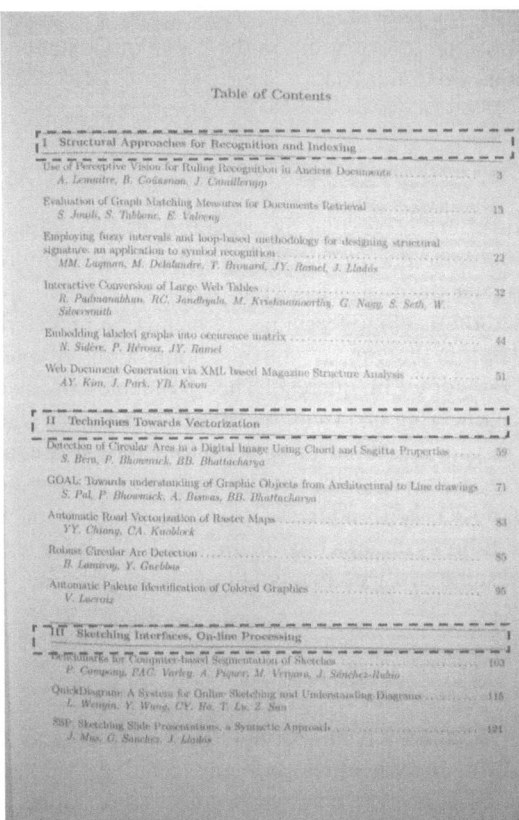

Fig. 8. Block segmentation of a page

5 The Table of Contents Generator

Having ways to fast navigate in digital libraries is mandatory. The blocks of interest spotted during the segmentation process shown above are transcribed via OCR. This information is used for indexing articles and volumes. The index generating module of LiveMemory takes the set the transcription of the Table of Contents pages from a volume and tries to match a formation rule of a regular expression to find the "page_number". The Java library for regular expression parsing was used to create the parser generator.

Each image that corresponds to a volume page is segmented to find its number and title blocks, which are transcribed using the Tesseract OCR. This information becomes attributes of the image. Figure 9 shows the results of block segmentation from the Table of Contents and of the paper title page for the article shown in Figure 8. As one may observe, the information in the two blocks are not the same. Even the title does not fully coincide as in the paper title there is a spelling mistake, corrected by the volume editor in the Table of Contents. Now, the system tries to unify the page_number information with the page attributes. The title pages are the key for the image and contents matching.

4.5 Dlslcell: A Software Package for the Analysis and Design of Cellular Radio Systems *P.73* D. Lara R., M. C. Ruiz S., G. Ramírez S., G. Hernández V., *Instituto Politécnico Nacional, Mexico*	DISICELL: A SOFTWARE PACKAGE FOR THE ANALYSIS AND DEIGN OF CELLULAR RADIO SYSTEMS[*]. Domingo Lara, Concepción Ruiz,Gustavo Ramírez and Genaro Hernánde Center for Research and Advanced Studies. Electrical Engineering Dept. Av. Instituto Politécnico Nacional #2508.México D.F. 07000. Phone: + 52 5 5861282 Fax: + 52 5 7520590.
4.5 Dlslcalls A Sollwara Package far lha Analysls and Daslgn al Câ€¢IIÃ¤Iar lalla lyslans P. 73 D. Lara R._ M. C. Ruiz S., G. Ramirez S., G. HcmÃ©ndcz V., Instituto PolitÃ©cnico Nacional, Mexico	DISICELL: A SOFTWARE PACKAGE FOR THE ANALYSIS AND DEIGN OF CELLULAR RADIO SYSTEMS*. Domingo Lara, Concepcion Ruiz,Gustavo Ramirez and Genaro Hernandez. Center for Research and Advanced Studies. Electrical Engineering Dept. Av. Instituto PolitÃ©cnico Nacional #2508.MÃ©xico D.F. 07000. Phone: + 52 5 5861282 Fax: + 52 5 7520590.

Fig. 9. Top: (Left) Information from Table of Contents; (Right) Paper Title; Bottom – Tesseract© transcriptions

The top-left part of Figure 9 presents an image block extracted from the Table of Contents of a volume. Its automatic transcription performed by the Tesseract OCR is shown immediately below. The use of regular expressions has enabled to spot the page number in the volume. That information was used to find the corresponding image file by offsetting the list_of_filenames. The segmentation process shown in the last section is able to find the image of the paper_title_block as shown in the top of right hand side.

Another aiding element is provided by the image filenames: they follow a strict numerical order. This means that the image filenames follow a pattern such as volume_year_page_number. For instance, the first image scanned of the 1991 volume is 1991_001, the second one is 1991_002, the third page is 1991_003, etc. Then, the problem of tying hyperlinks between the table_of_contents and images becomes finding the right offset in the two lists. Unfortunately, in some volumes of the SBrT proceedings there are missing and repeated pages. This may cause unrecoverable trouble to any automatic indexing system.

6 Conclusions and Lines for Further Work

This concluding section provides some evidence of the effectiveness of the index generation scheme presented herein. A total of 613 pages of 323 papers within several volumes of the SBrT proceedings with different page, table of contents, heading and footnote layouts, typesetting and printing technologies, paper color due to aging, etc. were tested. The title blocks were correctly recognized in 96.9% of the total number of pages, while page numbers were spotted in 97.1% of cases. The linking of the "Table of Contents" to page numbers was successful in 98.3 % of the total number of papers.

LiveMemory is no doubt a valuable and user friendly platform for the generation of digital libraries of event proceedings. Its use in the case of the SBrT digital library witnesses its usability.

The automatization of the detection procedure of the layout of the pages of the Table of Contents using a classification tool such as Weka [8] is under development. Other lines for further work include automatic author and keyword searching.

Acknowledgments. Research reported herein was partly sponsored by CNPq - Conselho Nacional de Pesquisas e Desenvolvimento Tecnológico and CAPES - Coordenação de Aperfeiçoamento de Pessoal de Nível Superior, Brazilian Government. The authors also express their gratitude to the SBrT – the Brazilian Telecommunications Society, for granting the permission to use the images from their proceedings.

References

1. Ávila, B.T., Lins, R.D.: A New Alg. for Removing Noisy Borders from Monochromatic Documents. In: ACM-SAC 2004, pp. 1219–1225. ACM Press, New York (2004)
2. Ávila, B.T., Lins, R.D.: A New and Fast Orientation and Skew Detection Algorithm for Monochromatic Document Images. In: ACM DocEng 2005. ACM Press, New York (2005)
3. Gonzalez, R.C., Woods, R.E.: Digital Image Processing, 3rd edn. Prentice-Hall, Englewood Cliffs (2007)
4. Lins, R.D., Ávila, B.T., de Araújo Formiga, A.: BigBatch: An Environment for Processing Monochromatic Documents. In: Campilho, A., Kamel, M.S. (eds.) ICIAR 2006. LNCS, vol. 4142, pp. 886–896. Springer, Heidelberg (2006)
5. Silva, J.M.M., da Lins, R.D., da Rocha Jr., V.C.: Binarizing and Filtering Historical Documents with Back-to-Front Interference. In: ACM Symposium on Applied Computing, Dijon. Proceedings of SAC 2006, pp. 853–858. ACM Press, New York (2006)
6. van Beusekom, J., et al.: Example-Based Logical Labelling of Document Title Page Images. In: ICDAR 2007, pp. 919–924. IEEE Press, Los Alamitos (2007)
7. Tesseract, http://code.google.com/p7tesseract-ocr/
8. Weka 3: Data Mining Software in Java, http://www.cs.waikato.ac.nz/ml/weka/
9. Larkey, L.S.: A patent search and classification system. In: Proceedings of DL 1999, 4th ACM Conference on Digital Libraries, Berkeley, CA, pp. 179–187 (1999)
10. Maron, M.: Automatic indexing: an experimental inquiry. Journal of the ACM (JACM) 8(3), 404–417 (1961)

Segmentation of Colour Layers in Historical Maps Based on Hierarchical Colour Sampling

Stefan Leyk

Department of Geography, University of Colorado, 260 UCB,
Boulder, CO 80309, USA
stefan.leyk@colorado.edu

Abstract. A colour image segmentation (CIS) process for scanned historical maps is presented to overcome common problems associated with segmentation of old documents such as (1) variation in colour values of the same colour layer within one map page, (2) differences in typical colour values between homogeneous areas and thin line-work, which belong both to the same colour layer, and (3) extensive parameterization that results in a lack of robustness. The described approach is based on a two-stage colour layer prototype search using a constrained sampling design. Global colour layer prototypes for the identification of homogeneous regions are derived based on colour similarity to the most extreme colour layer values identified in the map page. These global colour layer prototypes are continuously adjusted using relative distances between prototype positions in colour space until a reliable sample is collected. Based on this sample colour layer seeds and directly connected neighbors of the same colour layer are determined resulting in the extraction of homogeneous colour layer regions. In the next step the global colour layer prototypes are recomputed using a new sample of colour values along the margins of identified homogeneous coloured regions. This sampling step derives representative prototypes of map layer sections that deviate significantly from homogeneous regions of the same layers due to bleaching, mixed or false colouring and ageing of the original scanned documents. A spatial expansion process uses these adjusted prototypes as start criterion to assign the remaining colour layer parts. The approach shows high robustness for map documents that suffer from low graphical quality indicating some potential for general applicability due to its simplicity and the limited need for preliminary information. The only input required is the colours and number of colour layers present in the map.

Keywords: Colour image segmentation, two-stage colour sampling, historical maps, homogeneity, cartographic pattern recognition.

1 Introduction

The unique value of cartographic documents for the extraction of spatial information about the landscape has motivated a number of research efforts including the recognition of e.g., elevation contours [1], roads [2], symbol chains [3], forest area [4], or map text [5]. Topographic maps show the highest degree of complexity [6,7] and,

J.-M. Ogier, W. Liu, and J. Lladós (Eds.): GREC 2009, LNCS 6020, pp. 231–241, 2010.

with increasing age, the level of graphical quality decreases [8,9], which limits attempts of data capture and pattern recognition in such documents, to date. However, spatial representations in historical maps are invaluable projections of information about the landscape in the past over large areas. Large amounts of digital historical map series exist in national and international archives indicating the high demand for automated recognition processes to extract relevant information. One of the most important pre-processing steps in image analysis efforts is colour image segmentation (CIS) [10,11]. Subsequent procedures of pattern recognition and object extraction strongly depend on the results of the segmentation step. Due to the low graphical quality of scanned historical maps in which topographic and thematic information is represented as colour map layers CIS remains a challenging field of research. Different approaches to CIS have been described for well-conditioned maps [12-15]. For low-quality map documents only few such attempts can be found [9,16], which still suffer from limited applicability to other map products and the need for extensive parameterization, which results in a lack of robustness.

In this paper an approach for CIS is presented, which overcomes typical problems encountered in historical cartographic (and related) documents produced manually using engraving techniques and printing technologies of the 19th century. These problems include (1) varying colour values of the same colour layer within one map page, and (2) differences in colour values between area objects and line-work which belong to the same colour layer as a consequence of critical image resolution and bleaching effects. The process described herein retrieves colour values from a representative sub-area of the map based on similarity to extreme colour values in colour space. This first step derives robust global colour layer prototypes, which are used to identify homogeneous regions of each colour layer. To classify remaining areas, which deviate in colour from homogeneous areas, a re-sampling of colour values is conducted at the margins of the identified homogeneous areas and a spatial layer-specific expansion is carried out. This two-stage sampling approach compensates for variations in colour values and saturation in different parts of the same colour layer by adjusting the colour layer prototypes, which are used for segmentation.

2 Data and Methods

The CIS procedure is tested on different map pages of the historic topographic USGS map series, which were produced between 1899 and 1961. Problems include shape deformations, object interactions, as well as bleached, mixed and blurred colours due to ageing of the paper, manual reproduction and scanning parameters. Sample maps in the scale of 1:62,500 or 1:125,000 are available from the San Francisco Bay Area Regional Database (BARD) of the U.S. Geological Survey (http://bard.wr.usgs.gov/) [17] as scanned documents with a resolution of 400 dpi in 24 bit RGB colour (Fig. 1).

The CIS process presented here consists of two main steps: (1) the derivation of primary global colour layer prototypes and the extraction of homogeneous or well-saturated regions of different colour layers, and (2) re-sampling of colour values for the adjustment of prototypes to classify parts of map layers that significantly deviate in colour from identified homogeneous areas (Fig. 2).

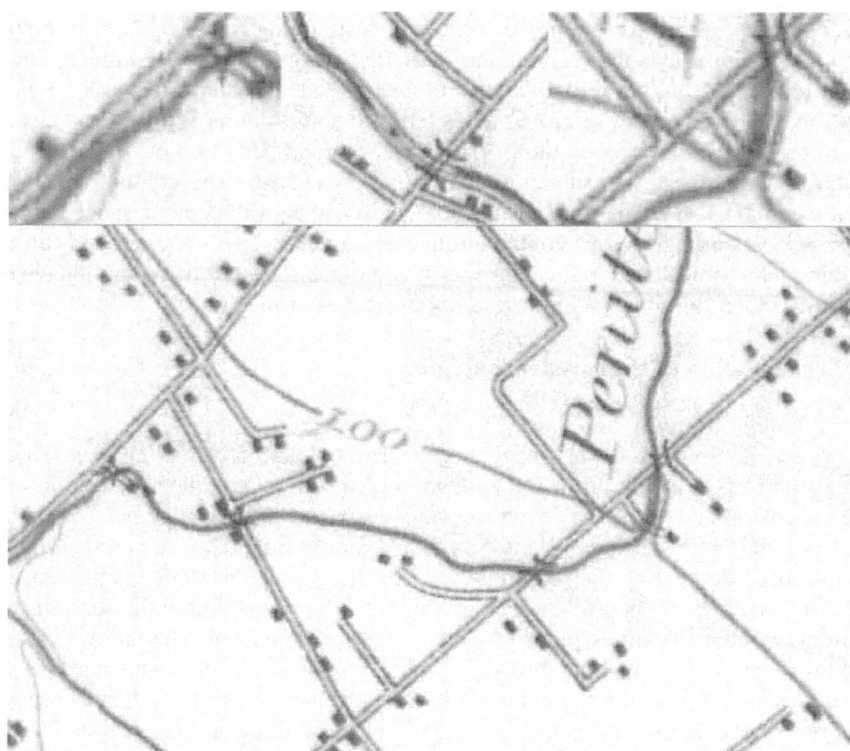

Fig. 1. Subsections of map page San Jose (1899) in the scale 1:62,500 from the San Francisco Bay Area Regional Database (BARD) (http://bard.wr.usgs.gov/)

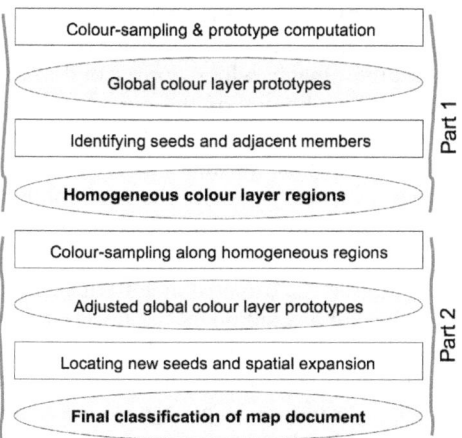

Fig. 2. Overview of the CIS process. Part 1 focuses on the colour sampling for prototype definition to identify homogeneous regions; part 2 includes the re-sampling of colour values and the adjustment of colour layer prototypes to find remaining parts of map layers.

In both parts iterative constrained colour value sampling is carried out for the computation of colour layer prototypes and their dispersion measures. Homogeneous or well-saturated regions identified in the first step and their prototypes are input to the second step where a re-sampling takes place to adjust colour layer prototypes. The only information, which is required a priori, is the number of colour layers and the colours used for these map layers in the topographic map. The colours of the map layers are quantified by theoretical full colour extremes of the colour space. The historic USGS quadrangle series contains four map layers: whitish background, blue for hydrography and wetlands, red for elevation and black for other thematic objects such as infrastructure, buildings, administrative boundaries or map text.

2.1 Segmentation of Homogeneous Regions

Colour Value Sampling and Initial Colour Layer Prototypes

One or more sub-regions of the image are selected and used as sampling area. The size of each sub-region should allow the collection of a sufficient number of colour values for each colour layer and thus a representative sample to derive global colour layer prototypes and layer-specific variation measures. Initial candidates of colour layers are determined by computing the minimum Euclidean distances in RGB colour space between the locations defined by the colour values of the pixels within the sampling area and the full colour locations for white, black, red and blue. This step allows the sampling procedure to start with more realistic initial colour candidates, which are found in the original image, and not theoretical full colour values. Instead candidates represent the most extreme colour values that occur within the sampling area defined by the proximity to full colour values in colour space.

For each pixel in the sampling area the Euclidean distances in colour space are computed between the positions defined by the colour values of this pixel and each of the initial colour layer candidates (white, black, red and blue). The computed distances are written into four new "colour distance grids", one for each colour layer. Next, the minimum distance values are identified in each of the four distance grids; the colour values in the original image, which correspond to these locations in colour space, define the new colour candidates of the corresponding colour layers. These new candidates are used for the next distance calculation in colour space and excluded from the following iteration in which further colour layer candidates are collected based on minimum distance in colour space to the new colour candidates. This iterative process continues until a robust set of initial colour layer candidates could be identified. The colour values of all collected colour layer candidates are written into data arrays, one for each colour space dimension and thus three arrays for each colour layer (R,G,B). These arrays serve as containers for storing point coordinates in colour space to define the point clouds of each colour layer as described below. This first sampling step ensures that colour layer candidates show representative colour values for the colour layers of the considered map page. During this iterative process the locations of these candidates in colour space move away from the initial extreme colour values.

The colour layer candidates that have been collected during the initial sampling are expected to represent typical colour values and to indicate the expected variability within well-saturated sections of the different colour layers. 3D point clouds are defined for each colour layer in colour space using the colour values of colour layer candidates, which have been stored in data arrays. The locations of the centroids of each of the point clouds are determined; these centroids represent the colour values of the first colour layer prototypes. Furthermore, the standard deviations of the distances in colour space between the individual colour layer candidates and the associated colour layer prototype (the centroid of the point cloud) are derived to quantify the expected variation in colour values within the same colour layer.

Defining Robust Colour Seeds and Homogeneous Regions of Colour Layers
Based on the identified prototypes the image is searched for robust colour seeds to be written into the image. First, the distances in colour space between each pair of colour layer prototypes are calculated (e.g., between black and red prototypes). Next for each pixel in the image it is tested whether its position in colour space is close to one of the prototypes and at the same time distant enough from all other prototypes. The procedure starts at a proportion of 1/10 (e.g., distance must be less than 1/10 of the RED-BLACK distance to prototype RED and farther than 9/10 away from prototype BLACK). The denominator is iteratively decreased (resulting in 1/9 and 8/9 of the inter-prototype distances, 1/8 and 7/8, etc.) and thus the range of positions in 3D colour space to be admitted is increased until a sufficient number of colour seeds is identified to shape a statistically robust sample. The colour values of the seeds identified are further collected in the data arrays to continuously refine the positions of colour layer prototypes in colour space. Based on these recalculated positions the distances between pixels and prototypes as well as between all pairs of prototypes in colour space are also recalculated before any change in distance proportions. If a sufficient number of colour seeds have been found the identified pixels are marked in the image object. A grey dummy layer prototype is determined by the midpoint in colour space between the black and white prototypes. This dummy layer is used to define pixels that cannot be unambiguously classified to one colour layer but are at risk to be misclassified.

Next, the physical neighbours of the colour seeds are identified using a simple test. If an adjacent pixel of a seed has the minimum distance in colour space to the same colour layer the pixel is immediately marked as part of the colour layer. The colour values of these new colour layer members are continuously collected to even further refine the colour layer prototypes. This member assignment is carried out until no new value is identified in the image. The strict constraint of colour similarity between adjacent pixels allows extraction of homogenous and well-saturated parts of the different colour layers (Figs. 3b,c).

2.2 Segmentation of Remaining Regions with Deviating Colour Properties

Repeated Colour Sampling and Prototype Re-computation
Typically, line-work such as elevation contours or river networks in topographic maps show only few homogeneous or well-saturated areas and can appear in light colours in some areas where line thickness becomes critical for the scanning resolution. Where colour layer elements include both areas and line-work, the colour layer prototypes can be very different between homogenous areas (saturated colours) and line-work

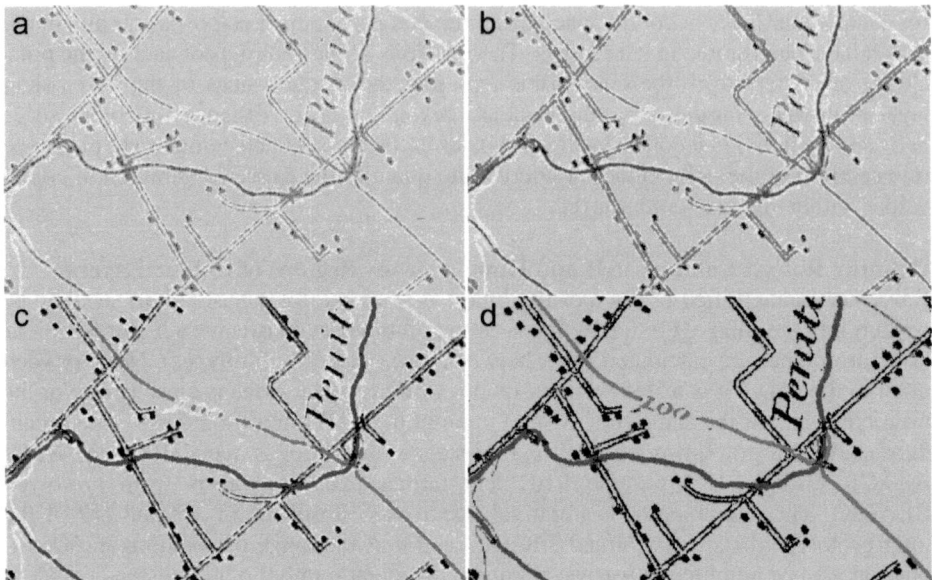

Fig. 3. Illustration of the single steps of the CIS process (the original map is shown in Fig. 1). Crème colour indicates background, blue, red and black are used for the corresponding colour layers, dark grey in (b) and (c) indicates the grey dummy colour layer): (a) Initial colour seeds based on global colour layer prototypes; (b) Homogeneous regions of the different colour layers; (c) Identified locations for colour sampling along the margins of homogeneous areas for prototype adjustment; (d) final segmentation after colour assignment, constraint filtering and expansion.

(less saturation, bleached colours), e.g., buildings and road lines both belong to the black map layer but appear in different tones in the historic USGS maps. Such variations represent particular problems since these bleached regions show high similarity to the background or could be easily misclassified as members of a different colour layer and cannot be detected based on primary colour layer prototypes. For this reason the second step of the CIS approach attempts to classify the remaining parts that suffer from colour deviation or lack in homogeneity.

First, a second sampling procedure is carried out to derive new colour layer prototypes for coloured foreground layers (i.e., red and blue). The rationale is to define prototypes that are representative for these bleached colour layers and thus would be located farther away in colour space from the black colour layer prototype. The sampling is conducted along the margins of the identified homogeneous colour regions of the blue and red foreground layers; only pixels in direct neighbourhood to these regions that are not yet classified and at the same time are not adjacent to homogeneous areas of any other foreground colour layer are sampled. The underlying assumption is that pixels of thin colour line-work appear very light and bleached similar to the colours of pixels at the margins of homogeneous colour regions of the same map layer. Sampling is carried out until a sufficient amount of sample pixels are collected for each colour layer. The following conditions have to be fulfilled: (a) The pixel has to be adjacent to one homogeneous region of a coloured foreground layer (red or blue), (b) The pixel must not be adjacent to any other homogeneous area that belongs to a foreground layer, and (c) the Euclidean distance in colour space between the colour point of this pixel and the colour layer prototype of the adjacent foreground layer has to be less than the distances to all other homogeneous colour layer prototypes (foreground). For each colour layer the collected sample is used to define adjusted global colour layer prototypes for blue and red by calculating centroids of layer specific 3D point clouds in colour space similar to the above described procedure (see 2.1).

After the new prototypes for the coloured foreground layers have been derived the black layer prototype can also be recomputed. Since the distances between the (old) black prototype and the new prototypes for blue and red increased in colour space this step can be done with looser constraints. All pixels that are not yet classified and not adjacent to a blue or red homogenous region are considered. If the distance in colour space between the considered pixel and the black layer prototype (or the grey prototype while the second-least distance must be to the black prototype) the pixel is marked as member of the black layer. Based on all identified pixels the black layer prototype is also recomputed by calculating the centroid of the 3D point cloud in colour space.

Assignment and Expansion of Colour Layers
Only non-classified pixels are considered in the final assignment and expansion step. Based on the distances in colour space between the pixel's colour values and the adjusted colour layer prototypes (colour similarity) and tests in the local environment of this pixel each pixel is assigned a map layer class. To overcome effects of false colouring and misclassification some additional tests for adjacency of each pixel are performed and constrained filters are applied to produce the final colour segmentation (Fig. 3d). This final result also includes a post-processing step to create a cartographic representation of road network similar to the original map. A filling step was carried out to re-label the background pixels to grey, which appear enclosed by grey and black layer elements. Next, the inner dimension of these filled compact regions could be determined and the interior parts were transformed back to background to create a double-line geometry. If the representation of a road changes to single lines this step simply does not find any locations with such inner dimensions of grey pixels.

3   Results and Discussion

First experimental segmentation results are illustrated in Figs. 4 and 5. The first step of the CIS approach proved a very high degree of robustness. Colour layer seeds (pixels) were misclassified in less than 3%. As can be seen in Figs. 3a-c many parts of well-saturated and homogeneous sections of the map layers (e.g., buildings, parts of map text or thicker line-work) as well as background were detected. Global colour layer prototypes as computed from 3D point clouds of sampled colour values represented suitable quantifiers for the identification of homogeneous colour layer regions. Consequently the segmentation of homogeneous regions was very reliable and represented an optimal basis for step two of the CIS approach.

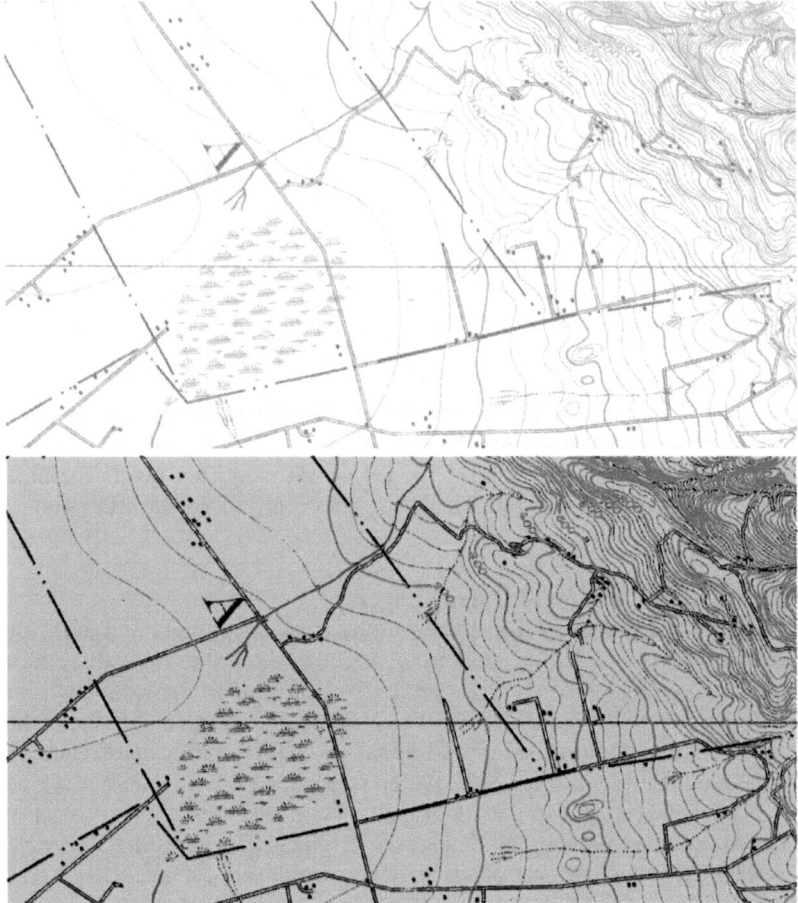

Fig. 4. Result of colour image segmentation shown on a larger subsection of map page San Jose (1899) in the scale 1:62,500

The sole use of such global layer prototypes did not allow classification of non-homogeneous regions or parts of colour layers whose colours significantly deviated from homogeneous parts. This high degree of within-layer colour variation would have resulted in considerable lack or misclassification of colour layer segments given traditional approaches.

The presented CIS approach could solve this problem to a great extent by performing a hierarchical segmentation approach. The colour value re-sampling step allowed the computation of new colour layer prototypes after the homogeneous regions were identified. These prototypes were more representative for the bleached parts or thin line-work of the different colour layers and allowed the successful assignment of non-classified locations (e.g., elevation contour line sections in the lower left and upper right corner in Fig. 3d). The original map (Figs. 1, 4 and 5) shows particular problems with regard to colour deviation within the red and blue colour layers but also due to differences between area features and line-work of the black layer. Thinner line-work is represented by colours of lower saturation most likely due to age-caused bleaching.

Some problems occurred if background colour was significantly darker (shadow effects), such as in urban areas where black layer objects are in close proximity; background area can be misclassified as black layer region. Similarly, if elevation contours became too dense no background could be identified between them (Fig. 4). In some instances thin and bleached parts of river networks were not identified if they significantly changed colour appearance due to colour mixing (Fig. 5). However some parts of such streams were classified correctly, which indicates some potential for post-processing steps using symbol chain tracking or gap filling techniques as well as cluster analysis for wetland symbols (Fig. 4). False colours impeded the final expansion and assignment of pixels in some cases, which required the definition of

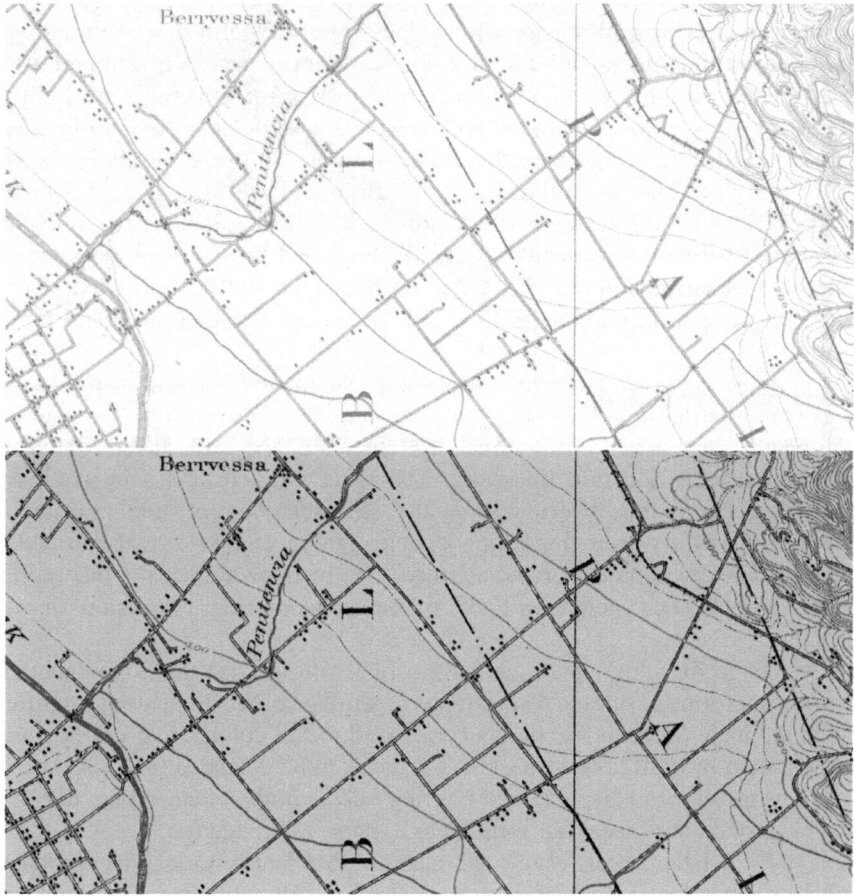

Fig. 5. Result of colour image segmentation shown on a larger subsection of map page San Jose (1899) in the scale 1:62,500

simple neighbourhood relationships during the expansion. Quantitative accuracy assessments are difficult to conduct and hard to interpret for colour image segmentation results in maps if both line-work and area features are involved [9,15]. However the results demonstrate a high degree of completeness and an accuracy assessment will be conducted based on vectorized spatial data in a GIS when larger areas can be examined based on visual inspections as demonstrated in [9].

4 Conclusions

The described procedure represents an attempt to conduct colour image segmentation of scanned historical USGS map documents with a minimum of parameterisation, which is the most significant contribution of this approach. The procedure relies only on the preliminary knowledge of the number and colours of map layers present in the document. This preliminary knowledge drives the definition of initial colour seeds based on theoretical full-colour values in colour space. The strength of this approach is its simplicity, which indicates potential for general applicability to similar documents. Various national archives of thousands of historical topographic maps exist in different countries that contain unique information about the landscape in the past. The hierarchical determination and adjustment of colour layer prototypes ensures the robustness of the approach against colour variation in colour layers and allows the segmentation of area features and line-work at the same time. Alternative segmentation approaches based on homograms [10], local window operations [15] or fuzzy c-means classification were tested and found unsuccessful for these kinds of problems.

The presented approach overcomes typical problems in historical map documents such as (1) variation in colour values of the same map layer within one map page, (2) the change in typical colour values if elements of the same layer contain both homogeneous areas and thin line-work. The latter problem arises in particular if the scanning resolution is critical for the representation of line objects or if the document suffers from bleaching due to ageing and historical reproduction technology. However, the described approach can be improved by the incorporation of more sophisticated neighbourhood relationships during expansion or shape descriptions.

The use of RGB colour space can be justified with the fact that topographic map production conforms to rules and constraints regarding colours to be used for different layers in the map. Since these colours correspond to full colours that can be numerically quantified in RGB colour space this choice can be defended [9]. Also since the main problems refer to bleaching and blurring effects in the scanned map documents employment of colour space transformations remains of limited use.

Next steps include the test of this approach for different historical map products to examine its robustness for different document types. Also the incorporation of topologic relationships, line continuation and smoothing, as well as gap filling criteria will be tested to improve the results of this pre-processing step.

References

[1] Khotanzad, A., Zink, E.: Contour line and geographic feature extraction from USGS colour topographical paper maps. IEEE Trans. Pattern Anal. Mach. Intell. 25(1), 18–31 (2003)

[2] Chiang, Y.-Y., Knoblock, C.A., Shahabi, C., Chen, C.-C.: Automatic and Accurate Extraction of Road Intersections from Raster. GeoInformatica 12(2), 121–157 (2009)

[3] Gamba, P., Mecocci, A.: Perceptual grouping for symbol chain tracking in digitized topographic maps. Pattern Recognit Lett. 20, 355–365 (1999)

[4] Leyk, S., Boesch, R., Weibel, R.: Saliency and semantic processing—extracting forest cover from historical topographic maps. Pattern Recognition 39(5), 953–968 (2006)

[5] Cao, R., Tan, C.: Text/graphics separation in maps. In: Blostein, D., Kwon, Y.-B. (eds.) GREC 2001. LNCS, vol. 2390, pp. 167–177. Springer, Heidelberg (2002)

[6] Llados, J., Valveny, E., Sanchez, G., Marti, E.: Symbol recognition: Current advances and perspectives. In: Blostein, D., Kwon, Y.-B. (eds.) GREC 2001. LNCS, vol. 2390, pp. 104–128. Springer, Heidelberg (2002)

[7] Watanabe, T.: Recognition in maps and geographic documents: Features and approach. In: Chhabra, A.K., Dori, D. (eds.) GREC 1999. LNCS, vol. 1941, pp. 39–49. Springer, Heidelberg (2000)

[8] Bajcsy, P.: Automatic Extraction Of Isocontours From Historical Maps. In: 7th World Multiconference on Systemics, Cybernetics and Informatics Proceedings (SCI 2003), Orlando, Florida, vol. 4, pp. 99–104 (2003)

[9] Leyk, S., Boesch, R.: Colours of the past: Colour Image Segmentation in Historical Topographic Maps Based on Homogeneity. GeoInformatica 14(1), 1–21 (2010)

[10] Cheng, H.D., Jiang, X.H., Sun, Y., Wang, J.: Colour image segmentation: advances and prospects. Pattern Recognit. 34, 2259–2281 (2001)

[11] Lucchese, L., Mitra, S.K.: Colour image segmentation: a state-of-the-art survey, Image processing, vision, and pattern recognition. Proc. of the Indian National Science Academy (INSA-A) 67A(2), 207–221 (2001)

[12] Centeno, J.: Segmentation of thematic maps using colour and spatial attributes. In: Chhabra, A.K., Tombre, K. (eds.) GREC 1997. LNCS, vol. 1389, pp. 221–230. Springer, Heidelberg (1998)

[13] den Hartog, J., ten Kate, T., Gebrands, J.: Knowledge based segmentation for automatic map interpretation. In: Kasturi, R., Tombre, K. (eds.) Graphics Recognition 1995. LNCS, vol. 1072, pp. 159–178. Springer, Heidelberg (1996)

[14] Santos, R., Ohashi, T., Yoshida, T., Ejima, T.: Filtering and segmentation of digitized land use map images. Int'l. J. Doc. Anal. Recognit. 1, 167–174 (1998)

[15] Chen, Y., Wang, R., Qian, J.: Extracting contour lines from common-conditioned topographic maps. IEEE Trans. Geosci. Rem. Sens. 44(4), 1048–1057 (2006)

[16] Leyk, S., Boesch, R.: Extracting Composite Cartographic Area Features in Low-Quality Maps. Cartography and Geographical Information Science 36(1), 71–79 (2009)

[17] U.S. Geological Survey, USGS 15 and 30 minute Historical Maps for San Francisco Bay Area. San Francisco Bay Area Regional Database (BARD), Menlo Park, CA (2004), http://bard.wr.usgs.gov/

A New Image Quality Measure Considering Perceptual Information and Local Spatial Feature

Nathalie Girard[1], Jean-Marc Ogier[1], and Étienne Baudrier[2]

[1] Laboratoire d'Informatique, Image et Interactions
University of La Rochelle, La Rochelle, France
nathalie.girard@univ-lr.fr, jean-marc.ogier@univ-lr.fr
[2] Laboratoire des Sciences de l'Image, de l'Informatique et de la Télédétection
University of Strasbourg, Strasbourg, France
baudrier@unistra.fr

Abstract. This paper presents a new comparative objective method for image quality evaluation. This method relies on two keys points: a local objective evaluation and a perceptual gathering. The local evaluation concerns the dissimilarities between the degraded image and the reference image; it is based on a gray-level local Hausdorff distance. This local Hausdorff distance uses a generalized distance transform which is studied here. The evaluation result is a local dissimilarity map (LDMap). In order to include perceptual information, a perceptual map based on the image properties is then proposed. The coefficients of this map are used to weight and to gather the LDMap measures into a single quality measure. The perceptual map is tunable and it gives encouraging quality measures even with naive parameters.

Keywords: Quality measure, gray-level image, image comparison, Hausdorff distance, distance transform, local dissimilarity measure.

1 Introduction

Image quality evaluation is a key point in several domains including image compression algorithm assessment or graphical image quality evaluation. Even if the best method is the subjective method MOS (Mean opinion score), which is based on observers' evaluation, it is not always possible to seek it: it is subject to variations and it involves many people and a lot of time. An alternative is to use an automatic quality evaluation. In this frame, the measure can be estimated just on the transformed image itself or in comparison with a reference image. We focus on the latter kind of methods so-called comparative objective methods. There exists a lot of well known comparative objective methods like the Mean Square error (MSE) or the Peak Signal to Noise Ratio (PSNR), ... But none of the current methods take in account a perceptual evaluation because they are often based on a pixel to pixel difference and perceptual information include both local and global aspects of the image. In order to move closer to the evaluation

J.-M. Ogier, W. Liu, and J. Lladós (Eds.): GREC 2009, LNCS 6020, pp. 242–250, 2010.

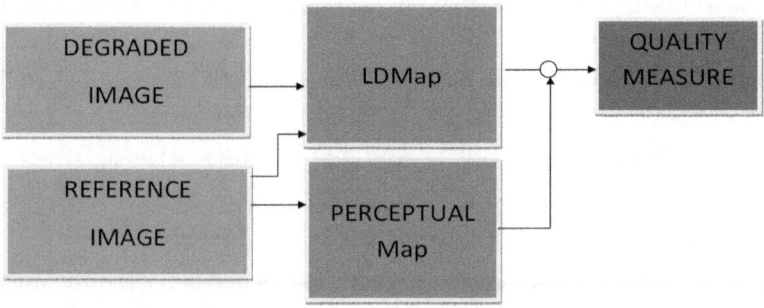

Fig. 1. Scheme of the quality measure method

of the final user, it is important to integrate a perceptual evaluation. We propose a new method which is based on two key points:

- a local objective evaluation
- a perceptual gathering.

Unlike to human vision, the pixel to pixel difference is very sensitive to small translations. A generalization of it to a less sensitive local measure has been developed for binary images in [1]: the so-called Local Dissimilarity Map (LDMap). It is based on the distance transform and several distance transform generalization to gray-level images are available. A comprehensive study of the LDMap generalization to gray-level images is presented here in quality measure aim.

Our method then exploits the spatial distribution of the dissimilarity measures gathered in the LDMap generalized to gray-level images thanks to a perceptual weighting that emphasizes image areas that are important for human vision. These weights, gathered in a so-called weighting map, are based on brightness, shape and texture; they can be tuned in function of the final application. Our method results in a single measure, the Figure 1 illustrates the measurement principle. The proposed measure has been compared to subjective evaluations on the test image base *IVC database* and gives encouraging results. The two key points are detailed in the following sections, the section 2 present the Local Dissimilarity Map - LDMap, the section 3 present the Perceptuel Map and the quality measure. The last section presents experiments and comparison with the LIVE base.

2   The Local Dissimilarity Map

2.1   Definition

The LDMap is based on a local and adaptative evaluation of dissimilarities between images. It has been first defined on binary images via a local measure of dissimilarities, and is based on a distance between a pixel and a set of pixel.

When the Hausdorff distance is used as a local measure, the formula of the LDMap is simple as follows

Definition 1 (LDMap). *Let A, B be two bounded sets of points in $\mathbb{R}^2$, and $x \in \mathbb{R}^2$ a pixel, the LDMap between A and B is defined as*

$$LDMap(x) = |ind_B(x) - ind_A(x)| \max(d(x, A), d(x, B)) \qquad (1)$$

where $ind_E(x) = 1$ if $x \in E$ and 0 otherwise, d is a distance transform (DT), measuring the distance between a point and a set of points.

The formula gives for each pixel x a value that depends on the distance transformation from the sets A and B. Figure 2 illustrates the notion of local dissimilarity. Each image contains two letters. The dissimilarities are quantified: big dissimilarities are represented in *dark* and small ones in *light*. Moreover, they are spatially localized: from the LDMap, one can see that the bigger dissimilarities are situated on the straight line of the "e" and on the top of the "t", so the corresponding dissimilarities are important. The ones between the left part of the "o" and the bottom of the "t" and between the loop of the "e" and the "c" are light, so they are small.

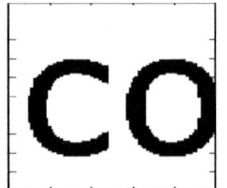  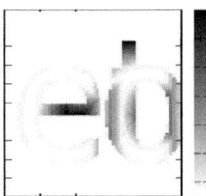

Fig. 2. Letters CO and ET and their LDMap illustrating their local dissimilarities

Remark 1. The DT : $x \mapsto d(x, A) \in \mathbb{R}^+$ is null for $x \in A$.

This definition 1 can be generalized to gray-level images:

- the term $|ind_B(x) - ind_A(x)|$ is changed in $\frac{|B(x) - A(x)|}{\max(|B(x) - A(x)|)}$ so as its result stays in $[0, 1]$
- the rest is a maximum between two DT. Several DT generalizations have been proposed in the literature, e.g. GRAYMAT[2], DTOCS[3]. These DT are generalizations in the sense that they hold the same result for gray-level version of binary images

Thus the LDMap for gray-level images can be defined as follows:

Definition 2 (Gray-level LDMap). *Let A, B $\mathbb{R}^2 \to \mathbb{R}$ be two bounded functions $\mathbb{R}^2$, and $x \in \mathbb{R}^2$ a pixel, the gray-level LDMap between A and B is defined as*

$$LDMap(x) = \frac{|B(x) - A(x)|}{\max(|B(x) - A(x)|)} \max(d(x, A), d(x, B)) \qquad (2)$$

where d is a generalized distance transform (DT).

These different generalizations to gray-level images lead to different dissimilarity measures. In the next section, we study the properties of these DT so as to choose the most suitable for our application.

2.2   Computation of DT in the Gray Level Case

The distance transform and its continuous generalizations

The DT are based on an underlying distance between pixels. The different DT come from distinct underlying distances. We focus here on the underlying distance computation so as to choose the most adapted DT to image processing.

For binary images, pixels can be seen as points of the plan. The distance transforms are then based on 2D mathematical spatial distances (e.g. the Euclidean distance, the Mahalanobis distance and so on). A first possibility for replacing the distance transform (in the binary case) is to define a new point set (in the gray-level image), that is the set of the points equal to the minimum or the maximum value and to measure the distances from this point set. This is the choice in all the articles but [4]. Under this hypothesis, there is still to compute the distance between pixels. For gray-level images, the gray-level dimension is added to the two plan dimensions. There are different ways of including this dimension in the distance computation, and they influence the produced dissimilarity measure. Considering a gray-level image as a set of points of $\mathbb{R}^3$, the image is a surface in $\mathbb{R}^3$. Then the distance between two points of this surface is the shortest path between these two points accorded to a path-length measure. There are two possibilities in the literature for this measure: the measure of the length of the path on the image surface or the measure of the area of the surface under the path on the image surface. A formal definition is given thereafter (Definition 3).

Definition 3 (path-length measure). *Let I be a continuous function defined on X the image support:*

$$I : \begin{cases} X \longrightarrow & \mathbb{R} \\ (x, y) \longmapsto & I(x, y) \end{cases}$$

and let $\pi : t \in [0, 1] \longmapsto \pi(t) \in \mathbb{R}^2$ be a continuous path between the pixels p and q of I. The length of the path π (so-called L_π) is given by

1. the length of the path on the image surface

$$L_\pi^1 = \int_0^1 |I'(\pi(t))| dt \tag{3}$$

2. the area of the surface under the path on the image surface

$$L_\pi^2 = \int_0^1 |I(\pi(t))\pi'(t)| dt \tag{4}$$

Both of these possibilities have been implemented in the discrete space. The discretization step implies also choices so the different versions of the DT underlying distances are briefly introduced in the next section. A last generalization so-called Continuous distance transform (CDT) is proposed by Arlandis and al [4]. The CDT is not based on a path-length measure but on a darkness saturation measure around the pixel on which the distance is computed. The CDT only has a algorithm, so it will not be detailed in the next section.

Discrete implementations of the distance transforms

The detailed presentation of the different DT is too long for this paper so we will only give a brief insight on them.

1. methods based on eq. 3 (introduced by P. Toivanen):
   - the Distance Transform On Curved Space [3,5] (DTOCS), the Weighted DTOCS (WDTOCS)
   - the method improvements: 3-4-DTOCS and Optimal WDTOCS (Opt-WDTOCS).
2. methods based on eq. 4
   - the Gray-Weighted Medial Axis Transform [2] (GRAYMAT) introduced by G. Levi and U. Montanari. Its underlying distance is based on the pixel difference value and as a consequence, it promotes low gray-level pixel paths. It is aimed at skeletization.
   - the Gray-Weighted Distance Transform [6] (GWDT) introduced by D. Rutovitz.

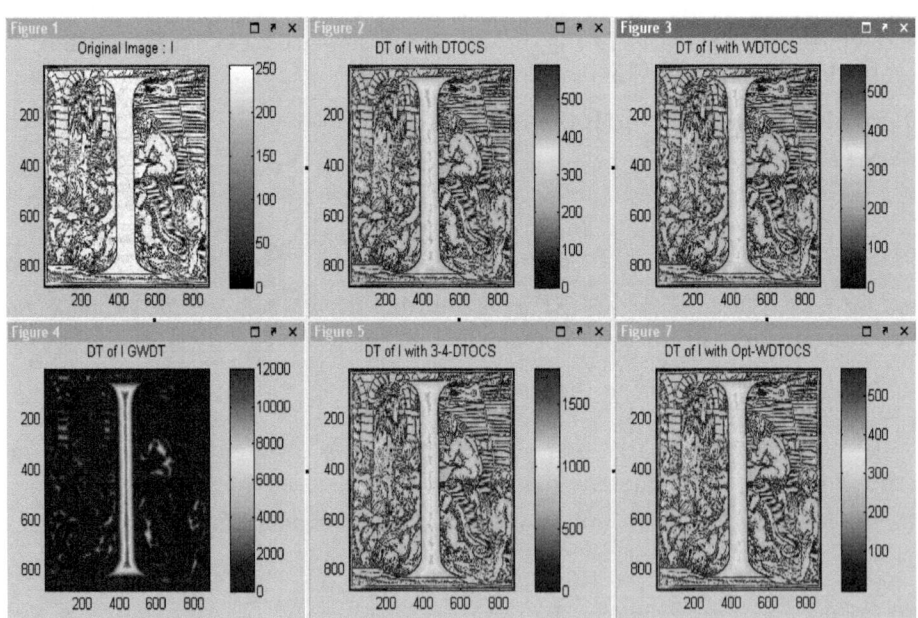

Fig. 3. An example of some DT

2.3   Comments and Choice

It is quite difficult to anticipate the qualities of the different DT from their definition, nevertheless, we have chosen the first family method because their underlying distance is closer to the notion of distance in the graylevel dimension. The first family indeed, takes into account differences of gray levels on a path and the spatial distance. It seems to us more appropriate than the second family which overwrites some spatial information for the benefit of gray level information (see Fig. 3). In the second family the differences between the versions are minor. Nevertheless the 3-4-DTOCS seems the most interesting to us because it does not underestimate the diagonal distance unlike to DTOCS. The CDT gives also good visual results. Our study on this point should be furthered to detail the choice.

For a $m \times n$ image, the LDMap always gives $m \times n$ dissimilarity measures for one image comparison, which represent to much information. So we want to used all these measures in order to define a single quality measure. This is the goal of the following section.

3   The Quality Measure

It is known that the visual perception do not take all the image into account [7], e.g. the perception of a compression artefact depends on the characteristics of the surrounding pixels [8]. As influencing characteristics, one can cite the texture, the luminance and the edges [9]. The LDMap measures the local dissimilarities but does not take into account these perceptual characteristics. Thus, we propose to defined a weighting map (WMap) where the weights model the perceptual sensitivity. A weight equal to 1 means that, according to its model, the eye will be very sensitive to perturbation on this pixel. On the contrary, a weight equal to 0 means that a perturbation there will not be perceptible. The sum of the LDMap values weighted with these weights is based on the objective dissimilarity measures and their perceptual importance. It is important to notice that the choice of the WMap computation that is given there after is only a proposition dependent on the application. The scheme of the process proposed Fig. 1 is a general one where the WMap computation can be adjusted to the application. This scheme shows that the WMap is computed from the original image and weights the LDMap obtained from the image comparison. We make this choice because we suppose that the protocol is to compare the original image with the transform one. Thus the reference image from which the defaults are evaluated is the original one.

The WMap combines the brightness, shape, texture attributes which are factors the eye is sensitive to. Each attribute is calculated on the reference image and for each one we define a map: for the brightness B, for the shape S and for the texture T. The WMap is defined as weighted sum (Definition 4) of B, S and T. Then the WMap is normalized in order to be a weighting Map. Figure 4 presents an example of WMap.

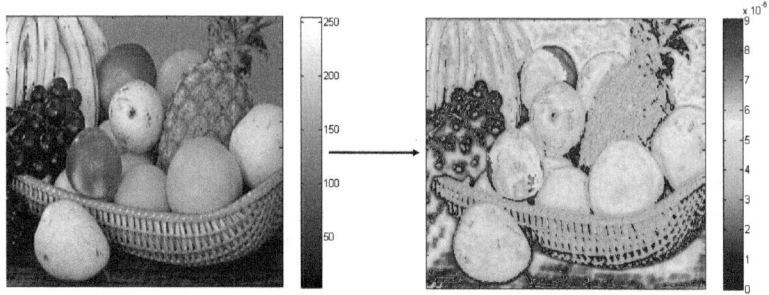

Fig. 4. WMap with p_1=2, p_2=1 and p_3=-2

Definition 4 (WMap). *Let A be an image, the WMap of A is defined by:*

$$WMap_A = \frac{p_1 \cdot B + p_2 \cdot S + p_3 \cdot T}{\max(p_1 \cdot B + p_2 \cdot S + p_3 \cdot T)} \tag{5}$$

Where p_1, p_2 and p_3 are the weights respectively of the brightness, the shape and the texture, and where B, S and T are positive.

By balancing the LDMap with the WMap as illustrated Fig. 1, we obtain the perceptual local dissimilarity map (PLDMap), containing local information on dissimilarities and visual perception. The quality of the transformed image can be measured by extracting the maximum value or the mean value of the PLDMap. We have study the two values and the maximum appears to be the best quality measure.

4   Results

The measure has been tested on the LIVE base [10], with the JPEG compressed images. The DT that have been tested are the 34-DTOCS and the CDT, which are the most relevant for the quality measure (cf sec.2). As we do not succeed in finding automatically the right parameters for the WMap, we have made a learning step. It consists in finding the parameters that minimize the variance of our quality measure against the MOS on a learning base of 40 images. The next step is a test that is made on a 50 image set totally disjoint from the learning set. The parameters values are those obtained thanks to the learning. The parameters obtained with the learning are

- for the 34-DTOCS: $p_1 = \frac{2}{15}, p_2 = \frac{1}{2}, p_3 = \frac{11}{30}$
- for the CDT: $p_1 = \frac{1}{4}, p_2 = \frac{1}{2}, p_3 = \frac{1}{4}$

As a result, we test the correlation between our measure and the MOS value for the 34-DTOCS and the CDT. The hypothesis of no correlation has a probability of 0.01 for the CDT measure and 0.17 for the 34-DTOCS measure. The threshold of rejection is commonly 0.05, which means that the CDT can not be seen as no correlated with the MOS, unlike to the 34-DTOCS measure.

5 Conclusion and Future Works

We present a generalization of the CDL to the graylevel images. The generalized CDL is improved with perceptual weights so as to give an objective quality measure. Our test shows that this measure based on the CDT is well correlated with the MOS on a test base of 50 images. First of all, the WMap for which the optimal parameters have been computed as an illustration of the method, could take in account other attributes. Obviously, weights of maps constituting the WMap can be refined to improve the measure. Then, the correlation coefficient of the measure should be compared with those of other quality measures. Finally, the LDMap could be developed to color images.

References

1. Baudrier, E., Morain-Nicolier, F., Millon, G., Ruan, S.: Binary-image comparison with local-dissimilarity quantification. Pattern Recognition 41(5), 1461–1478 (2008)
2. Levi, G., Montanari, U.: A grey-weighted skeleton. Inform. Control 17, 62–91 (1970)
3. Toivanen, P.: New geodesic distance transforms for gray scale images. Pattern Recognition Letters 17, 437–450 (1996)
4. Arlandis, J., Pérez, J.C.: The continuos distance transformation: A generalization of the distance transformation for continuos-valued images. In: Amsterdam, I. (ed.) Pattern Recognition & Applications (2000)
5. Toivanen, P., Elmongui, H.: Sequential local transform algorithms for gray-level distance transforms. In: Proc. of the 9th Eur. Sig. Proc. Conf. (1998)
6. Rutovitz, D.: Data structures for operations on digital images. In: Cheng, G.C., Ledley, R.S., Pollok, D.K., Rosenfeld, A. (eds.) Pictorial Pattern Recognition, pp. 105–133 (1968)
7. Lorenzetto, G.P.: Image comparison metrics: A review, July 25 (1998)
8. Wang, Z., Bovik, A.C.: A universal image quality index. IEEE Signal Processing Letters 9(3), 81–84 (2002)
9. Dinet, E., Bartholin, A.: A spatio-colorimetric model of visual attention. In: Proc. of the Expert Symp. on Visual Appearance, Paris, CIE, October 2006, pp. 97–105 (2006)
10. Sheikh, H.R., Wang, Z., Cormack, L., Bovik, A.C.: Live image quality assessment database release, Technical report, University of Texas (2005), http://live.ece.utexas.edu/research/quality

A   Appendix

- **LDMap** : Local Dissimilarity Map (section 2)
- **MOS** : Mean opinion score
- **MSE** : Mean Square error
- **PSNR** : Peak Signal to Noise Ratio
- **DT** : Distance Transform (section 2)
- **GRAYMAT** : Gray-Weighted Medial Axis Transform
- **DTOCS** : Distance Transform On Curved Space
- **CDT** : Continuous Distance Transform
- **WDTOCS** : Weighted Distance Transform On Curved Space
- **GWDT** : Gray-Weighted Distance Transform
- **WMap** : Weighting Map (section 3)
- **PLDMap** : Perceptual Local Dissimilarity Map (section 3)

GREC'09 Arc Segmentation Contest: Performance Evaluation on Old Documents

Hasan S.M. Al-Khaffaf, Abdullah Z. Talib, Mohd. Azam Osman, and Poh Lee Wong

School of Computer Sciences, Universiti Sains Malaysia, 11800 USM Penang, Malaysia
{hasan,azht,azam}@cs.usm.my, wongpohlee@hotmail.com

Abstract. Empirical performance evaluation of raster to vector methods is an important topic in the area of graphics recognition. By studying automatic vectorization methods we can reveal the maturity of the tested methods whether as a research prototype or a commercial software. Arc Segmentation Contest held in conjunction with the eighth IAPR International Workshop on Graphics Recognition (GREC09) is an excellent opportunity for researchers to present the results of their proposed raster to vector methods. The contest provides a uniform platform where the output of different methods can be analyzed. The relevance of the contest is further revealed by the creation of new test images with their ground truth data. Old documents were used in this contest. Five methods participated (two research prototypes and three commercial software). Two tests were performed namely between-methods test (participated by all methods) and within-method test (participated by only one method). This paper presents the results of the contest.

Keywords: Performance Evaluation, Graphics Recognition, Raster to Vector Conversion Methods, Line Drawings.

1 Introduction

The Arc Segmentation Contest 2009 is the fifth in the series of contests and was held during GREC'09 at the City University of La Rochelle, France, in July 2009. It was organized by the School of Computer Sciences, Universiti Sains Malaysia, Malaysia. The previous contest was held during GREC'07 in Curitiba, Brazil, in September 2007 and was organized by Image Understanding and Pattern Recognition (IUPR) Research Group, University of Kaiserslautern, Germany.

In the next paragraphs a short synopsis for each of the previous reports of the arc segmentation contests are presented. Note that, the first two contests were called dashed-line detection contests rather than arc segmentation contests.

GREC'95: Bin Kong et al. [1] performed dashed-line detection contest on a set of synthetic images.

GREC'97: Chhabra and Ihsin [2] evaluated the performance of participating methods in dashed-line detection contest. Solid and dashed lines, solid and dashed circular arcs, and segmentation of text were evaluated in the contest. Only synthesized images were

J.-M. Ogier, W. Liu, and J. Lladós (Eds.): GREC 2009, LNCS 6020, pp. 251–259, 2010.

used. Many performance metrices were used to judge the performance of the segmentation/detection.

GREC'99: Chhabra and Ihsin [3][4] evaluated the performance of four vectorization software: one research prototype (VrLiu) and three commercial software (Scan2CAD, Vectory, TracTrix). Ten real scanned images were used in the test (five mechanical engineering drawings and five architectural drawings). *EditCost Index* [5] was used as the criterion of the performance. Many metrices (false alarms, miss detection, one-to-many matches, and many-to-one matches) are combined into one index (*EditCost Index*). Continuous straight lines, circular arcs, and text regions were evaluated.

GREC'01: Liu et al. [6] evaluated the performance of two research prototypes (TIF2VEC and RANVEC). Seven images were tested (four synthesized and three real scanned). Different types of noise were used (Gaussian noise, high frequency, hard pencil, geometrical noise). Vector Recovery Index (*VRI*) was used to measure the performance of the methods. Only solid circular arcs were included in the test.

GREC'03: Liu [7] evaluated two research prototypes (Elliman's and JiQiang's algorithms). Twelve images were used (eight real scanned images, four synthesized images created by adding noise to the real scanned images). *VRI* was used to measure the performance of the methods. Only solid circular arcs were tested.

GREC'05: Liu [8] evaluated three research prototypes (TIF2VEC, RANVEC, and Keysers & Breuel method). Eighteen test images were used in the test (six real scanned and twelve synthesized images created by adding salt and pepper noise to the six real images). The *VRI* was also used but with slightly modified formula ($VRI = \sqrt{D_v * (1 - F_v)}$) in order to avoid the case of scoring when no (or very small) arcs could be detected. Only continuous circular arcs were considered in this contest.

GREC'07: Shafait et al. [9] evaluated the performance of four vectorization methods: one research prototype (Wenyin's method) and three commercial software (VPstudio, Scan2CAD, Vectory). Five real scanned images (with no artificial noise) were used in the contest. A recent performance evaluation method was the criteria (*vectorial scores*) used to judge the quality of vector detection. The calculation of *vectorial scores* [10] requires the ground truth data to be represented in color pixels. *Vectorial scores* are a set of metrices that represents the errors made by the vectorization algorithm. The metrices include total oversegmentation, total undersegmentation, oversegmented components, undersegmented components, missed components, and false alarms. Only solid circular arcs were considered in this contest.

This paper presents the results of the Arc Segmentation Contest at GREC'09. The purpose of this contest is to study the current status of raster to vector methods and to generate new test images with their ground truth data available to the public. For these reasons there was no attempt to find a winner in this contest. The test data in this contest were selected from two old books. This would ensure that a level of complexity was already added to the data in order to make the correct detection of the graphical elements more challenging and it would also serve to avoid the use of degradation model (which could be controversial) since some kind of degradation was already embedded in the original scanned raster image due to low quality of image preparation. One element of *VRI* index (namely D_v) was used as the acceptable performance evaluation criterion in this contest. D_v is the vector detection rate obtained by calculating the length-weighted sum of all vector detection quality of all ground truth lines [11]. In this contest a within-method comparison was performed. The comparison involves comparing different versions of a raster to vector method developed over time. Also, a research prototype called Qgar-Lamiroy [16] participated for the first time in GREC Arc Segmentation Contests.

The rest of the paper is organized as follows. Section 2 shows how test images and ground truth images are created. In Section 3, the participating methods and the two test methods performed in this contest are provided. Section 4 shows the criterion to be used for performance evaluation. Section 5 shows the results of the contest. Section 6 presents the summary.

2 Test Images and Generating Ground Truth Data

Ground truth data were generated for the contest by selecting a set of mechanical engineering drawings from two text books [12][13]. Color images were obtained by scanning the paper drawings with 450 DPI each. The images were then converted to grey level images using ImLab imaging software [14]. A total of twelve binary images were obtained by thresholding the grey images. Seven images were selected as test images (shown with their ground truth data in Figs. 1 and 2), and five images were used as training images. Figure 1 shows the test images selected from an old book published in 1960s [12]. The drawings seem to be prepared by hand with the same graphical element having different thickness at different places. Figure 2 shows the test images selected from a more recent text book [13]. The quality of the drawings is better than those of Fig. 1 but the drawings were less complex. The ground truth data is superimposed on the original scanned images using a web-based service[1].

The ground truth images for test and training images were created by manual measurement of the primitives' parameters. An engineer generated the ground truth data manually by drawing, resizing, positioning the vector data over a displayed raster image using a vector editor. The first author then checked the generated data visually and randomly verified some of the real numerical data. The widths of the graphical elements were generated by measuring it using a raster image editor. Only circles and circular arcs were included in the ground truth files (shown in Figs. 1 and 2 in green color [electronic form]). The attributes (centers and radii) of the arcs are measured in pixels. The images are diverse in terms of graphics complexity as well as in the number of graphical entities. However, images with higher complexity were selected as test images. No vector representation is generated for any arc-like or circle-like characters such as the letters 'c', 'O' and others; and symbols such as '(' and ')'. In other words, all text strings in the images were ignored during ground truthing. This is true for all training images and test images. For this reason, in this contest we have focused our attention on D_v rather than the combined index (*VRI*). A website (http://www.cs.usm.my/arcseg2009/) was made online for the contest. The training images and the test images generated in this contest were made available for upload by third party.

---

[1] http://demo.iupr.org/vec2img/

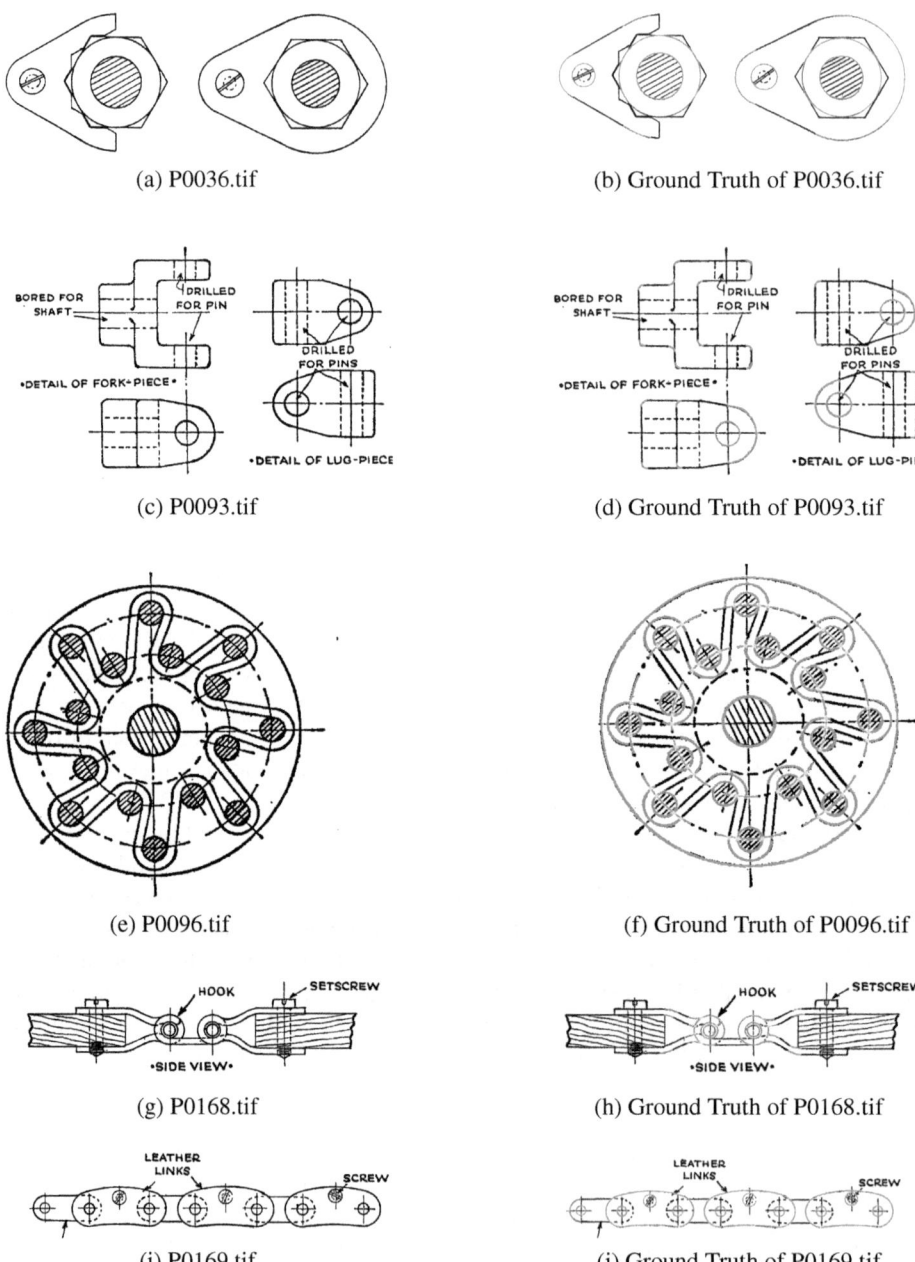

(a) P0036.tif

(b) Ground Truth of P0036.tif

(c) P0093.tif

(d) Ground Truth of P0093.tif

(e) P0096.tif

(f) Ground Truth of P0096.tif

(g) P0168.tif

(h) Ground Truth of P0168.tif

(i) P0169.tif

(j) Ground Truth of P0169.tif

Fig. 1. Test images (Scanned from [12])

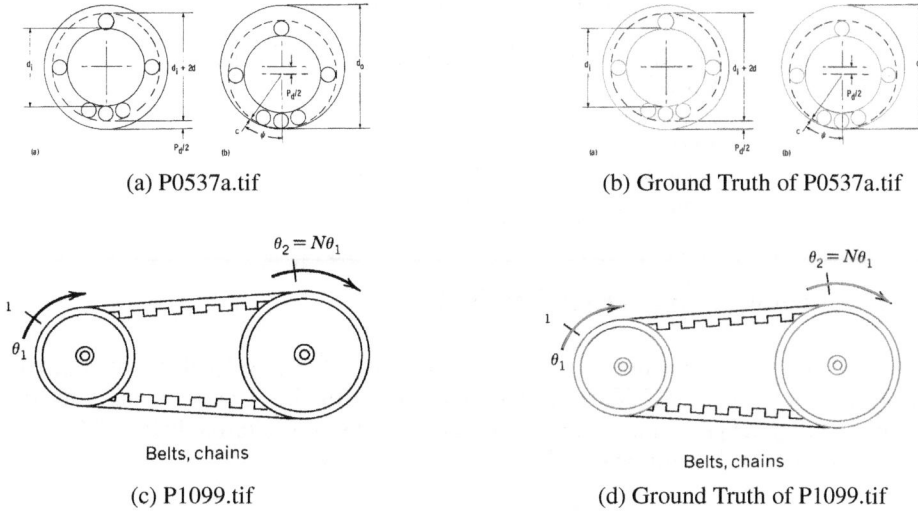

(a) P0537a.tif

(b) Ground Truth of P0537a.tif

Belts, chains

(c) P1099.tif

Belts, chains

(d) Ground Truth of P1099.tif

Fig. 2. Test images (Scanned from [13])

3   Participating Methods and Tests Performed

Research prototypes as well as commercial software were accepted in this contest. We have two research prototypes VrLiu [15] and Qgar-Lamiroy [16] as well as five different trial versions of three commercial software (VPstudio 8, 9, and 10 [17], Scan2CAD 7.5d [18], Vectory 5.0 [19]). All the software that we have used in the contest are trial versions. However, they are fully functional. The limitation is only on the size of the images and the duration in which the software can be used. Both limits have no effect on the outcomes of the contest. Some software parameters were preset while running the experiment in order to ensure consistency between the outputs of all the methods. To be fair to all software, no attempt has been made to fine tune software parameters for the purpose of getting better results. Getting the most performance of a software package requires a person experienced with the package and there were no seasoned operator on hand for the contest.

We were particularly interested in performing within-method comparison i.e. studying the performance of different versions of the same method (developed over time) to gauge the improvement in the performance of line detection. However, only one method (VPstudio) participated because of the unavailability of earlier versions of other methods.

Two different kinds of tests (Table 1) were performed in this contest. In the first test (between-method) different raster to vector methods were used to vectorize the raster images. *VRI* was used to judge the quality of arc detection. In the second test (within-method) different versions of the same method developed over time were used to vectorize the test images in an endeavor to reveal any improvements to the method (only VPstudio participated in this case) as to its ability at detecting arcs.

Table 1. Comparisons performed

Method	Between-methods	Within-method
VrLiu	Y	-
Qgar-Lamiroy	Y	-
VPstudio	Y	Y
Vectory	Y	-
Scan2CAD	Y	-

4 Performance Evaluation Method

The *VRI* index [11] was the performance evaluation of choice used in previous Arc Segmentation Contests [6,7,8]. It was also the criterion of choice used by many authors when reporting the performance of their proposed raster to vector methods [20,21].

VRI is calculated as follows:

$$VRI = \sqrt{D_v * (1 - F_v)} \tag{1}$$

where D_v is the detection rate and F_v is the false alarm rate. As noted in Section 2 all text strings were ignored during ground truthing including the arc-like symbols. Some systems will try to detect arc-like symbols as real arcs which may increase the false alarm rate thus decreasing the *VRI* score. Hence, we used only detection rate D_v rather than the combined index (*VRI*) in order to focus on the detection ability of the participating methods and to cancel the effect of recognizing some symbols as real arcs.

5 Results

The contest was performed offsite after the workshop. The participants were provided with the test images through e-mail. They have provided us with the results of their methods in VEC format. The other results were prepared by the authors of this report. The scores of the first test which include two research prototypes and three commercial software are shown in Table 2. The numbers in bold highlight the highest D_v scores for that image. From the D_v column it is noted that VrLiu's method has the largest D_v score (on average) and hence it is the best performer in this contest. The second best performer is VPstudio followed by Qgar-Lamiroy at third place. VPstudio scored higher than VrLiu for the image P0096.tif which is the most complex image in the set of test images. All VEC files (except for VrLiu) were prepared or provided to us with arc/circle widths set to 1. This may have resulted in lower D_v scores for these methods.

The scores of the second test are shown in Table 3. The numbers in bold highlight the highest D_v scores for that image. From the D_v column of Table 3 it is shown that version 9 scores slightly better than version 8. Version 10 performs better (on average) than the older versions. The differences within these three versions could be visualized by looking at Fig. 3.

Table 2. Performances of different systems

Image	VrLiu D_v	F_v	VRI	Qgar-Lamiroy D_v	F_v	VRI	VPstudio 10 D_v	F_v	VRI	Vectory 5 D_v	F_v	VRI	Scan2CAD 7 D_v	F_v	VRI
P0036	**.740**	.287	.726	.546	.264	.634	.584	.398	.593	.155	.831	.162	.250	.586	.322
P0093	**.483**	.715	.371	.459	.774	.322	.477	.435	.519	.443	.341	.540	.441	.554	.444
P0096	.563	.344	.608	.356	.335	.487	**.700**	.413	.641	.003	.994	.004	.299	.434	.411
P0168	**.696**	.748	.419	.629	.739	.405	.534	.467	.534	.421	.343	.526	.437	.650	.391
P0169	**.788**	.245	.772	.417	.424	.490	.595	.467	.563	.298	.504	.384	.417	.429	.488
P0537a	**.811**	.205	.803	.426	.318	.539	.413	.649	.381	.063	.937	.063	.382	.576	.403
P1099	.755	.283	.736	.693	.377	.657	**.767**	.175	.795	.198	.776	.211	.540	.450	.545
Avg	**.691**	.404	.634	.504	.462	.505	.581	.429	.575	.226	.675	.270	.395	.526	.429

Table 3. Performances of different versions of VPstudio software

Image	VPstudio 8 D_v	F_v	VRI	VPstudio 9 D_v	F_v	VRI	VPstudio 10 D_v	F_v	VRI
P0036	0.572	0.353	0.608	0.572	0.353	0.608	**0.584**	0.398	0.593
P0093	0.477	0.444	0.515	0.477	0.444	0.515	0.477	0.435	0.519
P0096	0.682	0.413	0.633	0.699	0.413	0.641	**0.700**	0.413	0.641
P0168	0.534	0.489	0.522	0.534	0.467	0.534	0.534	0.467	0.534
P0169	**0.599**	0.448	0.575	0.592	0.470	0.560	0.595	0.467	0.563
P0537a	0.414	0.647	0.382	0.414	0.647	0.382	0.413	0.649	0.381
P1099	0.767	0.175	0.795	0.760	0.175	0.795	0.767	0.175	0.795
Avg	0.578	0.424	0.576	0.578	0.424	0.576	**0.581**	0.429	0.575

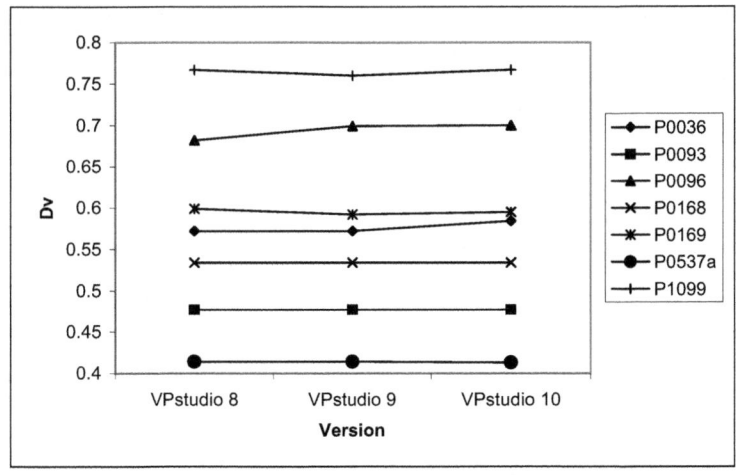

Fig. 3. Performances (D_V) of three versions of VPstudio software

6   Summary

In this paper we have described the detail of the Arc Segmentation Contest 2009 held in conjunction with GREC'09 that includes preparation of ground truth data, the tests performed, the participating methods, and the performance evaluation method. Real scanned images scanned from two old text books were used as test data. No artificial noise and no degradation model were used in the contest. However, since most drawings were scanned from an old text book, some distortions and noise are inherited in the scanned images. The ground truth data were created by manual measurement of arcs/circles attributes. Five methods participated in this contest; two research prototypes and three commercial software. To focus on the detection ability of the software D_v score was used as performance evaluation criterion. Two tests were performed: A test between different methods and a test between different versions of one of the methods. The former test shows that Liu's method outperforms the other methods. The latter test shows that the recent version of VPstudio is slightly better than the previous two versions.

Acknowledgment

We would like to express our gratitude to Abdul Halim Ghazali for his contribution in preparing the ground truth images. The first author is supported by USM Fellowship.

References

1. Kong, B., Phillips, I., Haralick, R., Prasad, A., Kasturi, R.: A benchmark: Performance evaluation of dashed-line detection algorithms. In: Kasturi, R., Tombre, K. (eds.) Graphics Recognition 1995. LNCS, vol. 1072, pp. 270–285. Springer, Heidelberg (1996)
2. Chhabra, A.K., Phillips, I.T.: The second international graphics recognition contest - raster to vector conversion: A report. In: Chhabra, A.K., Tombre, K. (eds.) GREC 1997. LNCS, vol. 1389, pp. 390–410. Springer, Heidelberg (1998)
3. Chhabra, A.K., Phillips, I.T.: Performance evaluation of line drawing recognition systems. In: Proc. 15th International Conference on Pattern Recognition, Barcelona, vol. 4, pp. 864–869 (2000)
4. Tombre, K.: International workshop on graphics recognition - dashed-line detection contest, http://www.loria.fr/~tombre/contest.html (accessed in June 2009)
5. Phillips, I.T., Chhabra, A.K.: Empirical performance evaluation of graphics recognition systems. IEEE Transactions on Pattern Analysis and Machine Intelligence 21(9), 849–870 (1999)
6. Liu, W.Y., Zhai, J., Dori, D.: Extended summary of the arc segmentation contest. In: Blostein, D., Kwon, Y.-B. (eds.) GREC 2001. LNCS, vol. 2390, pp. 343–349. Springer, Heidelberg (2002)
7. Liu, W.: Report of the arc segmentation contest. In: Lladós, J., Kwon, Y.-B. (eds.) GREC 2003. LNCS, vol. 3088, pp. 364–367. Springer, Heidelberg (2004)
8. Wenyin, L.: The third report of the arc segmentation contest. In: Liu, W., Lladós, J. (eds.) GREC 2005. LNCS, vol. 3926, pp. 358–361. Springer, Heidelberg (2006)
9. Shafait, F., Keysers, D., Breucl, T.M.: GREC 2007 arc segmentation contest: Evaluation of four participating algorithms. In: Liu, W., Lladós, J., Ogier, J.-M. (eds.) GREC 2007. LNCS, vol. 5046, pp. 310–320. Springer, Heidelberg (2008)

10. Shafait, F., Keysers, D., Breuel, T.M.: Pixel-accurate representation and evaluation of page segmentation in document images. In: Proceedings of the 18th International Conference on Pattern Recognition, vol. 1, pp. 872–875 (2006)
11. Liu, W.Y., Dori, D.: A protocol for performance evaluation of line detection algorithms. Machine Vision and Applications 9(5-6), 240–250 (1997)
12. Peatfield, A.E.: Teach Yourself Mechanical Engineering. E.L.B.S. and English Universities Press, London (1965)
13. Kurtz, M.: Mechanical Engineers' Handbook. John Wiley & Sons, Chichester (1986)
14. ImLab: Free image processing system, http://imlab.sourceforge.net/
15. Liu, W.Y., Dori, D.: Incremental arc segmentation algorithm and its evaluation. IEEE Transactions on Pattern Analysis and Machine Intelligence 20(4), 424–431 (1998)
16. Lamiroy, B., Guebbas, Y.: Robust Circular Arc Detection. In: Ogier, J.-M., Liu, W., Lladós, J. (eds.) GREC 2009. LNCS, vol. 6020. Springer, Heidelberg (2009)
17. VPstudio: Raster to vector conversion software, softelec, Munich, Germany, http://www.softelec.com and http://www.hybridcad.com
18. Scan2CAD: Raster to vector conversion software, softcover international limited, Cambridge, England, http://www.softcover.com
19. Vectory: Vectorization software, graphikon gmbh, Berlin, Germany, http://www.graphikon.de
20. Hilaire, X., Tombre, K.: Robust and accurate vectorization of line drawings. IEEE Transactions on Pattern Analysis and Machine Intelligence 28(6), 890–904 (2006)
21. Jiqiang, S., Feng, S., Chiew-Lan, T., Shijie, C.: An object-oriented progressive-simplification-based vectorization system for engineering drawings: model, algorithm, and performance. IEEE Transactions on Pattern Analysis and Machine Intelligence 24(8), 1048–1060 (2002)

A Performance Characterization Algorithm for Symbol Localization

Mathieu Delalandre[1,2], Jean-Yves Ramel[1], Ernest Valveny[2],
and Muhammad Muzzamil Luqman[1,2]

[1] Laboratoire d'Informatique (LI), 37200 Tours, France
mathieu.delalandre@univ-tours.fr,
jean-yves.ramel@univ-tours.fr,
muhammadmuzzamil.luqman@etu.univ-tours.fr
[2] Computer Vision Center (CVC), 08193 Bellaterra (Barcelona), Spain
ernest@cvc.uab.es

Abstract. In this paper we present an algorithm for performance characterization of symbol localization systems. This algorithm is aimed to be a more "reliable" and "open" solution to characterize the performance. To achieve that, it exploits only single points as the result of localization and offers the possibility to reconsider the localization results provided by a system. We use the information about context in groundtruth, and overall localization results, to detect the ambiguous localization results. A probability score is computed for each matching between a localization point and a groundtruth region, depending on the spatial distribution of the other regions in the groundtruth. Final characterization is given with detection rate/probability score plots, describing the sets of possible interpretations of the localization results, according to a given confidence rate. We present experimentation details along with the results for the symbol localization system of [1], exploiting a synthetic dataset of architectural floorplans and electrical diagrams (composed of 200 images and 3861 symbols).

Keywords: symbol localization, groundtruth, performance characterization.

1 Introduction

In recent years there has been a noticeable shift of attention, within the graphics recognition community, towards performance evaluation of symbol recognition systems. This interest has led to the organization of several international contests and development of performance evaluation frameworks [2]. However, to date, this work has been focussed on recognition of isolated symbols. It didn't take into account the localization of symbols in real documents. Indeed, symbol localization constitutes a hard research gap, both for recognition and performance evaluation tasks.

Different research works have been recently undertaken to fill this gap [3]. Groundtruthing frameworks for complete documents and datasets have been proposed in [4,5], and different systems working at localization level in [6,1,4]. The key problem is now to define characterization algorithms working in a localization context. Indeed the characterization of localization in complete documents is harder, as comparison of results with

J.-M. Ogier, W. Liu, and J. Lladós (Eds.): GREC 2009, LNCS 6020, pp. 260–271, 2010.
© Springer-Verlag Berlin Heidelberg 2010

groundtruth needs to be done between sets of symbols. These sets could be of different size, and significant differences could appear between the localizations of symbols provided by a given system, and the corresponding ones in the groundtruth. Characterization metrics must then be reformulated to take these specificities into account.

In the rest of the paper, we will first introduce in section 2 related work on this topic. We will present next in section 3 the approach we propose. Section 4 will give experiments and results we have obtained with our algorithm. Conclusion and perspectives arising from this work will be presented in section 5.

2   Related Work

Performance characterization, in the specific context of localization, is a well known problem in some research topics such as computer vision [7], handwriting segmentation [8], layout analysis [9], text/graphics separation [10], etc. Concerning symbol localization, at the best of our knowledge only the work of [4] has been proposed to date. Performance characterization algorithms (in the specific context of localization) aim to detect possible matching cases between groundtruth and localization results, as detailed in Table 1. They determine about true or false localization results, without considering the class of objects. It is therefore a two-class recognition problem, to separate objects from background. Once objects are correctly located/segmented, we could proceed the evaluation of recognition. This is known as whitebox evaluation in the literature [11], the goal is to characterize the performance of individual submodules of a complete system and to see how they interact each other.

Table 1. Matching cases between groundtruth and localization results

single	an object in the results matches with a single object in the groundtruth
misses	an object in the groundtruth doesn't match with any object in the results
false alarm	an object in the results doesn't match with any object in the groundtruth
multiple	an object in the results matches with multiple objects in the groundtruth (merge case) or an object in the groundtruth matches with multiple objects in the results (split case)

The key point when developing such a characterization algorithm, is to decide about representations to be used, both in results and groundtruth. Two kinds of approach exist in the literature [9], exploiting pixel-based and geometry-based representations.

In a pixel-based representation, results and groundtruth correspond to sets of pixels. For that reason, algorithms exploiting such a representation are very accurate. They are usually employed to evaluate segmentation tasks in computer vision [7] or handwriting recognition [8]. However, groundtruth creation is more cumbersome and requires a lot more storage. Comparison of groundtruth with the results is also time-consuming.

In a geometry-based representation, algorithms employ geometric shapes to describe the regions (in results and groundtruth). The type of geometric shapes depend on the application: bounding boxes - text/graphics separation [10], isothetic polygons – layout analysis [9], convex hulls - symbol spotting [4], etc. Comparison of the groundtruth with the results is time-efficient, and the corresponding groundtruth straightforward to produce. Such a representation is commonly used in the document analysis field, as it is more focused on semantic [10,9,4]. Because the main goal of the systems is recognition, evaluation could be limited to detection aspects only (i.e. to decide about a wrong or a correct localization without evaluation of the segmentation accuracy).

In both cases, characterization algorithms exploit information about regions in localization results, and compare them to groundtruth. This results in boolean decisions about positive/negative detections, which raises several open problems:

Homogeneity of results: Regions provided as localization results could present a huge variability (set of pixels, bounding boxes, convex hulls, ellipsis, etc.). This variability disturbs the comparison of systems. A characterization algorithm should take these differences into account, and put the results of all the systems at a same level.

Reliability of results: Large differences could appear between the size of regions in the results and the groundtruth. These differences correspond to over or under segmentation cases. This results in aberrant positive matching cases between the groundtruth and the detection results, when large regions in the results intersects smallest ones in groundtruth. To be able to detect these ambiguous cases a characterization algorithm should be able to reconsider the localization results provided by a system.

Time complexity: Complex regions, such as symbols, must be represented by pixel sets or polygons to obtain a correct localization precision. However, their comparison is time-consuming both for geometry-based and pixel-based representations [9,8]. This involves to use specific approaches to limit the complexity of the algorithms [9].

In this paper we propose an alternative approach to region-based characterization, to solve these problems. We present this approach in the next section 3.

3   Our Approach

3.1   Introduction

Our objective in this work is to provide a "reliable" and "open" solution to characterize the performance of symbol localization systems. To achieve that, we propose an algorithm exploiting only single points as the result of localization and offering the possibility to reconsider the localization results provided by a system. It uses the information about context in groundtruth, and overall localization results, to detect the ambiguous localization cases. A probability score is computed for each matching between a localization point and a groundtruth region, depending on the spatial distribution of the other regions in the groundtruth. Final characterization is given with detection rate/probability score plots, describing the sets of possible interpretations of the localization results, according to a given confidence rate.

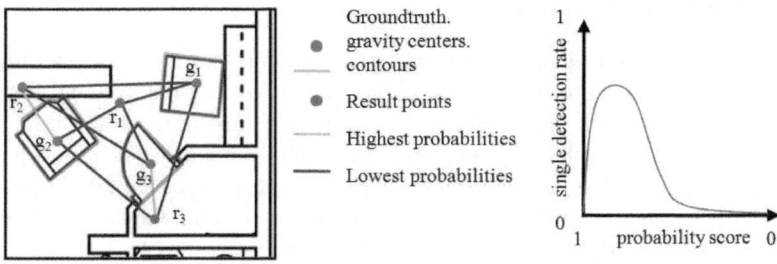

Fig. 1. Our approach

Fig.1 illustrates our approach. For each result point r_i, probability scores are computed with each symbol g_i in the groundtruth. These probability scores will depend on the spatial distribution of the symbols g_i in the groundtruth, they will change locally for each result point r_i. In Fig. 1, r_1, r_2 and r_3 are located at similar distances of symbols g_1, g_2 and g_3 in groundtruth. However, r_2 and r_3 present highest probability scores to be matched with g_2 g_3 respectively, but not r_1. Local distribution of symbols g_i in groundtruth around r_1 makes ambiguous the characterization of this localization result. For that reason, a positive matching between r_1 and any symbols g_i will be considered with a low probability. Final characterization is given with a detection rate/probability score plot, describing the sets of possible interpretations of the localization results according to their probability scores.

In the rest of this section we describe our characterization algorithm. We exploit three main steps to perform the characterization: (1) in a first step we use a method to compare the localization of a result point to a given symbol in groundtruth (2) exploiting this comparison method, we compute next for each result point its probability scores with all the symbols in groundtruth (3) at last, we employ a matching algorithm to identify the correct detection cases, and draw the detection rate/probability score plots. We will detail each of these steps in next subsections 3.2, 3.3 and 3.4 respectively. Table 2. provides the list of mathematical symbols that we have used for detailing the different steps of our proposed algorithm.

Table 2. Table of mathematical symbols

i	index value	j	index value
r_i	result object	g_i	groundtruth object
θ	direction between result and groundtruth objects	s_i	scaling factor
q	number of result objects	n	number of groundtruth objects
p_i	probability score	$f(s_i/s_j)$	probability score function
ε	score error ($\varepsilon = 1 - p_i$)		
s	single detection	T_s	single detection rate (=s/n)
f	false detection	T_f	false alarm rate (=f/q)
m	multiple detection	T_m	multiple detection rate (=m/q)

3.2 Localization Comparison

In our approach, groundtruth is provided as regions (contours with corresponding gravity centers) and localization results as points (Fig. 1). We ask the systems to provide points that should be in regions near the location of symbols in groundtruth. These points could be gravity centers, key-points of interest or junctions. The use of points as results makes impossible to compare them directly with groundtruth, due to the scaling variations. Indeed, symbols appear at different scales in drawings. To address this problem, we compare the result points with groundtruth regions by computing scale factors (Fig. 2). In geometrical terms, a factor s specifies how to scale a region in the groundtruth so that its contour fits with a given result point. Thus, result points inside and outside a symbol will have respectively scale factors of $s \leq 1$ and $s > 1$.

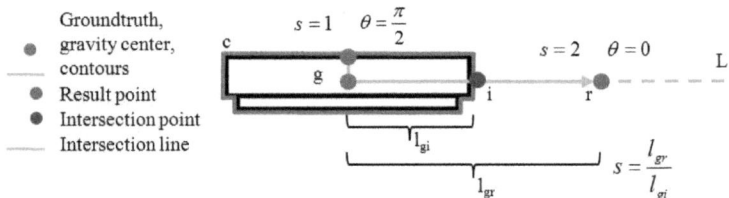

Fig. 2. Scale factor

The factor s is computed from the line L of direction θ, joining the gravity center g defined in groundtruth to a result point r. On this line L, s corresponds to the ratio of lengths l_{gr} and l_{gi}. l_{gr} is the Euclidean distance between the gravity center g and the result point r. l_{gi} is computed in the same way, but with the intersection point i of L with contours c of symbol. This intersection point i is detected using standard line intersection methods [12]. If several intersections exist (i.e. cases of concave contours or holes in symbol), the farthest one from g is selected.

3.3 Probability Scores

In a second step, we compute probability scores between result points and groundtruth (Fig. 3). For each result point r, relations with groundtruth are expressed by a set of n points $S = \bigcup_{i=1}^{n} g_i(\theta, s)$ (Fig. 3 (a)), where θ and s represent respectively the direction and scale factor between the result point and a symbol i. We define next the probability scores between the result point r and symbols $\bigcup_{i=1}^{n} g_i$ as detailed in Fig. 3 (b). When all symbols $\bigcup_{i=1}^{n} g_i$ are equally distant (i.e $s_1 = s_2 = ...s_i.. = s_n$), thus $\bigcup_{i=1}^{n} p_i = 0$. In the case of a $s_i = 0$, thus the corresponding $p_i = 1$. Otherwise, any other cases $0 < s_i < s_j$ will correspond to probability scores $0 < p_j < p_i < 1$.

The equations (1) (2) below give the mathematical details we employ to compute the probability scores. For each g_i, we compute the probability score p_i to the result point r as detailed in (1). To do it, we exploit the other symbols $\bigcup_{j=1, j \neq i}^{n} g_j$ in the groundtruth. We mean the values $f\left(\frac{s_i}{s_j}\right)$ corresponding to local probabilities g_i to r, regarding g_j.

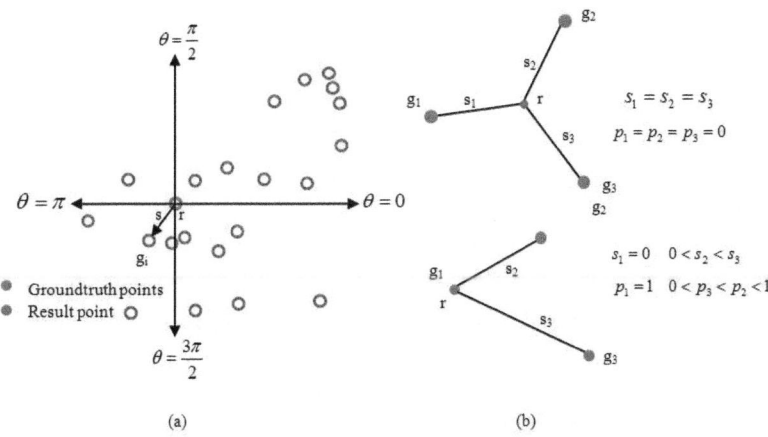

Fig. 3. (a) plot (θ, s) (b) probability scores

The function $f\left(\frac{s_i}{s_j}\right)$ must respect the conditions given in Table 3. In fact, there exists different mathematical ways to define such a function (based on inverse, linear, cosinus functions, etc.). We have employed here a gaussian function (2), as it is a common way to represent random distributions. This function is set using a variance $\sigma^2 = 1$, and normalization parameters k_y and k_x. These parameters are defined in order to obtain the key values $f(x = 0) = 1$ and $f(x = 1) \rightarrow 0$. Fig. 4 gives plots of gaussian and probability score functions to illustrate how kx and ky impact the values. The value of k_y is defined as $\sqrt{2\pi}$, to bound $f(x = 0)$ at 1 (e^0). Concerning k_x, we determine it using the approximation into a taylor serie of the cumulative distribution function $\Phi_0(x) = \int_0^x \phi(x)$. We define k_x when $\Phi_0(x) = 1 - \lambda$, with λ a precision defined manually. In our experiments, we have fixed $\lambda = 10^{-6}$ corresponding to $kx = 3.9$.

$$p_i = \sum_{j=1, j \neq i}^{n} \frac{f\left(\frac{s_i}{s_j}\right)}{n - 1} \tag{1}$$

$$f\left(x = \frac{s_i}{s_j}\right) = \frac{k_y}{\sqrt{2\pi}} \times e^{-\frac{(k_x x)^2}{2}} \tag{2}$$

Table 3. Table of function $f(x = \frac{s_i}{s_j})$

$f\left(x = \frac{s_i}{s_j}\right)$	0		1	$+\infty$	
	$s_i = 0$	$s_i \rightarrow +\infty$	$s_i = s_i$	$s_i \rightarrow +\infty$	$s_i \rightarrow 0$

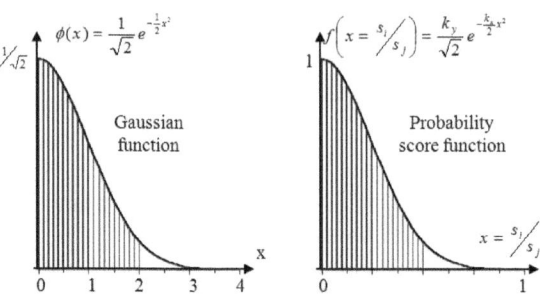

Fig. 4. Gaussian and probability score functions

3.4   Matching Algorithm

Once probability scores are computed, we look for relevant matchings between ground-truth and results of a system. In order to make open interpretation of localization of symbols, we provide characterization results using distribution plots. We compute different sets of matching results, according to the ranges of the probability scores. Final results are displayed into distribution plots (Fig. 5 (a)), where the x axis corresponds to score error ε (i.e. inverse of probability score $\varepsilon = 1 - p$), and the y axis to performance rates.

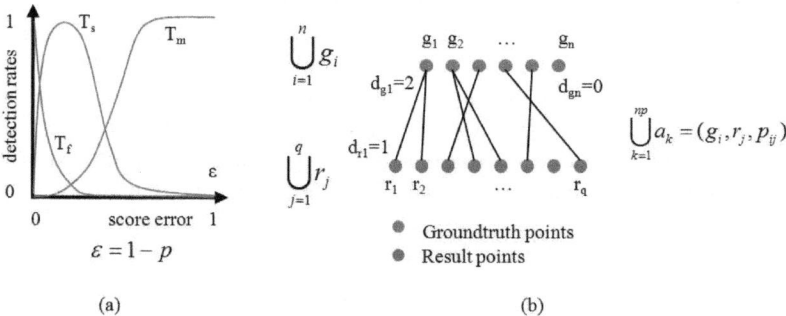

Fig. 5. (a) result plot (b) correspondence list

We compute different performance rates, T_s, T_f and T_m corresponding respectively to single detections, false alarms and multiple detections (see Table 1.). The rate of "misses cases" corresponds to $1 - T_s$. To do it, we build-up a correspondence list between results and groundtruth as detailed in Fig. 5 (b). This correspondence list is bipartite, composed of nodes corresponding to the groundtruth $\bigcup_{i=1}^{n} g_i$ and the results $\bigcup_{j=1}^{q} r_j$. Our probability scores are given as undirected arcs $\bigcup_{k=1}^{nq} a_k = (g_i, r_j, p_{ij})$ of nq size. We use these arcs to make the correspondences in the list in an incremental way, by shifting the ε value from 0 to 1. An arc a_k is added to the list when its $p_{ij} \geq 1 - \varepsilon$.

For each ε value, the T_s and T_f rates are computed by browsing the list, and checking the degrees of nodes $\bigcup_{i=1}^{n} d_{gi}$ and $\bigcup_{j=1}^{q} d_{rj}$, as detailed in (3), (4) and (5).

$$\forall g_i \leftrightarrow r_j, d_{gi} = d_{rj} = 1 \rightarrow s = s + 1 \tag{3}$$

$$T_s = \frac{s}{n}$$

$$\forall r_j, d_{rj} = 0 \rightarrow f = f + 1 \tag{4}$$

$$T_f = \frac{f}{q}$$

$$\forall r_j \leftrightarrow g_i, d_{rj} > 1 \vee d_{gi} > 1 \rightarrow m = m + 1 \tag{5}$$

$$T_m = \frac{m}{q}$$

4 Experiments and Results

In this section, we present experiments and results obtained using our algorithm. We have applied it to evaluate the symbol localization system of [1]. This system relies on a structural approach, using a two-step process.

First, it extracts topological and geometric features from a given image, and represents it using an ARG[1] (Fig. 6). The image is preliminary vectorized into a set of quadrilateral primitives. These primitives become nodes in the ARG (labels 1, 2, 3, 4), and connections between become arcs. Nodes have, as attributes, relative lengths (normalized between 0 and 1) whereas arcs have connection-type (*L junction, T junction, X junction*, etc.) and relative angle (normalized between 0° and 90°).

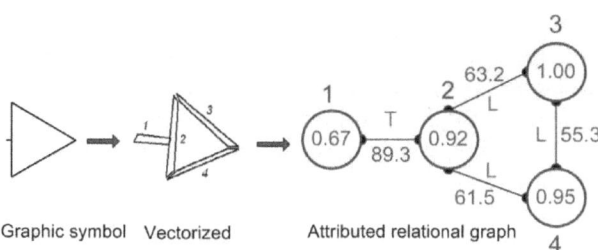

Graphic symbol   Vectorized   Attributed relational graph

Fig. 6. Representation phase of [1]

In the second step, the system looks for potential regions of interest corresponding to symbols. It detects parts of the ARG that may correspond to symbols i.e. symbol seeds. Scores, corresponding to probabilities of being part of a symbol, are computed for all edges and nodes of the ARG. They are based on features such as lengths of segments, perpendicular and parallel angular relations, degrees of nodes, etc. The symbol seeds are detected next during a score propagation process. This process seeks and analyzes

---

[1] Attributed Relational Graph.

Table 4. Dataset used for experiments

		Drawing level		Symbol level	
floorplans	Setting	backgrounds	5	models	16
	Dataset	images	100	symbols	2521
diagrams	Setting	backgrounds	5	models	17
	Dataset	images	100	symbols	1340

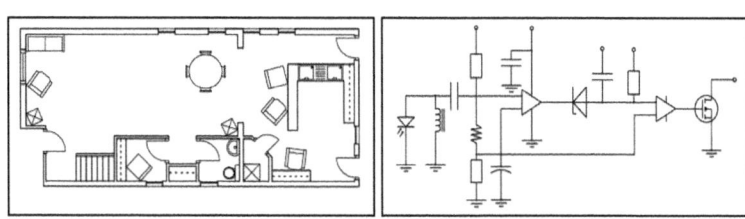

Fig. 7. Examples of test documents

the different shortest paths and loops between nodes in the ARG. The scores of all the nodes belonging to a detected path are homogenized (propagation of the maximum score to all the nodes in the path) until convergence, to obtain the seeds.

To test this system, we have employed datasets coming from the SESYD database[2]. This database is composed of synthetic document images, with the corresponding ground-truth, produced using the system described in [5]. This system allows the generation of synthetic graphic documents, containing non-isolated symbols in a real context. It is based on the definition of a set of constraints, that permit to place the symbols on a predefined background, according to the properties of a particular domain (architecture, electronics, etc.). The SESYD database is composed of different collections, including architectural floorplans, electrical diagrams, geographic maps, etc. In this work, we have limited our experiments to subsets of this database, including documents from the electrical and architectural domains. Table 4. gives details of the dataset we use, and Fig. 7 gives some examples of test documents.

Fig. 8 presents the characterization results we have obtained on floorplans, with variation of $\{T_s, T_f, T_m\}$ rates. The system presents in A a good confidence for the detection results $T_s \leq 0.50$, with an score error $\varepsilon \leq 0.05$ and nearly none multiple detections $T_m \leq 0.03$. The best detection results is obtained in B with a $T_s = 0.57$, corresponding to an score error of $\varepsilon = 0.11$. However, these detection results are joined to a $T_f = 0.31$, highlighting a bad precision of the system. In addition, at this state confusions appear in localization results with a $T_m \geq 0.20$. This rate results in the merging of false alarms with less confident results, as the false alarm rate goes down to $T_f = 0.31$. Up to this

---

[2] http://mathieu.delalandre.free.fr/projects/sesyd/

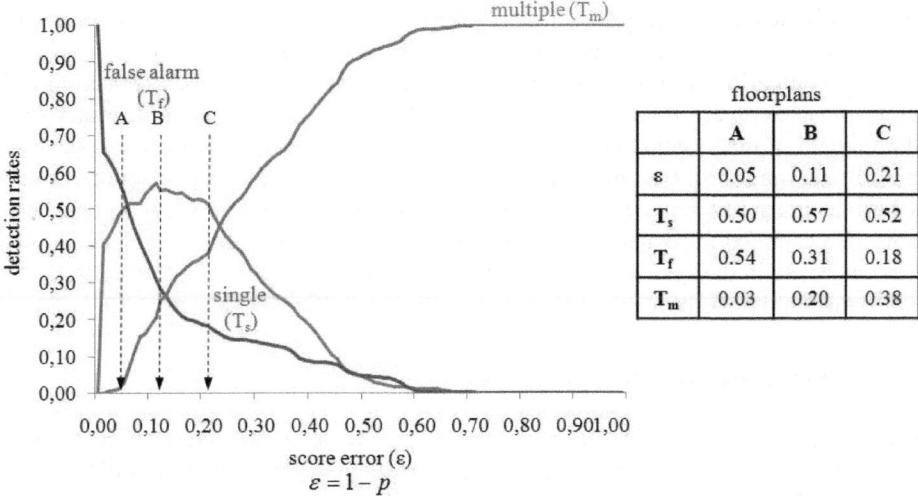

Fig. 8. Characterization results on floorplans

point, T_s dies down slowly, and for score error $\varepsilon \geq 0.21$ in C the T_s and T_m curves start to be diametrically opposite.

Fig. 9 gives results concerning electrical diagrams. The system presents a good confidence in A, for $\varepsilon \leq 0.04$ and $T_s = T_f = 0.45$. However, at this point the system already does multiple detections with a $T_m = 0.10$. The best localization score is obtained at B with $\varepsilon = 0.13$ and $T_s = 0.62$. Therefore, the best localization score is better for electrical diagrams than floorplans. In addition, the system doesn't make a

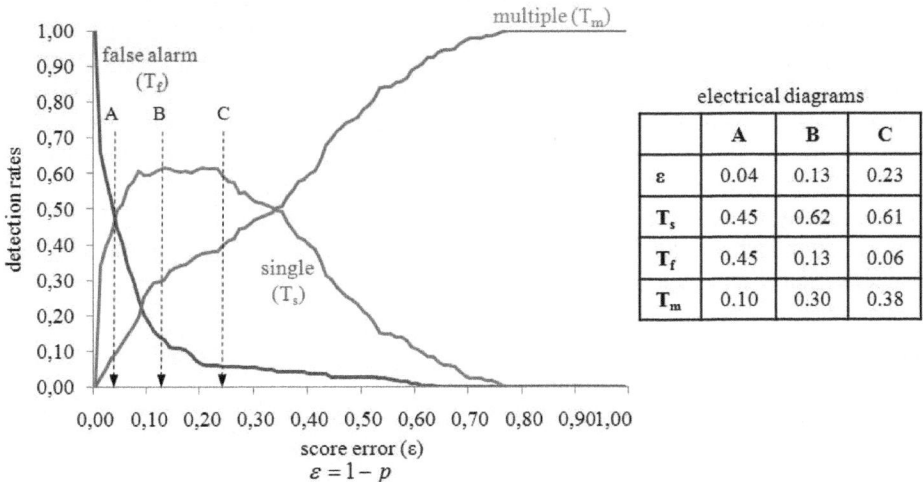

Fig. 9. Characterization results on electrical diagrams

lot of false detections with a $T_f = 0.13$. However, multiple detections stay higher with $T_m = 0.30$. Up to C, T_s decreases in linear way for score errors $\varepsilon \geq 0.23$.

When comparing results obtained by the system [1] on these two application domains, the ones on electrical diagrams are best. However, each of them illustrates specific failures of the system. The first is the high level of multiple detection appearing on electrical diagrams. Certainly the systems could increase a lot its detection results, by introducing split / merge procedures of detected regions of interest. The second concerns the generated false alarms on floorplans. In order to reduce this problem, a point to be improved in the system is to introduce a checking procedure of detected ROIs, to reduce the hypothesis of localization.

5   Conclusions and Perspectives

In this paper we have presented an algorithm for performance characterization of object localization systems. This algorithm has been applied in the context of symbol localization, but we believe it could be applied to other problems (medical image segmentation, mathematical formula recognition . . .). It aims to propose a more "reliable" and "open" solution to characterize the performance of systems, by offering the possibility to reconsider the localization results. It exploits only single points as the results of localization. Then a probability score is computed for each matching between a localization point and a groundtruth region, depending on the spatial distribution of the other regions in the groundtruth. They will change locally for each result point. Characterization results are given with detection rate/probability score plots, describing the sets of possible interpretations of the localization results, according to a given confidence rate. We present experiments and results obtained using our algorithm, to evaluate the symbol localization system of [1]. These experiments have been done on a dataset of synthetic images (with the corresponding groundtruth), composed of 200 document images and around 3861 symbols from electrical and architectural domains. We conclude about the performance of this system, in terms of localization accuracy, precision level (false alarms and multiple detections) on both datasets.

In future, we aim to take-forward our experimentations to evaluate the scalability (large number of symbol models) and robustness (noisy images [13]) of systems. We also plan to perform experiments with real datasets [4], and to compare the obtained results with the synthetic ones. And, our final goal is the comparison of different symbol localization systems. We plan to take benefit of the work done around the EPEIRES project[3], to look for potential participants interested in testing their systems with this characterization approach.

Acknowledgements

This work has been partially supported by the Spanish projects CONSOLIDER-INGENIO 2010 (CSD2007-00018), TIN2008-04998 and TIN2009-14633-C03-03. The authors wish to thank Hervé Locteau (LORIA institute, Nancy, France) for his comments and corrections about the paper.

---

[3] http://epeires.loria.fr/

References

1. Qureshi, R., Ramel, J., Barret, D., Cardot, H.: Symbol spotting in graphical documents using graph representations. In: Liu, W., Lladós, J., Ogier, J.-M. (eds.) GREC 2007. LNCS, vol. 5046, pp. 91–103. Springer, Heidelberg (2008)
2. Valveny, E., Tabbone, S., Ramos, O., Philippot, E.: Performance characterization of shape descriptors for symbol representation. In: Liu, W., Lladós, J., Ogier, J.-M. (eds.) GREC 2007. LNCS, vol. 5046, pp. 278–287. Springer, Heidelberg (2008)
3. Delalandre, M., Valveny, E., Lladós, J.: Performance evaluation of symbol recognition and spotting systems: An overview. In: Workshop on Document Analysis Systems (DAS), pp. 497–505 (2008)
4. Rusiñol, M., Lladós, J.: A performance evaluation protocol for symbol spotting systems in terms of recognition and location indices. International Journal on Document Analysis and Recognition (IJDAR) 12(2), 83–96 (2009)
5. Delalandre, M., Pridmore, T., Valveny, E., Trupin, E., Locteau, H.: Building synthetic graphical documents for performance evaluation. In: Liu, W., Lladós, J., Ogier, J.-M. (eds.) GREC 2007. LNCS, vol. 5046, pp. 288–298. Springer, Heidelberg (2008)
6. Locteau, H., Adam, S., Trupin, E., Labiche, J., Heroux, P.: Symbol spotting using full visibility graph representation. In: Liu, W., Lladós, J., Ogier, J.-M. (eds.) GREC 2007. LNCS, vol. 5046, pp. 49–50. Springer, Heidelberg (2008)
7. Unnikrishnan, R., Pantofaru, C., Hebert, M.: Toward objective evaluation of image segmentation algorithms. Pattern Analysis and Machine Intelligence (PAMI) 29(6), 929–944 (2007)
8. Breuel, T.: Representations and metrics for off-line handwriting segmentation. In: International Workshop on Frontiers in Handwriting Recognition (IWFHR), pp. 428–433 (2002)
9. Bridson, D., Antonacopoulos, A.: A geometric approach for accurate and efficient performance evaluation. In: International Conference on Pattern Recognition (ICPR), pp. 1–4 (2008)
10. Wenyin, L., Dori, D.: A proposed scheme for performance evaluation of graphics/text separation algorithms. In: Chhabra, A.K., Tombre, K. (eds.) GREC 1997. LNCS, vol. 1389, pp. 335–346. Springer, Heidelberg (1998)
11. Kanungo, T., Resnik, P.: The bible, truth, and multilingual ocr evaluation. In: Document Recognition and Retrieval (DRR). SPIE Proceedings, vol. 3651, pp. 86–96 (1999)
12. Balaban, I.: An optimal algorithm for finding segments intersections. In: Symposium on Computational Geometry (SGC), pp. 211–219 (1995)
13. Kanungo, T., Haralick, R.M., Phillips, I.: Non-linear local and global document degradation models. International Journal of Imaging Systems and Technology (IJIST) 5(3), 220–230 (1994)

Graphics Recognition—What Else?

Karl Tombre[1,2,3]

[1] INRIA, 615 rue du jardin botanique, 54600 Villers-lès-Nancy, France
[2] Université de Lorraine, INPL, LORIA, B.P. 239, 54506 Vandœuvre-lès-Nancy
CEDEX, France
[3] CNRS, LORIA, B.P. 239, 54506 Vandœuvre-lès-Nancy CEDEX, France
Karl.Tombre@inria.fr

Abstract. This paper tries to sum up the discussions held during the
sessions of GREC'09, as well as at the final panel session. As it is always
good to know where you are coming from, the paper briefly takes a look
back at the discussions held two years earlier, before looking ahead at
the future challenges for our research community. A number of points
raised two years ago remain very much valid, but we also try to identify
some new grand challenges for the field of graphics recognition.

1   Introduction

In this paper, we will not repeat the brief summary of the history of graphics
recognition, which was presented in the report from the GREC'07 panel discussion [1]. The historical view of the reasons for gathering a specific document
analysis community, and within that a graphics recognition sub-group, remains
valid and the reader should keep it in mind in the thoughts raised here.

At the end of the GREC'09 workshop, a panel session was organized to wrap
up the numerous and lively discussions held throughout the sessions of the workshop, and try to come up with some more general conclusions. The panel members were Young-Bin Kwon, George Nagy, Sitaram Ramachandrula, and Karl
Tombre. A number of workshop participants also contributed to the debate.

This article presents the main discussions and conclusions of the panel.

2   Looking Back: Some of Our 2007 Conclusions

> *"The only person who likes change is a baby with a wet diaper."*
> *Mark Twain*

Traditionnally, we start by taking a look back at the "hot" topics discussed at
the 2007 panel. At that time, we identified the following categories:

2.1   Features

It becomes increasingly difficult to answer the question: which features distinguish graphics recognition from general pattern recognition problems? This

J.-M. Ogier, W. Liu, and J. Lladós (Eds.): GREC 2009, LNCS 6020, pp. 272–277, 2010.
© Springer-Verlag Berlin Heidelberg 2010

stems among other causes from the fact that we experience a deep convergence with the methods used in content-based image retrieval. The specificities of black-and-white images and of graphical information tend to become a detail in this broader context.

We pointed out that one interesting contribution would be to work on the characterization of various features for shape representation and recognition. In that sense, graphics recognition can be at the forefront of putting together large-scale repositories of features, so as to avoid the recurring appearance of "new" features which are actually minor variations on old ideas.

2.2 User Interaction

We pointed out that little work had been done, in the area of user interaction, on modeling the user, despite the fact that there are not many common features between a general, low-technicity user and a highly specialized technician, mastering the knowledge specific to a given application. This becomes a crucial problem when we produce applications aimed at the general public, but with complex user interaction which becomes accessible only after months of training.

2.3 Large-Scale Applications

A lot of discussions addressed the specific challenges of building large-scale applications, i.e. scaling from an academic problem to a really useful system. It was felt that this was not only an engineering problem but a cultural question and one of scaling our approaches. The need for a true policy on software development firmly rooted in the scientific achievements and tested on large datasets becomes crucial; we have to include the composition with reusable sofware and stress the building of production-quality code.

We also agreed that graphics recognition was still looking for its "killer application". One idea put forward was that a general sketching interface could be the answer, or maybe a combined sketching/retrieval/recognition system, making it possible to navigate in documentation by sketching simple examples of what is being searched for. It was pointed out that the GREC community does not seem to be very interested in dealing with digital documents such as PDF documents or web graphics. This may stem from a lack of good opportunities, or be an illustration of Mark Twain's quote cited above...

2.4 Performance Evaluation

We were disappointed by the low number of participants in our contests, even the more because this is an area where our community has often been showing the path to the image analysis community at large. We agreed that beyond the contests, we need to have open-source, robust benchmarking tools available online, with a sufficient amount of ground-truthed data. The question remains to know whether we have access to benchmarking data covering all our needs.

We suggested to make the methods available as web services, so as to be able to test the limits of new methods as soon as they are developed. Another idea vented two years ago was to announce in advance a grand challenge for the commmunity to work on.

3   Topics Discussed during the GREC'09 Workshop

> *"All progress is precarious, and the solution of one problem*
> *brings us face to face with another problem."*
> *Martin Luther King Jr.*

The workshop was a good opportunity to explore the variety of topics addressed nowadays by the graphics recognition community. The format of the workshop, the fact that it (still) gathers a significant part of the research teams active in the area, makes it a good observatory for the state of the art.

3.1   Technical / Methodological Aspects

Segmentation of graphics document is the historical theme of GREC, maybe also the only one for which you can be reasonably confident that you will see the most advanced results at GREC! We had several nice presentations at GREC'09; some of them dealt with *vectorization,* and especially the problem of robust arc detection, as well as specific ad hoc improvements, including domain knowledge, for various applications. Still, a very interesting question was raised during the workshop: why do we insist on vectorizing our raw data? Is it just something we have inherited from the "David Marr paradigm" or is it really necessary for the problems we have to deal with? Another question was whether it still makes sense to work mostly on binary images, whereas our raw data are often grey-level or color images.

Another typical segmentation topic of our community is *text-graphics separation.* We had several presentations dealing directly or indirectly with this question, including some nice applications such as the analysis of business cards on mobile phones. Somebody pointed out in a very interesting way that we improve our text-graphics separation methods because we have changed our focus: we now do text detection, not image segmentation!

Symbol recognition and spotting has also been a very active domain for many years already. Several papers addressed that area and questions asked included the way of dealing with complex symbols made of simpler symbols (we probably do not have robust approaches to that yet), the nature and the use of the available context (which is often difficult to capture), and the recurring search for genericity, which may be a kind of unreachable "Holy Grail" as it makes much more sense to build ad hoc systems for specific application areas.

Due to the very nature of the data we have to deal with, *structural pattern recognition* approaches are often a straightforward choice for many problems we deal with. This workshop was no exception and we saw a number of contributions using structural signatures, numeric signatures containing embedded graphs with the usual questions about graph edit distances, graph classification problems, adjacency grammars used in sketching applications, spatial relations represented in various ways. Using graphs necessarily leads to the general problems of graph matching, sub-graph matching, and the numerous embeddings and/or heuristics which aim at making these operations more efficient and applicable to large-scale problems. Note that these are not problems specific to graphics recognition.

It was pointed out that despite the known limitations of structural methods, they will remain useful for our problems for many reasons, including the structural and spatial nature of a lot of information we have to deal with, and the fact that we often do not have sufficient learning data to be able to perform statistical learning of our recognition methods.

Any recognition problem, be it solved by structural or statistical methods, has to deal with a representation of the shapes to be recognized through appropriate *features*. With respect to our 2007 conclusions, we must acknowledge that at GREC'09, we saw some very nice new results on features, signatures and descriptors used in graphics recognition problems. But the question pointed out two years ago remains very much valid, maybe even more so than two years ago: which features distinguish graphics recognition from general pattern recognition problems?

This question is also at the core of work on *content-based indexing*, where we have some exciting results on our graphics-specific problems, but no real specific methodology, compared to the large area of information indexing and spotting in image databases.

Finally, our community remains active in **performance evaluation and contests**, but we still have problems gathering participants, as pointed out two years ago. On the positive side, our GREC databases are used as reference in many papers. We still need to foster discussions about a broader policy of sharing data used in our publications.

3.2 Applications

We had the opportunity to see a lot of applications during these workshop days. Here is an overview of the variety encountered:

- Engineering drawings, architectural drawings, etc. do not seem to attract a lot of interest these days.
- Specific diagrams or notations such as chemical diagrams or music notation were more present; maybe this is a trend to move to very specialized areas, far from the main trend where a lot of effort has already been done (including with manual labor).
- There is a lot of interest in historical archives, i.e. legacy documents, often drawn and/or written by hand, with the purpose of archival, indexing and retrieval tasks.
- Ad hoc tasks such as identification and recognition of tables are addressed more easily than full-fledged, large-scale applications.
- The interest for sketching interfaces, already largely present two years ago, was confirmed at this workshop.

4   Some Hot Topics

> *"It's a long way to Tipperary, It's a long way to go.*
> *It's a long way to Tipperary, To the sweetest girl I know!"*
> British music hall and marching song by Jack Judge

During the closing panel, we discussed several possible hot topics for the coming years.

A first item was the recurring wish for methods capable of efficiently combining structural and statistical methods. Of course, this is not specific to graphics recognition, but as said before, the very structural and spatial nature of the information we work with makes structural methods quite natural in the community. Their efficient integration into methods which also take full advantage of statistical learning and classification is certainly the right path to take.

As much as it was two years earlier, the need for the development of large-scale applications remains a strong incentive. We need toolboxes of robust document image processing algorithms. We need to make code and test data available. But we often end up with the dilemma of the cost we are ready to pay, as academic researchers, in order to develop professional-quality code.

Questions were also raised about the usefulness of the contest model for performance evaluation. Wouldn't it be desirable to stabilize noise models and evaluation metrics, make test databases available, maybe even have a consortium in charge of maintaining them and delivering the service of performance evaluation throughout the year.

Historical documents seem to become a major issue, but this is not specific to GREC and we even wondered whether it was the right place to deal with the issues which had been presented. The problems to be solved go all the way from image processing for restoration purposes to large-scale indexing and retrieval, based on the right features and descriptors, computed both from images, graphics and text.

5   Food for Thought

> *Will the last person to leave graphics recognition*
> *please turn off the lights?*

Let us conclude this report by mentioning some broader questions which were discussed at the panel:

– Is there life outside Google? Services such as those delivered by Google and other major industry players have drastically changed the way we deal with digitized information. How can we be reasonaly confident that we will not discover one day that we discuss at GREC "new" research trends which are already out there, available as a web service by one of these players? Said in other words, how can an academic community interested in graphics recognition have a real impact on large-scale applications for collections of historical documents, for instance?

- Can we define a Grand Challenge? In other fields (speech recognition, autonomous vehicles, etc.) the scientific community has worked towards solving a Grand Challenge, involving cooperation between different teams. This has been a driving force towards progress in these fields. How could this kind of higher goals be set in an area like graphics recognition? Would it make sense for it to be specific to graphics recognition, or should it be a broader problem?
- What is the value to end-users or to customers? More precisely, what are the difficult questions for today's users (i.e. those who use information), and what partial answers do we have to these questions? Are we willing and ready to consolidate these answers to bring real value, so as to have a real impact?

Of course, these questions look more like open and general problems than like a conclusion, but maybe the best way of concluding a panel is actually to leave the audience with some food for future thought...

Reference

1. Tombre, K.: Is Graphics Recognition an Unidentified Scientific Object? In: Liu, W., Lladós, J., Ogier, J.-M. (eds.) GREC 2007. LNCS, vol. 5046, pp. 329–334. Springer, Heidelberg (2008); Report from final panel discussion at GREC 2007

Author Index